Understanding Past Earthquakes

Austin Elliott · Christoph Gruetzner
Editors

Understanding Past Earthquakes

Springer

Editors
Austin Elliott
United States Geological Survey
Moffett Field, USA

Christoph Gruetzner
Institute for Geological Sciences
Friedrich Schiller University Jena
Jena, Germany

ISBN 978-3-031-73582-0 ISBN 978-3-031-73580-6 (eBook)
https://doi.org/10.1007/978-3-031-73580-6

This work was supported by Innsbruck University, Friedrich-Schiller University Jena, USGS Earthquake Science Center and KAUST.

This Springer imprint is published by the registered company Springer Nature Switzerland AG
The registered company address is: Gewerbestrasse 11, 6330 Cham, Switzerland

Contents

Introduction to Understanding Past Earthquakes

1

Austin Elliott and Christoph Grützner

1.1 Purpose of This Book

This book presents overviews of modern, twenty-first century methods that are used to investigate the physical characteristics of *past* earthquakes. Where did earthquakes occur? How large were they? Which tectonic faults caused them? How strong were the ground motions and where? These are fundamental questions that we ask about any seismic event, yet only in the late 20th and early twenty-first century have they been operationalized to regularly obtain immediate answers for new earthquakes that occur (e.g., Hayes et al., 2011). While advances in recording, delivering, and analyzing earthquake data have made these processes virtually immediate for contemporary earthquakes, there are many pre-digital, pre-instrumental, and pre-historic earthquakes for which physical aspects of source and effects may yet be quantified. By extending our documentation of earthquake phenomena deeper into the past with a modern lens, we enhance understanding of their causative structures, resulting ground-motions, and lingering physical impacts, incrementally filling knowledge gaps within a natural system that operates on timescales that far exceed those of human study. We learn more about fault rupture in earthquakes by building a much larger database than possible with modern events alone.

Though earthquakes occur with regularity in every continent and beneath every ocean, the human record of major seismic events is relatively sparse, considering

A. Elliott
USGS Earthquake Science Center, PO Box 158, Moffett Field, CA 94035, USA
e-mail: ajelliott@usgs.gov

C. Grützner (✉)
Institute for Geosciences, Friedrich Schiller University Jena, Burgweg 11, 07749 Jena, Germany
e-mail: christoph.gruetzner@uni-jena.de

A. Elliott and C. Gruetzner (eds.), *Understanding Past Earthquakes*,
https://doi.org/10.1007/978-3-031-73580-6_1

the thousands of seismogenic tectonic faults that we know about on Earth that have *not* produced major earthquakes in modern times (e.g., Yeats, 2012). Our modern scientific understanding of earthquakes is scarcely more than a century old and our comprehensive recording and analysis of them with modern digital instruments is far younger than that. The Global Seismographic Network was established in 1986 (Butler et al., 2004), and so our routine identification and measurement of earthquakes has been comprehensively global and digital for less than half a century.

While oceanic subduction zones are particularly prolific in producing earthquakes, faults within the continents–where they are most readily observed and their direct effects most proximal to human settlements–generally produce major earthquakes at longer intervals of hundreds to thousands of years. This means that the vast majority of the thousands of hazardous active faults on the planet have *not yet* produced a major earthquake in modern human history, let alone during our brief span of contemporary instrumental recording with seismometers, dedicated survey teams, and now Earth-imaging satellites (England & Jackson, 2011).

Despite the youthfulness of modern earthquake recording technology and practice, the impacts of earthquakes prior to the twentieth century are neither unrecorded nor unknowable. Written and oral records of seismic events date back millennia in some parts of the world (e.g., Ambraseys, 2009; Cheng et al., 2017; Lee et al., 1976; Ludwin et al., 2007; Sbeinati et al., 2005; Zare et al., 2014); archeological evidence of their effects survives even longer than the direct accounts of their eyewitnesses (e.g., Sintubin, 2011); and the landscape and rocks themselves retain the signatures of past seismic upheaval until they are eroded away across the eons or buried by deposits of sediments where we must use excavation, drilling, or seismic imaging to uncover them (e.g., McCalpin, 2009).

Recent scientific and technological advances permit us now to identify, characterize, and analyze the phenomena that accompanied individual past earthquakes. Furthermore, the wealth of data that record earthquakes today have revealed immense complexity in their source processes and effects, elucidating through each unique seismic event the physical processes and properties by which earthquakes initiate, propagate, and terminate, and by which their seismic waves radiate and attenuate. The richness of modern recorded earthquakes, while commonly surprising earthquake scientists and emergency managers with their novelty or complexity, suggests that prior historical and prehistoric earthquakes shared similar complexities or novelties that were previously unrecognized due to sparse recording or limited investigation. Indeed, digging back to investigate earthquakes of the past and revisit past analyses with an updated understanding and set of scientific questions has yielded crucial evidence about seismic history and potential. For most hazardous faults, the most significant hints about future earthquake behavior come from records of the past.

Practitioners from the diverse earthquake-researching sub-disciplines of seismology, geology, geodesy, archeology, and history have been taking renewed looks at past seismic events, devising constructive approaches that reanalyze or synthesize previously made observations. Deriving a more complete picture of a particular seismic event often requires multiple lines of evidence, for example various combinations of measurements of slip magnitude, rupture length, shaking intensity, and spatial distribution of impacts. None of these alone is uniquely sufficient to define the size of a past earthquake and the complete extent of its impacts, but together they give an understanding of a past seismic event that approximates the information we would glean from a modern one.

Disparate practitioners, however, may attempt to appropriate data sources and measurements from one discipline into another without due consideration of the uncertainties, errors, and nuanced implications that dedicated researchers have been carefully constraining. For example, while a single surface offset measurement of ~5 m may conform with the magnitude roughly estimated from the distribution of shaking effects, the relatively large uncertainties in magnitude-slip scaling relationships (e.g., Stirling et al., 2013; Wells & Coppersmith, 1994) and the varying reliability of single offset measurements (e.g., Salisbury et al., 2015; Zielke et al., 2015) make this far from a robust corroboration. Likewise for similar reasons an apparent shaking intensity of VII on the Modified Mercalli scale at a point near observable surface rupture may carry sufficient uncertainty, without further contextual analysis and calibration, to form only a weak corroboration with geomorphically estimated event magnitudes.

Scientists, researchers, and historians attempting to constrain the size, location, and impacts of past events by synthesizing multiple lines of evidence should be aware of the measurement methods, approaches, uncertainties, and caveats in the fields whose data they are synthesizing. The chapters of this book present synoptic views of the history, methods, uncertainties, considerations, and caveats that have recently been demonstrated and explored in a wide suite of disciplines that are now commonly used to assess past earthquakes.

This book is addressed to both beginners in the field of earthquake research and those who wish to explore new methods for their studies. Beginners may find it a useful overview of how to study past earthquake events; advanced users can benefit from in-depth discussions of pitfalls and methodological details. Practitioners and those working on seismic hazard will find information on the potential and the limits of different techniques to ascertain certain parameters of past earthquakes. Specialized scientists already familiar with one or more of the methods discussed here can get an overview on which other techniques from neighboring disciplines are possible to use for additional constraint. Reviewers of projects and scientific papers will find in these chapters a brief overview of the dos and don'ts of each method in case they are not intimately familiar with the techniques. Journalists might use it as a reference to look up methodological details of new studies that have made or should make it to the headlines.

1.2 Organization of This Book

The measurable physical phenomena of earthquakes can be grouped into two categories: the effects of transient, dynamic elastic wave propagation through the solid Earth–**ground motion** otherwise known as **shaking**–and the semi-permanent net displacement of solid Earth along tectonic faults–**ground deformation**. The techniques described in the chapters of this book are thus grouped into two sections accordingly. For contemporary earthquakes, ground motion effects are typically the domain of seismologists and earthquake engineers, while ground deformation effects are studied by geodesists, geologists, and geotechnical engineers. A similar distinction between these two categories was adopted for the Environmental Seismic Intensity Scale ESI-2007, which differentiates between primary effects (fault offset and tectonic subsidence/uplift) and secondary effects due to shaking (Michetti et al., 2007).

The first chapter in the ground motion category, on **analog seismology,** is about the wealth of data that can be gathered from analog seismograms (Kanamori, 1988). These old data can extend the instrumental record back to the early 20th Century, or even the late 19th Century (e.g., Krüger et al., 2017). Extracting information on location, mechanism, and magnitude of earthquakes is possible at least for strong events, but needs meticulous work and a good understanding of how seismic waves were measured in the nascent days of instrumental seismic recording. Naturally, event coverage declines the further back in time one wants to look. In their contribution to this volume, *Ou* et al. give an overview of what type of information can be expected from analog seismograms and how to locate, convert, process, and analyze them. They discuss published examples of such analyses for events dating back as far as 1902.

In the second chapter, *Hough* describes how **macroseismic** data, i.e. reports of earthquake damage and shaking intensity, can characterize earthquakes beyond the timescale covered by the instrumental record, and are often available even for events that happened hundreds of years ago (e.g., Caracciolo et al., 2021). However, recent research has not only shown the usefulness of these data, but also highlighted the limitations and discrepancies that arise when dealing with varied and subjective reporting of shaking effects in all manner of different environments. In this chapter, *Hough* describes in detail the various methodologies for analyzing historical accounts of shaking with their pros and cons, providing a historical perspective on this widely used approach.

Chapter three takes earthquake analyses even further back in time, showing how **archaeoseismology** can help understand past seismic events that are even older than written records by inspecting the damage to ancient sites (Galadini et al., 2006; Noller, 2001; Rodríguez-Pascua et al., 2011; Sintubin, 2011; Stiros, 1996). A major problem that archaeoseismology must confront is the attribution of damage to seismic shaking rather than having been caused by other processes such as abandonment, warfare, or other environmental phenomena. *Sintubin* summarizes how this disambiguation may be accomplished and lays out the potential of archaeoseismological techniques to define various parameters of past earthquakes.

In the last chapter on ground motions, "**Lacustrine Records of Past Seismic Shaking**," *Moernaut* et al. illustrate one of the more recent advancements in studies of past earthquakes providing some of our oldest records of ground shaking, preserved geologically. Lakebottom sedimentary records of earthquakes may be used to constrain the epicentral location and shaking distribution, and one of their major advantages is that their sedimentary record often spans many seismic cycles. *Moernaut* et al. detail the methods that relate the types, extents, and thicknesses of lake records to ground shaking parameters and discuss the advantages of multi-lake studies. Three published case studies illustrate the technique in intraplate, transform, and subduction zone settings.

The ground deformation section of this book opens with a chapter on measuring **terrestrial ruptures**. *Zielke & Klinger* address the measurement of surface offsets for constraining the along-fault distribution of surface displacement from individual earthquakes. These data are essential not only for assessing paleoearthquake magnitude, but also for informing modern estimates of shaking hazard and surface displacement hazard. Perhaps counterintuitively, measuring the offset of geomorphic markers is far from trivial. In this chapter the authors provide guidance on which types of data provide suitable imaging of these features in the landscape and discuss the challenges of what and how to measure. An additional challenge is posed by the distribution of deformation away from a main identifiable fault trace (e.g., Antoine et al., 2022), a problem which can be addressed for earthquakes that occurred within the era of aerial and satellite imaging but presents an unresolved obstacle for slip measurements in earlier quakes.

The final chapter ventures into the offshore environment; "**Paleoseismology and Paleogeodesy Using Coral Microatolls**" deals with the near-coastal record of large earthquakes as captured by the measurable uplift and subsidence of shallow-water biogenic landforms. Studies of offshore faults including subduction zones generally contend with the challenge that direct access to the fault and its surface exposure is difficult, to say the least. Thus, researchers rely on alternative indicators of paleoshaking and paleogeodesy. In this chapter, *Philibosian* explains how coral microatolls record tectonic uplift and subsidence, and hence allow extraction of parameters needed to quantify the slip, rupture extent, and timing of past events. The chapter describes how to obtain time series of relative sea level and vertical tectonic motion from coral growth records, and how this translates into estimates of paleo-fault slip and fault coupling.

These six chapters represent state-of-the-science benchmark reviews of a variety of methods for evaluating the physical parameters of past earthquakes. Each of these techniques is either recently developed, or a commonly used method in which technical, computational, or conceptual advancements have recently been realized that now permit careful interrogation of interpretation of past quakes based on these fundamental data sources. Additional methods of investigating pre-modern, pre-instrumental, and pre-historic earthquakes exist and their number is growing along with the tools we use to measure contemporary earthquakes. Many of these techniques have recently been covered by comprehensive review articles or dedicated books. For example, classical paleoseismology with all its underlying principles, its pitfalls, and its potential has been extensively covered by McCalpin's *Paleoseismology*, which has also been translated into Russian and Chinese (McCalpin, 2009), and by recent reviews (Kondo et al., 2022; McCalpin et al., 2023). Similarly, the use of environmental earthquake effects and the ESI-2007 is comprehensively covered in guidebooks, review papers, and web applications (Michetti et al., 2007; Audemard et al., 2015; Serva et al., 2015; Quigley et al., 2016; ISPRA, 2019; Ferrario et al., 2022). The topic of coseismic landslides has been detailed by Fan et al. (2019); soft-sediment deformation has been analyzed by Allen (1986) and Alsop et al. (2022), just to name a few. Studies on tsunamis as earthquake proxies constitute a subject on their own. The fields of coastal marsh paleogeodesy and paleoseismology as well as turbidite paleoseimology offer great promise in defining chronologies and effects of the largest and rarest earthquakes which occur in subduction zones, covered in recent reviews by Goldfinger (2011), Dura et al. (2016), and Sawai (2020).

Acknowledgements We thank all authors for their contributions, their dedication, and their patience. All reviewers are thanked for their work. The editors would like to thank SPRINGER for making this book project happen and countless colleagues for discussions, recommendations, and support.

References

Allen, J. R. L. (1986). Earthquake magnitude-frequency, epicentral distance, and soft-sediment deformation in sedimentary basins. *Sedimentary Geology, 46*(1–2), 67–75. https://doi.org/10.1016/0037-0738(86)90006-0

Alsop, G. I., Marco, S., & Levi, T. (2022). Recognising surface versus sub-surface deformation of soft-sediments: Consequences and considerations for palaeoseismic studies. *Journal of Structural Geology, 154*, 104493. https://doi.org/10.1016/j.jsg.2021.104493

Ambraseys, N. (2009). *Earthquakes in the Mediterranean and Middle East: A multidisciplinary study of seismicity up to 1900*. Cambridge University Press. https://doi.org/10.1017/CBO9781139195430

Antoine, S. L., Klinger, Y., Delorme, A., & Gold, R. D. (2022). Off-fault deformation in regions of complex fault geometries: The 2013, Mw7.7, Baluchistan rupture (Pakistan). *Journal of Geophysical Research: Solid Earth, 127,* e2022JB024480. https://doi.org/10.1029/2022JB024480

Audemard, F., Azuma, T., Baiocco, F., & Baize, S. (2015). Earthquake Environmental Effects for seismic hazard assessment: The ESI intensity scale and the EEE Catalogue; ISPRA: Rome, Italy, Vol. 97.

Butler, R., Lay, T., Creager, K., Earl, P., Fischer, K., Gaherty, J., Laske, G., Leith, B., Park, J., Ritzwolle, M., Tromp, J., & Wen, L. (2004). The global seismographic network surpasses its design goal. *Eos, Transactions American Geophysical Union, 85*(23), 225–229. https://doi.org/10.1029/2004EO230001

Caracciolo, C. H., Slejko, D., Camassi, R., & Castelli, V. (2021). The eastern Alps earthquake of 25 January 1348: New insights from old sources. *Bulletin of Geophysics and Oceanography, 63*(3), 335–364. https://doi.org/10.4430/bgo00364

Cheng, J., Rong, Y.-F., Magistrale, H., Chen, G.-H., & Xu, X.-W. (2017). An Mw-based historical earthquake catalog for mainland China. *Bulletin of the Seismological Society of America, 107*(5), 2490–2500. https://doi.org/10.1785/0120170102

Dura, T., Hemphill-Haley, E., Sawai, Y., & Horton, B. P. (2016). The application of diatoms to reconstruct the history of subduction zone earthquakes and tsunamis. *Earth-Science Reviews, 152*(1), 181–197. https://doi.org/10.1016/j.earscirev.2015.11.017

England, P., & Jackson, J. (2011). Uncharted seismic risk. *Nature Geoscience, 4*(6), 348–349. https://doi.org/10.1038/ngeo1168

Fan, X., Scaringi, G., Korup, O., West, A. J., van Westen, C. J., Tanyas, H., Hovius, N., Hales, T. C., Jibson, R. W., Allstadt, K. E., & Zhang, L. (2019). Earthquake-induced chains of geologic hazards: Patterns, mechanisms, and impacts. *Reviews of Geophysics, 57*(2), 421–503. https://agupubs.onlinelibrary.wiley.com/doi/full/10.1029/2018RG000626.

Ferrario, M. F., Livio, F., & Michetti, A. M. (2022). Fifteen years of environmental seismic intensity (ESI-07) scale: Dataset compilation and insights from empirical regressions. *Quaternary International, 625*, 107–119. https://doi.org/10.1016/j.quaint.2022.04.011

Galadini, F., Hinzen, K. G., & Stiros, S. (2006). Archaeoseismology: Methodological issues and procedure. *Journal of Seismology, 10*(4), 395–414. https://doi.org/10.1007/s10950-006-9027-x

Goldfinger, C. (2011). Submarine paleoseismology based on turbidite records. *Annual Review of Marine Science, 3*(1), 35–66. https://doi.org/10.1146/annurev-marine-120709-142852

Hayes, G. P., Earle, P. S., Benz, H. M., Wald, D. J., & Briggs, R. W., the USGS/NEIC Earthquake Response Team (2011). 88 Hours: The U.S. Geological Survey National Earthquake Information Center Response to the 11 March 2011 Mw 9.0 Tohoku Earthquake. *Seismological Research Letters, 82*(4), 481–493. https://doi.org/10.1785/gssrl.82.4.481.

ISPRA. The Earthquake Enviromental Effects (EEE) Catalogue (2019). Retrieved 15 June, 2022, from http://193.206.192.211/wfd/eee_catalog/viewer.php.

Kanamori, H. (1988). Importance of historical seismograms for geophysical research. *Historical seismograms and earthquakes of the world* (pp. 16–33). Academic Press.

Kondo, H., Owens, L. A., & Figueiredo, P. M. (2022). Paleoseismological studies. *Treatise on Geomorphology (Second Edition)*. Academic Press, pp. 495–562. https://www.sciencedirect.com/science/article/pii/B9780128182345001565.

Krüger, F., Kulikova, G., & Landgraf, A. (2017). Instrumental magnitude constraints for the 11 July 1889, Chilik earthquake. *Geological Society, London, Special Publications, 432*(1), 41–72. https://doi.org/10.1144/SP432.8

Lee, W. H. K., Wu, F. T., & Jacobsen, C. (1976). A catalog of historical earthquakes in China compiled from recent Chinese publications. *Bulletin of the Seismological Society of America, 66*(6), 2003–2016. https://doi.org/10.1785/BSSA0660062003

Ludwin, R. S., Smits, G. J., Carver, D., James, K., Jonientz-Trisler, C., McMillan, A. D., Losey, R., Dennis, R., Rasmussen, J., De Los Angeles, A., Buerge, D., Thrush, C. P., Clague, J., Bowechop, J., & Wray, J. (2007). Folklore and earthquakes: Native American oral traditions from Cascadia compared with written traditions from Japan. *Geological Society, London, Special Publications, 273*(1), 67–94. https://doi.org/10.1144/GSL.SP.2007.273.01.07

McCalpin, J. P. (2009). Paleoseismology. 2nd Ed. *International Geophysics, 95.*

McCalpin, J., Ferrario, F., Figueiredo, P., Livio, F., Gruetzner, C., Pisarska-Jamroży, M., Quigley, M., Reicherter, K., Rockwell, T., Štěpančíková, P., & Tábořík, P. (2023). New developments in onshore paleoseismic methods, and their impact on Quaternary tectonic studies. *Quaternary International, 10*, 59–76. https://doi.org/10.1016/j.quaint.2023.03.008

Michetti, A. M., Esposito, E., Guerrieri, L., Porfido, S., Serva, L., Tatevossian, R., Vittori, E., Audemard, F., Azuma, T., Clague, J., et al. (2007). Intensity scale ESI 2007. In Memorie Descrittive Carta Geologica d'Italia. Servizio Geologico d'Italia; Guerrieri, L., Vittori, E., Eds.; 74. Dipartimento Difesa del Suolo, APAT: Rome, Italy; Vol. 74, p. 53.

Noller, J. S. (2001). Archaeoseismology. In *Earth sciences and archaeology* (pp. 143–170). Springer, Boston, MA. https://doi.org/10.1007/978-1-4615-1183-0_6

Quigley, M. C., Hughes, M. W., Bradley, B. A., van Ballegooy, S., Reid, C., Morgenroth, J., Horton, T., Duffy, B., & Pettinga, J. R. (2016). The 2010–2011 Canterbury earthquake sequence: Environmental effects, seismic triggering thresholds and geologic legacy. *Tectonophysics, 672*, 228–274. https://www.sciencedirect.com/science/article/abs/pii/S0040195116000962.

Rodríguez-Pascua, M. A., Pérez-López, R., Giner-Robles, J. L., Silva, P. G., Garduño-Monroy, V. H., & Reicherter, K. (2011). A comprehensive classification of Earthquake Archaeological Effects (EAE) in archaeoseismology: Application to ancient remains of Roman and Mesoamerican cultures. *Quaternary International, 242*(1), 20–30. https://doi.org/10.1016/j.quaint.2011.04.044

Salisbury, J. B., Haddad, D. E., Rockwell, T., Arrowsmith, J. R., Madugo, C., Zielke, O., & Scharer, K. (2015). Validation of meter-scale surface faulting offset measurements from high-resolution topographic data. *Geosphere, 11*(6), 1884–1901. https://doi.org/10.1130/GES01197.1

Sawai, Y. (2020). Subduction zone paleoseismology along the Pacific coast of northeast Japan—Progress and remaining problems. *Earth-Science Reviews, 208*, 103261. https://doi.org/10.1016/j.earscirev.2020.103261

Sbeinati, M. R., Darawcheh, R., & Mouty, M. (2005). The historical earthquakes of Syria: An analysis of large and moderate earthquakes from 1365 B.C. to 1900 A.D. *Annals of Geophysics, 48*(3). https://doi.org/10.4401/ag-3206

Serva, L., Vittori, E., Comerci, V., Esposito, E., Guerrieri, L., Michetti, A.M., Mohammadioun, B., Mohammadioun, G. C., Porfido, S., & Tatevossian, R. E. (2015). Earthquake Hazard and the Environmental Seismic Intensity (ESI) Scale. *Pure Appl. Geophys., 173,* 1479–1515. https://doi.org/10.1007/s00024-015-1177-8

Sintubin, M. (2011). Archaeoseismology: Past, present and future. *Quaternary International, 242*(1), 4–10. https://doi.org/10.1016/j.quaint.2011.03.056

Stirling, M., Goded, T., Berryman, K., & Litchfield, N. (2013). Selection of earthquake scaling relationships for seismic-hazard analysis. *Bulletin of the Seismological Society of America, 103*(6), 2993–3011. https://doi.org/10.1785/0120130052

Stiros, S. C. (1996). Identification of earthquakes from archaeological data: Methodology, criteria and limitations. *Archaeoseismology, 7*, 129–152.

Wells, D. L., & Coppersmith, K. J. (1994). New empirical relationships among magnitude, rupture length, rupture width, rupture area, and surface displacement. *Bulletin of the Seismological Society of America, 84*(4), 974–1002. https://doi.org/10.1785/BSSA0840040974

Yeats, R. (2012). *Active faults of the world*. Cambridge University Press. https://doi.org/10.1017/CBO9781139035644

Zare, M., Amini, H., Yazdi, P., et al. (2014). Recent developments of the Middle East catalog. *Journal of Seismology, 18*, 749–772. https://doi.org/10.1007/s10950-014-9444-1

Zielke, O., Klinger, Y., & Arrowsmith, J. R. (2015). Fault slip and earthquake recurrence along strike-slip faults—Contributions of high-resolution geomorphic data. *Tectonophysics, 638*, 43–62. https://doi.org/10.1016/j.tecto.2014.11.004

Earthquake Source Parameters Determination Using Analog Seismic Records

2

Qi Ou, Galina Kulikova and Frank Krüger

2.1 Introduction

Formal earthquake observation can be traced back at least to 132 AD in ancient China where Chang Heng invented seismoscopes to detect earthquake occurrence and estimate direction of approach. Yet, it was not until the advent of continuous seismic recording in the mid-to-late 19th century that seismology started to transform into a modern observational science. As a result, earthquakes are often classified as either historical or instrumental, with the former only evidenced in historic writings, landscapes or excavated trenches and the latter with their ground motion recorded by seismic instruments. Nevertheless, not all instrumental earthquakes are equal. The range and quality of earthquake source parameters determined through time are greatly influenced by the historical stages of seismic instrumentation and accompanied theoretical advancements.

Early instrumental earthquakes (1889—mid-to-late 20th century) were recorded on paper by analog seismographs of a variety of instrumental designs, often with low

Q. Ou (✉)
School of GeoSciences The University of Edinburgh, Drummond Street, EH8 9XP Edinburgh, Scotland
e-mail: qi.ou@ed.ac.uk

G. Kulikova
Institute of Geociences University of Potsdam, Karl-Liebknecht-Strasse 24, 14476 Potsdam-Golm, Germany
e-mail: Galina.Kulikova@geo.uni-potsdam.de

F. Krüger
Institute of Geociences University of Potsdam, Karl-Liebknecht-Strasse 24, 14476 Potsdam-Golm, Germany
e-mail: Frank.Krueger@geo.uni-potsdam.de

A. Elliott and C. Gruetzner (eds.), *Understanding Past Earthquakes*,
https://doi.org/10.1007/978-3-031-73580-6_2

damping, low dynamics, and limited bandwidth; studies were often based on limited datasets measured by hand. After nearly a century of instrumental developments, the first seismometer network with unified instrumental parameters, the World-Wide Standardized Seismographic Network (WWSSN), was installed in the 1960s; the paper records were regularly copied onto microfilm and microfiche for effective data dissemination. The most revolutionary change came in the mid-to-late 1970s when digital broadband seismic stations started operating worldwide. The digital transformation drastically increased the dynamic ranges of seismological recording, from what was limited by the width of the paper and the tip of the recorder to ranges covered by 8-bit, 16-bit and nowadays 24-bit digital counts. The invention of force-feedback broadband seismometers enabled the entire teleseismic spectrum to be recorded by a single instrument (Ackerley 2014). Since then, the global digital records, in the form of time series disseminated via repositories such as IRIS, GEOSCOPE, and GEOFON, enabled rapid determination of earthquake focal mechanisms and source time functions through robust computational analysis. Today, rupture processes and fault structures as illuminated by dense seismic arrays are playing an increasingly important role in our understanding of earthquake sources, mechanics, and kinematics, giving us information unimaginable in the early instrumental era.

Over the 130 years of the instrumental period, however, considerably more (about 2/3 of the total number of) large magnitude earthquakes ($M \geq 7.5$) were recorded by analog instruments only (Fig. 2.1). In regions like California, the Caribbean, west of Portugal, Central Asia, India, and Mongolia (blue rectangles), large earthquakes that happened in analog recording times were not followed by any modern earthquakes of such magnitudes in the digital era. In China and Alaska, modern earthquakes recorded were much smaller than their analog counterparts. The two largest earthquakes of the world (1960-Valdivia (Chile) Earthquake (M9.5) and 1964 Great Alaska Earthquake (M9.2) U.S. Geological Survey 2017) also occurred prior to the digital era. Reanalysing these analog records with modern computational tools can greatly refine and add to our knowledge about these significant events, increase the sampling of the seismic cycle, and improve the seismic hazard assessments in these regions.

Earthquakes in the early instrumental period also form an important bridge between the modern seismological and pre-instrumental eras as they permit direct correlation of early instrumental seismic data, historical documentation, and field investigations. Such cross-comparison of independent paleoseismological, geomorphic, historical and seismological datasets commonly presents great opportunities for identifying causative faults for historical events and resolving contradictions in existing literature (e.g., Kulikova and Krüger 2015; Ou et al. 2020). By reconciling source parameters derived from different research methods, earthquakes in the turn of the 20th century can help calibrate and connect source parameters obtained for events before and after the instrumental period, extending our knowledge about events further back in time in a homogeneous manner.

Because of the clear benefits of analysis using this underutilized resource, this chapter gives an overview on methodologies for extracting earthquake source parameters from legacy analog seismograms, with the aim to help interested readers strategize for and plan historical seismological projects. We start with a description of

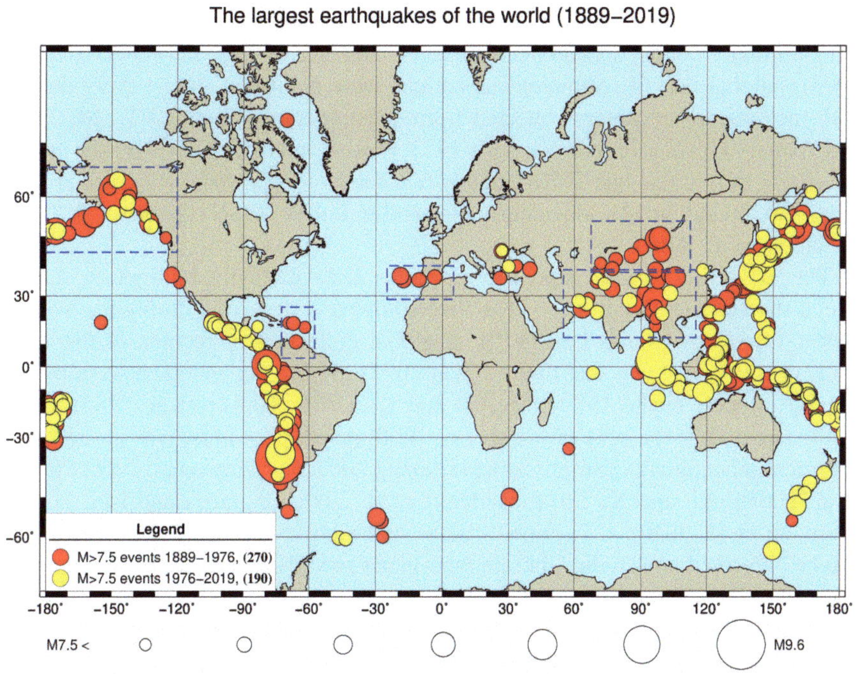

Fig. 2.1 The largest (magnitude 7.5 and larger) earthquakes of the world which occurred between 1889 (the first successful teleseismic earthquake records, Japanese earthquake recorded in Potsdam, Germany) and 2019. Rectangular frames show the regions where large earthquakes occurred only at the analog recording times or the regions where very few smaller magnitude modern earthquakes exist (blue rectangles)

the existence and availability of analog data recorded in space and time (Sect. 2.2), followed by a summary of the essential requirements that make an analog record suitable for determining earthquake source parameters (Sect. 2.3.1). Then, we delve into ways to implement or tailor modern methods to suit the characteristics of analog data (Sects. 2.3.2–2.3.5), before highlighting data sources and existing tools for faster and more productive usage of analog seismic data (Sect. 2.4). In the final section, we present some select interesting earthquake examples from our practice.

2.2 Data Availability

At the end of 19th and beginning of the 20th century—following the ground-breaking instrument developments by seismologists John Milne, Emil Wiechert and Fusakichi Omori then working in Germany and Japan (Dewey and Byerly 1969; Howell 1990; Ferrari 1992; Batlló 2014)—a global network of seismic stations already existed.

By the year 1920, at least 245 seismic stations were operational worldwide, covering all major continents and many oceanic islands (Fig. 2.2, (Schweitzer and Lee 2003)). Global lists of seismological stations were published almost every decade with instrument parameters collected from questionnaires (Wood 1921; McComb and West 1931; Charlier and Van Gils 1953; Powell and Fries 1964). By the end of the analog period (ca. late 1970s), nearly 600 seismological stations existed around the world, including 121 added during the roll-out of WWSSN (Glover et al. 1985; Alejandro et al. 2019).

The fate of these analog paper records varies. Data at Manila, Philippines, station which started operating in the 1880s were completely destroyed during the Second World War (Meyers and Lee 1979). A global collection of records of the 1960 Chilean earthquake was lost to fire (Okal 2005). Before the historical seismogram filming projects in the 1980s (Meyers and Lee 1979; Glover et al. 1985), those lost data could be the only physical copy. With recent pilot initiatives in digitising historical seismograms (Bromirski and Chuang 2003; Michelini et al. 2005; Pintore et al. 2005a; Xu and Xu 2014; Paulescu et al. 2016; Bogiatzis and Ishii 2016a; Murotani et al. 2020), it is hoped that as many legacy seismograms as possible can be preserved and studied, before their paper records as well as the decades-old microfilms deteriorate beyond usefulness.

Here we present a non-exhaustive list (white triangles in Fig. 2.2) of global, regional, national, network or station archives where historical seismograms are known to be available. Their contact details and inventory summaries can be found in Table 2.3.

2.3 Earthquake Source Parameters Determination

2.3.1 Data Requirements

As one might expect, analog seismograms are not famous for their convenience of use or completeness of metadata. Therefore, analysing them is an art of getting the most value out of an imperfect dataset, or finding methods that demand the least from the input data. To help interested readers strategize their data acquisition and digitisation efforts, we propose the following list of minimal and essential data requirements for each of the earthquake source parameters to be determined (Table 2.1). These data requirements are summarised from methods to be covered in the sections below and could serve as a check list for data preparation before the actual analysis.

Good azimuthal coverage of the stations helps constrain the earthquake start time, epicenter, and focal mechanism. Equal-distant stations in opposite azimuths are telling of rupture directivity. Dense seismic array in one azimuth helps reveal rupture subevents. While absolute timing (time labeling) is essential for calculating the event origin time and for aligning waveforms from seismic arrays, it is the relative timing (clear hour and minute marks) that is more important for epicenter relocation and period measurements for magnitude calculation. *Where waveform*

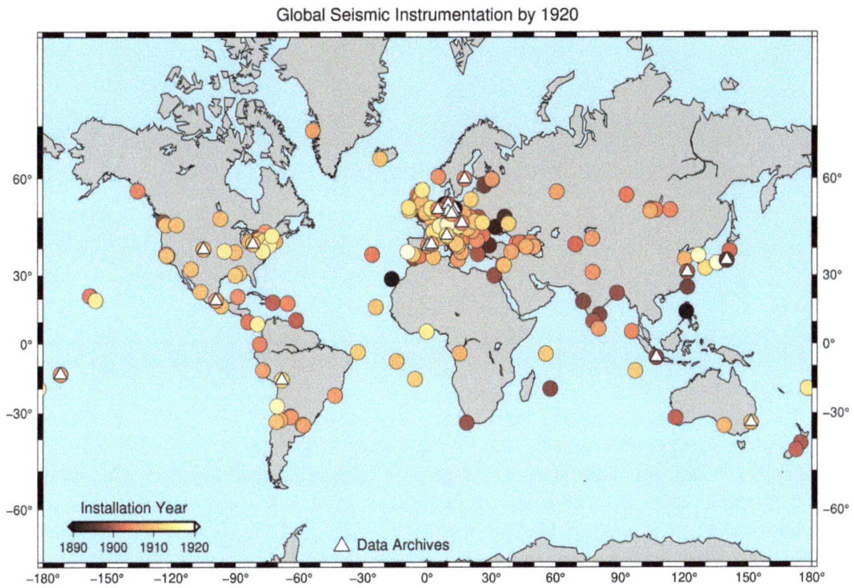

Fig. 2.2 Spatial and temporal roll-out of seismic instrumentation prior to 1920, based on bulletin data in (Schweitzer and Lee 2003). White triangles highlight archives hosting historical seismograms with contact details listed in Table 2.3

Table 2.1 Features required for determining each of the source parameters

Essential features	Origin Time	Epicenter	Magnitude	Mechanism	Kinematics
Station distribution	√	√		√	√
Absolute timing	√				√
Relative timing		√	√		√
True scale			√		
Instrument response			√	Either/	√
Orientation				Or	√

amplitude is required, the true scale of the scanned seismogram becomes an essential feature, which can be lost if the data is scanned with unknown resolution or without evidence of scale. Without knowing the instrumental responses, the amplitude information becomes meaningless and waveform modelling cannot be done. Without confirmed orientation of the record, the inferred mechanism and rupture kinematics can be wrong, although, as to be explained later, there are ways for deducing record orientations from other evidence.

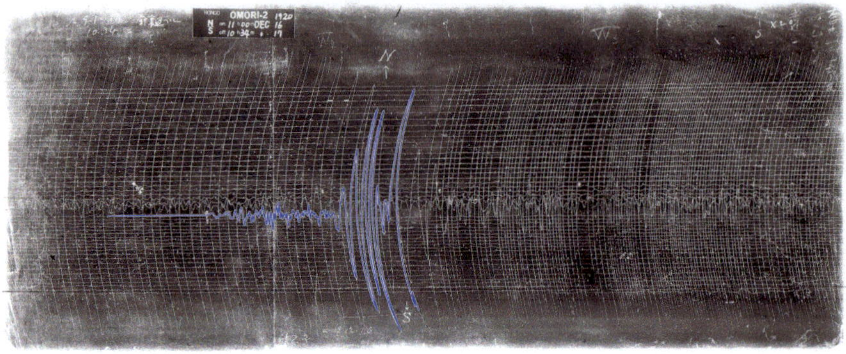

Fig. 2.3 An example smoked-paper horizontal seismogram of the December 16, 1920 Haiyuan (Gansu, China) earthquake recorded at the Hongo station (Tokyo, Japan) by an Omori type seismometer. This record was microfilmed by Earthquake Research Institute (ERI), The University of Tokyo during the Historical Seismogram Filming Project in the 1980s, and scanned from a microfilm reader at the USGS in 2017. The Japanese handwriting on the top left corner noted "needle flipped upside-down", referring to the trace discontinuity where we stopped digitising (in blue). Notes on the top right says "teleseismic" for this event happened out of Japan

In the example shown in Fig. 2.3, absolute timing is given on the plate at the top and relative timing is indicated by the white dots. The orientation of the record is up to the north and down to the south, as indicated on the plate and labeled next to the waveforms with arrows. The "× 20" scribble on the top right gives clue to the magnification of this Omori-2 seismograph. As this record was secondarily scanned from a microfilm instead of the original copy, and there is no evidence for scale on the image, the dimension of the seismogram cannot be derived from the image alone. Therefore, this record could be used for epicenter, focal mechanism and kinematic inversions (with other instrument parameters collected from Wood 1921), but is not suitable for magnitude determination.

The general procedure for converting such an analog into digital time series involves manually or semi-automatically tracing the useful portion of the waveform, correcting geometric distortions (e.g., slant and curvature typical of records from mechanical seismographs, Fig. 2.3), converting units from scan pixels into displacement or velocity versus time and finally assigning absolute record start time. More details of the conversion process between analog seismogram and digital ground motion time series will be covered in Sect. 2.4. Here we first proceed to how these data can be analysed to derive desired earthquake source parameters.

2.3.2 Epicenter Relocation

The epicenter or hypocenter locations of earthquakes play an important role in identifying responsible faults and deducing rupture directivity. The earliest earthquake

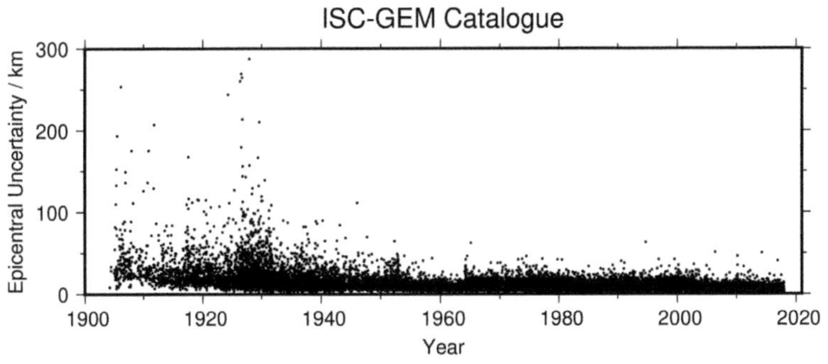

Fig. 2.4 Major axis of error ellipses of relocated epicenters in the International Seismological Centre—Global Earthquake Model (ISC-GEM) catalogue (International Seismological Centre 2022) showing large location uncertainties in the early instrumental period

locations were determined using a physical globe with station bearings and with very limited reported seismic arrival times. Early epicentral locations were often reported without uncertainties, though realistically uncertainties could be >100 km. With growing numbers of seismic stations worldwide, more refined velocity models, and more advanced algorithms for location determination, earthquakes are located increasingly accurately. Modern-day earthquakes are now located with uncertainties between 100 s of meters to 10 s of kilometers depending on the density of seismic stations surrounding the earthquakes (U.S. Geological Survey). However, for early instrumental earthquakes (pre-1960s), even after relocation with the latest velocity model and location procedure (Engdah et al. 1998; Bondár and Storchak 2011; Bondár et al. 2015) and with newly digitised phase arrival times from resources available to the International Seismological Centre (ISC) (Villasenor and Engdahl 2005; Villaseñor and Engdahl 2007; Di Giacomo et al. 2015; Bondár et al. 2015), the earthquake location uncertainties are still up to 10 s to 100 s of kilometers due to small dataset, incomplete azimuthal coverage, and timing errors (Fig. 2.4) (International Seismological Centre 2022). Therefore, the hypocenter locations of earthquakes before the 1960s, especially those before the 1940s, can still be improved further with phase arrival times from additional station bulletins and new phase time readings from newly recovered seismograms.

To determine the source location, the essential information includes (1) the times at which different phases of the wave energy are received at the stations, and (2) the latitudes and longitudes of the stations at the time. Seismic stations may have moved through history (e.g., the Cape Town Station moved in 1934 and again in 1950) (Okal 2015), so it is important to look for their historical locations in station bulletins, station books, and global station lists compiled quasi-decadally (e.g. Wood 1921; McComb and West 1931; Charlier and Van Gils 1953; Powell and Fries 1964). Station distribution plays an important role in constraining the earthquake epicenter. As there were more seismic stations in Europe and Japan in the early instrumental period (Fig. 2.5a–c), earthquake epicenters are usually better constrained in the direction between these two recording clusters. This was illustrated by the error ellipses of

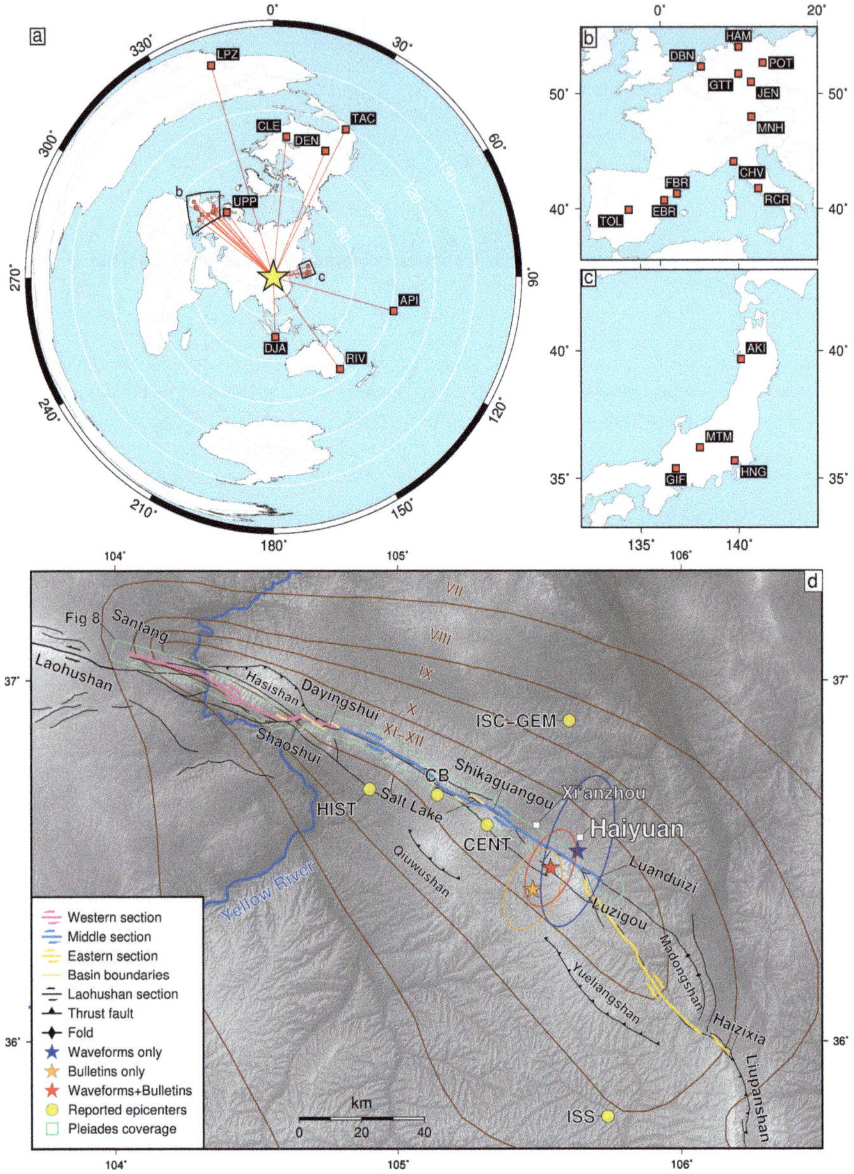

Fig. 2.5 **a** Station with collected waveforms for the 1920 Haiyuan earthquake with stations in Europe and Japan blown up in **b** and **c**, respectively. **d** Epicenter relocation for the 1920 earthquake, with stars with error ellipses showing results from different datasets and yellow dots showing previously published epicenters. Reproduced from Ou et al. (2020), licensed under CC-BY 4.0

the relocated epicenter of the 1920 Haiyuan Earthquake which happened on a 110° trending fault in central west China (Fig. 2.5d, (Ou et al. 2020)). As the earthquake was almost at the middle between these two clusters, the uncertainty in the along-strike direction is much smaller than that in the fault perpendicular direction.

As facilities for effective data transfer and sharing were not established until the WWSSN era, it is not uncommon to find previously unpublished seismograms or undigitised station bulletins and hence obtain station arrival time data that were not included in the ISS or ISC bulletins. The bulletins sometimes report different arrival times associated with the same phase name for the same station, possibly due to the ambiguity of the phase onset time on the seismograms or ambiguity as to which phase name a phase pick should be attributed to. Therefore, it is beneficial to perform phase picking on the original seismograms, and check out the additional notes in the station logbook and station bulletins which did not enter the compiled phase times in the global catalogues and bulletins.

Timing and clock information can be found in a range of sources from station bulletins to the logbooks and to the seismograms themselves. Most seismograms have a header at the top with stations and component names, and possibly paper start and end dates and times (Fig. 2.3). At most stations, seismogram papers were replaced every day such that all lines on a seismogram total approximately 24 h. Depending on the rotating speed of the drums, typically one round every hour or every half an hour, each line on the seismogram lasts approximately half or an hour. This information can be verified through line counts on the paper as well as the hour labels which are sometimes written on the paper at the start of each hour (Fig. 2.6). Different seismographs record units of time (minutes and hours) on the paper in different ways, some with gaps (Fig. 2.7), some with spikes, some with squiggles and some with separate strokes (Fig. 2.6). Markers of different types or sizes were used to represent a minute, half an hour and an hour. Clues can be found in the parts of the seismogram where the trace is in equilibrium and registering the background noise without the disruption of an earthquake, in which case the timing markers become obvious.

Apart from the markers, scribbles related to phase time, phase labels and hour labels on the seismograms can also help us match the waveforms in print onto the time axis (Fig. 2.6). It is worth noting that occasionally clock adjustments are written on the seismograms (Fig. 2.6). They are there because before radio time transfer became available in the 1920s, most analog seismic stations ran their own clocks and could only calibrate local timing against a central clock from time to time (Agnew 2020; Lee et al. 2020). Apart from incorporating the time shifts according to any written time adjustments, Lee et al. (2020) also developed a method to correct relative timing between seismograms from different stations using the ambient noise correction function.

Sometimes the spacing between minute marks on the image varies across the seismogram, due to either the unstable velocity of the drum rotation or artifacts from the scanning/microfilming process. A bespoke digitising software, DigitSeis (Bogiatzis and Ishii 2016b), digitises the stroke-type minute marks along with the waveform. The digitised minute marks evenly stretch each minute of the record to

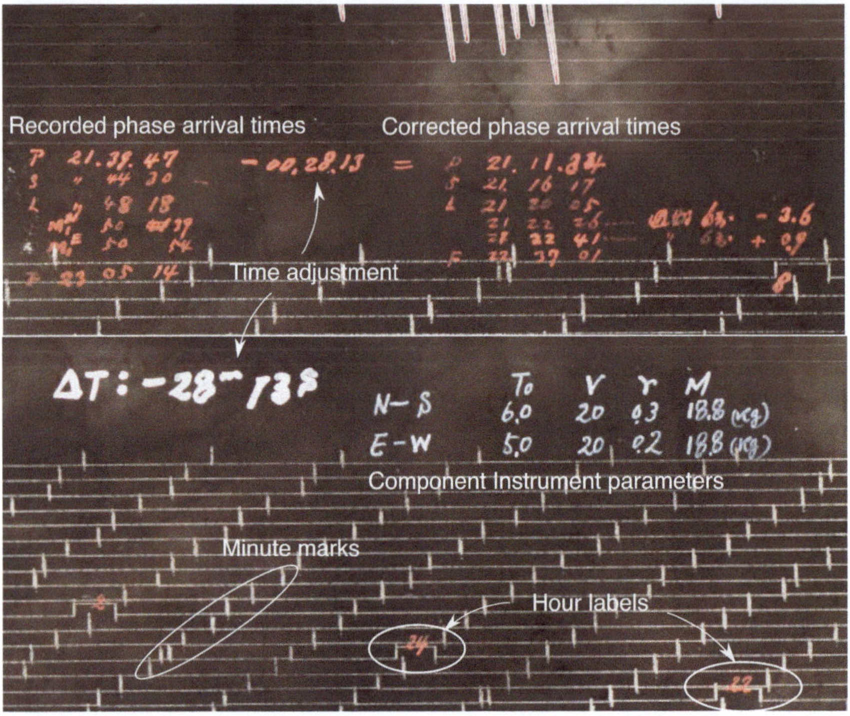

Fig. 2.6 Portions of a seismogram recorded at the MTM station showing minute marks, hour marks and the clock adjustment. This is the only record out of all 50 seismograms collected for the 1920 Haiyuan Earthquake that had a clock adjustments written on the seismogram. The component instrument parameters are also labeled

equal length in the time series. However, digitising minute marks in the forms of squiggles or gaps can be more challenging. For those seismograms that cannot have time marks easily digitised, one simplistic treatment is to measure the minute marks as closely to the earthquake waveform as possible and measure the lengths of a few minutes to take an average. The focus should be on the length of the waveforms that are actually digitised, which is typically the continuous signal from minutes before the first arrival to just before the waveform becomes saturated or broken (Fig. 2.3).

It is important to digitize a sufficient (several minutes) stretch of the record before the onset of the first seismic wave arrival in order to record the general trend of the trace and retain information about the background noise. Setting the record to begin at a whole minute also makes it easier to assign record time based on a clear reference point as compared to calculating an artificially precise second of the onset time. Tools and methods for digitising seismograms also typically correct for the gradient of the digitised trace to account for the helical record on the seismogram when wrapped around the drum, such that the x-axis of the digitised waveform can be converted to the time axis with a simple unit conversion.

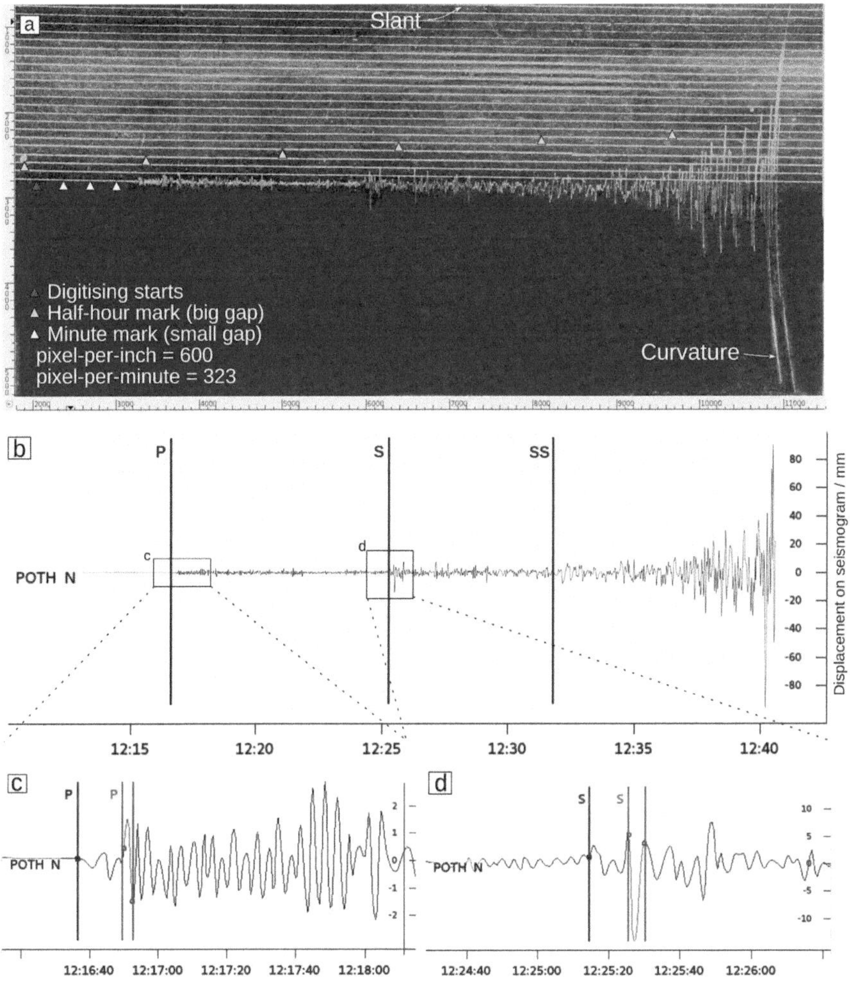

Fig. 2.7 **a** Scanned seismogram from Potsdam Station in pixel coordinates. **b** Digitised waveform in mm as a function of time after curvature correction. Red markers with P, S and SS labels are the picked phase arrival times in the Snuffler browser of the Pyrocko toolbox (Heimann et al. 2017). **c**, **d** Enlargements of the boxes in **b**. Green markers highlight the amplitude (measured from zero) and period (crest-to-crest or trough-to-trough) measurements for m_B calculation in Sect. 2.3.3.1. Reproduced from Ou et al. (2020), licensed under CC-BY 4.0

Once the waveforms are digitised, phase picking for epicenter location can be performed in much the same ways as in modern seismology (Fig. 2.7). As stations at different epicentral distances expect different sets of phases, the reported epicenter in the catalogues can be used to calculate an estimate of the epicentral distance for each station. The distances can be used to estimate the approximate arrival times of the expected phases for each record. Nevertheless, it is important to pick the phase arrival times independently from the calculated expected phase arrival times, as otherwise, the same epicenter used only as a guide will be obtained from the newly picked phase arrival times. Standard procedure for phase picking can be found in (Bormann et al. 2014). In addition, due to the aforementioned potential clock errors at the stations, the *relative times* between different phase times are more useful than the *absolute time* in constraining the location of the epicenters. Therefore, we pick arrival times of multiple body wave phases (P, PP, S, SS, SH, SV, SP etc.) when appropriate and visible rather than just the initial first arrivals. Then, there are many choices which epicenter relocation algorithm and velocity model to use, just as in modern seismology.

2.3.3 Magnitude Recalculation

There are a variety of ways to measure earthquake magnitude, which depends on different elements of a seismic record. Because the quality and availability of these elements differ amongst early instrumental records, some magnitude types might be more suitable or measurable than others. The magnitude of an earthquake can be measured in terms of local magnitude, M_L (Richter 1935), teleseismic surface wave magnitude M (later called M_S) and body wave magnitude m (later called m_B) (Gutenberg 1945b, a, c), as well as moment magnitude, M_W (Hanks and Kanamori 1979; Kanamori 1977). Gutenberg's surface wave magnitude scale underwent multiple revisions through history, with the formulation by Vanek et al. (1962) becoming the modern day standard. Gutenberg's broadband body-wave magnitude m_B was adapted to become the short-period narrow-band body-wave magnitude m_b which better suits the World-Wide Standardized Seismograph Network Short Period (WWSSN-SP) instruments installed in the 1960s.

Given the long-term development of the various magnitude scales, one can often find multiple disparate magnitudes assigned to a given earthquake in the early years of the instrumental period (Ou et al. 2020). Some of the differences arise from the choice of magnitude scales, whereas others reflect real errors due to incorrect instrument parameters or shortcomings of preliminary techniques. In this section, we first summarise the various classical magnitude formulae and how such magnitudes can be re-measured from historical seismograms. Then, we discuss ways moment magnitudes can be estimated for historical earthquakes.

2.3.3.1 Empirical Magnitude Formulae

The various empirical magnitude formulae share a common form of $M = log_{10}X + Y + a$ (Table 2.2), suggesting they are based on a similar physics about the waveform property. The X term is related to phase amplitude A and period T measured from the seismograms, Y is related to the distance Δ between the epicenter and the seismic station, and a is a calibration constant. The exact expression for X varies between the magnitude scales, and can be one of $\left(\frac{A_{max}}{T}\right)$, $\left(\frac{A}{T}\right)_{max}$, A_{max} and $\left(\frac{V_{max}}{2\pi}\right)$. Detailed standard procedures can be found in Bormann et al. (2013a) for modern records and in Bormann (2012) for early records.

For a simple harmonic oscillator, the displacement over time is $x(t) = Acos(\omega t)$, and the velocity over time is the derivative of $x(t)$, so $v(t) = -A\omega sin(\omega t)$. With $\omega = \frac{2\pi}{T}$, $\frac{V_{max}}{2\pi}$ is equivalent to $\frac{A}{T}_{max}$ by definition. This is also equivalent to $\frac{A_{max}}{T}$, if the recorded period T is filtered down to within a narrow desired range like in the case of the long-period and short-period seismographs of the WWSSN. This quantity can be further simplified to A_{max} if the period T is a known constant which can be absorbed into the calibration constant b. As such, these formulae are all trying to measure energy released far away through the maximum ground velocity experienced at the station. Which formula is applicable is a choice we would have to make in view of the historical data at hand. Here we aim to assist the readers to better understand the differences between classical and modern approaches and make decisions based on the events they are interested in and the data available.

First of all, it is important to note the type of seismograph that recorded the waveform. Early historical seismograms were typically registered by two main categories of seismographs, mechanical and electromagnetic. Mechanical seismographs, such as the Wiechert, Bosch-Omori and Mainka instruments among others (Reid 1912), amplify ground motion through levers that push a stylus to inscribe white lines on a black smoked paper wrapped on a rotating drum. Their waveforms tend not to be purely sinusoidal, but slightly slanted depending on the angle between the stylus and the drum and with curved swings most evident in the large-amplitude surface waves. These seismographs record amplified ground displacements, hence should apply the formulae with the $\frac{A}{T}$ term. Electromagnetic seismographs, represented by the Galitzin instrument, record ground velocity as black lines on white photographic paper based on the principle of induced current in an electrical coil when the coil moves in a magnetic field. These records should use the formulae with the $\frac{V_{max}}{2\pi}$ term instead.

Secondly, unlike the modern standard magnitude formulae which take measurements exclusively from the vertical components (Bormann et al. 2013a), their classical counterparts permit measurements from the horizontal components (Bormann and Dewey 2012; Bormann 2012). This is because very few stations in the early instrumental period had seismographs that recorded vertical motions. Fortunately, the agreement between surface wave magnitudes derived from horizontal and vertical components (Hunter 1972) means no correction is needed between them. For the classical broadband body-wave magnitudes m_B, tables of different calibration constants exist for measurements from horizontal and from vertical components respectively to take care of the effect of the polarization angle of the incoming ray

Table 2.2 (a) Modern standard magnitude formulae recommended by the Working Group on Magnitudes of the International Association of Seismology and Physics of the Earth's Interior (IASPEI) Commission on Seismological Observation and Interpretation (CoSOI) (Bormann et al. 2013a), where A is ground displacement amplitude in nm and V is ground velocity in nm/s measured from the vertical component, Δ is epicentral distance in km for ML and in arc-degree for teleseismic magnitudes, h is the hypocentral depth, Q is a attenuation function the values of which can be read from Table 2.2 in Bormann et al. (2013a) for measurements from the vertical component. Other equivalent nomenclatures used by various agencies and publishers are compiled in Table 2.3 of Bormann et al. (2013a). (b) Historically used classical magnitude formulae listed in Bormann et al. (2013b), where A is ground displacement amplitude in μm, with A_H denoting horizontal amplitude, Δ is epicentral distance in km for ML and in arc-degree for teleseismic magnitudes, Q is a attenuation function the values of which can be read from Table 7 in Bormann 2012 for measurements from P, PP and S waves from both the vertical and horizontal components

(a) Modern Standard Formulae	T/s	Δ	Remarks
ML = $\log(A) + 1.11 \log \Delta + 0.00189\Delta - 2.09$		< 1000 km	Southern California local
ML = $\log(A) + C(\Delta) + D$		< 1000 km	Elsewhere local
mB = $\log(V_{max}/2\pi) + Q(\Delta, h)$	0.2–30	20° – 100°	Broadband body wave
mb = $\log(A_{max}/T) + Q(\Delta, h)$	< 3	20° – 100°	Narrow band body wave
Ms_BB = $\log(V_{max}/2\pi) + 1.66 \log \Delta + 0.3$	3–60	2° – 160°	Broadband surface wave
Ms_20 = $\log(A_{max}/T) + 1.66 \log \Delta + 0.3$	18–22	20° – 160°	Narrow band surface wave
(b) Classical Formulae			
mB = $\log(A/T)_{max} + Q(\Delta, h)$	2–20	0° – 180°	Gutenberg (1945a)
Ms = $\log(A_H/T)_{max} + 1.66 \log \Delta + 3.3$	2–30	1° – 160°	Vanek et al. (1962)
Ms = $\log(A_H)_{max} + 1.66 \log \Delta + 3.3$	18–22	20° – 130°	Gutenberg (1945b)

path on the partitioning between displacements in the vertical and horizontal directions (Gutenberg 1956; Bormann 2012). It is important to note that, as there are two orthogonal horizontal components that partition the horizontal displacement, it is the vectorial sum of the horizontal amplitudes measured from the two components at the same time that should enter in the horizontal formulae. Gutenberg's initial approach of vectorially combining the highest amplitudes from the two horizontal components regardless of timing (Gutenberg 1945b) contributed to one of the reasons that many early surface wave magnitudes were overestimated (Bormann 2012; Geller and Kanamori 1977; Abe 1981).

The situation is trickier when only one horizontal component is available. The common, but less satisfying, approach for the surface wave magnitude is to multiply the maximum $\frac{A}{T}$ from the available component by $\sqrt{2}$, assuming equal amplitude at the same time on the orthogonal component (Gutenberg 1945b; Storchak et al. 2012). However, this likely results in overestimation as surface wave maxima in the two orthogonal horizontal components, typically corresponding to Love and Rayleigh waves respectively, tend not to occur at the same time (Bormann 2012). Nevertheless, the single horizontal component surface wave magnitude, with and without the $\sqrt{2}$, can be regarded as the upper and lower bounds respectively from the station. When ample data are available, better results can come from stations with vertical and/or both horizontal components. For the broadband body wave magnitude m_B, this problem can be handled with more confidence through the back-projection of the single component amplitude to the expected vectorial horizontal amplitude based on the back-azimuth of the epicenter relative to the station. This works particularly well for the P-wave amplitude if the available component is close to the radial direction. For the S phase, the horizontal components contain the S_H signal in full amplitude and a fraction of the S_V signal perpendicular to the S_H in the horizontal plane. *We consider the magnitude back-projected from the component closer to the transverse direction, where the S_H oscillations lie, more reliable than the one back-projected from a component closer to the radial direction.* This is because in the latter case, the signal-to-noise ratio is relatively smaller and the background noise, if not negligible, would also be magnified through the trigonometry, giving an over-estimated body-wave magnitude. Therefore, when only one horizontal component is available, preference should be given to the body wave magnitude measured from the P wave if the component is closer to the radial direction, and from the S wave if the component is closer to the transverse direction (Fig. 2.8).

Thirdly, it is important to note that these formulae require displacements or velocities of the ground motion, rather the waveform amplitude on the seismogram which is the ground motion magnified by a factor of tens to thousands depending on the instrumental gain. Therefore, accurate determination of the magnitudes, be it via the formulae discussed in this section or via waveform modelling in the next section, requires knowledge of the natural period, T_o, magnification factor, V_o, and damping constant, h of the seismograph operating at the time, in addition to knowing the physical scale of the seismogram, especially for recreations via scan or film. These parameters are used for calculating the transfer function $V(T)$ that determines the period-dependent magnification factor (Scherbaum 2001):

$$
\begin{aligned}
V(T) &= V_o \frac{\omega^2}{(P_1 - i\omega)(P_2 - i\omega)}, \quad \text{where} \quad \omega = \frac{2\pi}{T}, \\
P_{1,2} &= -\omega_o(h \pm i\sqrt{1-h^2}), \quad \text{and} \quad \omega_o = \frac{2\pi}{T_o}
\end{aligned}
\tag{2.1}
$$

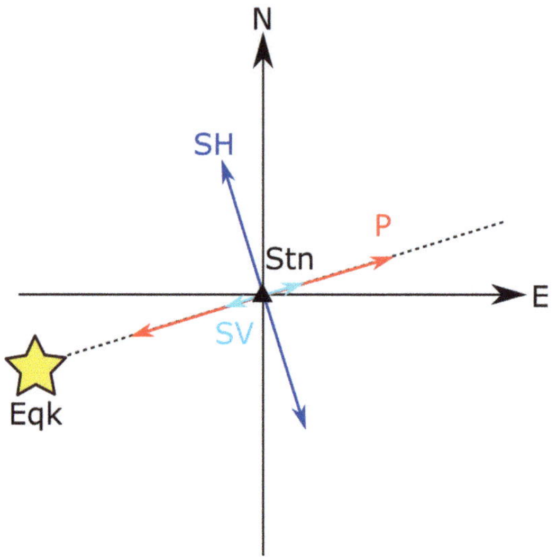

Fig. 2.8 Schematic example of back-projecting amplitude measured from one available horizontal component for the broad-band body-wave magnitude m_B. If only the N-S component is available, the back-projected horizontal amplitude from the S-wave is preferred. If only the E-W component is available, the back-projected horizontal amplitude from the P-wave is preferred

where $P_{1,2}$ are the poles of the low-pass filter that describes the frequency response of the seismograph. The most trusted instrumental parameters come from the seismograms themselves. If not labeled on the seismograms, sources to look for these parameters include station bulletins and log books as well as globally compiled station lists such as Wood (1921) and McComb and West (1931). Additional methods can also be employed to make sensible estimates subject to certain conditions (Sect. 2.4.3). When all these methods fail, one could potentially take similar values of other instruments of the same type as an estimate but the uncertainty would be unquantified. This is why it is beneficial to estimate magnitudes from data from as many stations as possible, be it reported amplitude and periods in the bulletins or new measurements from acquired seismograms. Sometimes, the median value can be less biased than the average, which is the approach taken by the International Seismological Centre—Global Earthquake Model (ISC-GEM) catalogue (Storchak et al. 2012).

2.3.3.2 Moment Magnitude Estimation

Since its invention in the 1970s (Kanamori 1977), the moment magnitude M_W became the most popular magnitude scale as it does not saturate at higher magnitude and is directly relatable to physical properties of the finite source fault. In modern seismology, seismic moment release from earthquakes is estimated from long (T>45 s) or very long (T>135 s) period components of the waveform for their ability to capture the entire rupture process that can be minutes long in large earthquakes (Ekström et al. 2012). The natural frequency of historical instruments was often too high and bandwidths too narrow to record such long-period signals (Kulikova and

Krüger 2015; Kulikova et al. 2016). Therefore, alternative approaches have to be sought to estimate M_W from historical waveforms.

The first method is through forward modelling and comparing the synthetic and recorded waveforms. To forward model a waveform, we need to have prior knowledge of the focal mechanism, rupture history, instrument response and the polarity of the seismogram. In Sect. 2.3.4, we discuss ways to determine the focal mechanism from historical data using first arrival polarity and amplitude ratio methods. Once the focal mechanism is well-constrained, we can test the rupture extent and directivity using waveforms with reliable instrument responses and polarities. A best-fit magnitude can then be found by scaling the synthetic waveforms and comparing with the historical waveforms (Ou et al. 2020).

The second approach is through comparing waveforms of a historical and a modern earthquake with a similar focal mechanism and location and registered by the same long-running instrument. The choice of a modern earthquake is typically backed by knowledge of the regional tectonics and confirmed by the similar shapes of the historical and modern waveforms. Using this approach, Rivera et al. (2002) selected clearly identifiable phases from waveforms of the 1935 and 1984 Sumatra subduction earthquakes and deconvolved the waveform of the smaller 1984 event from that of the 1935 event (Fig. 2.9). The resultant pulse from the deconvolution can be interpreted as the relative source-time function. The area under the pulse can be interpreted as the moment ratio between the two events. This then allows deduction of the seismic moment and M_W of the historical event from those of the modern event which should be better constrained. Successful deconvolution requires the historical instrument to have a wide dynamic range such that it is sensitive to oscillations of different frequencies and amplitude ranges, which was rather uncommon in the analog era.

2.3.3.3 Mantle Magnitude, Mm

Apart from the above mentioned IASPEI standard magnitude scales, Okal and Talandier (1989) developed a mantle magnitude scale, Mm, in a similar form as the empirical formulae, but using the amplitude of very long period Rayleigh waves, those that are sampling the mantle. This is useful for records with body-wave too weak (if the seismograph has small magnification) and saturated surface waves as the Rayleigh waves are often of intermediate amplitudes and more likely measurable from analog seismograms. This method has been systematically applied to 44 shallow earthquakes (Okal 1992a) and 41 intermediate-depth and deep earthquakes (Okal 1992b), with the Mm magnitude scale not saturating at the exceptionally large sizes of the 1960 Chilean and 1964 Alaskan earthquakes.

2.3.4 Focal Mechanism and Scalar Seismic Moment Determination

2.3.4.1 P/S Amplitude Ratios

A number of methods exist in modern seismology for focal mechanisms determination. From our experience, the most reliable and versatile method for dealing with historical data is the amplitude ratio method. The amplitude ratio method relies on

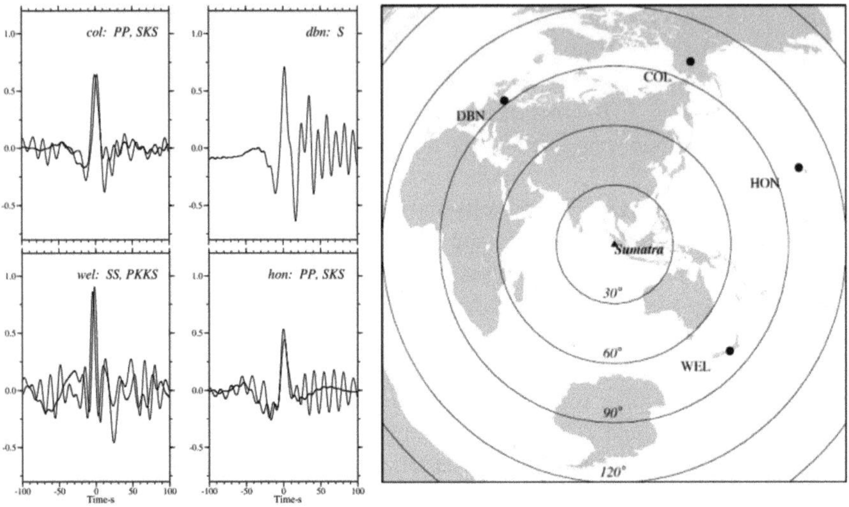

Fig. 2.9 Deconvolutions (the result of deconvolving the traces of 1984 from those of 1935) from the four stations shown in the map (Honolulu, De Bilt, College, and Wellington). Reproduced from Rivera et al. (2002)

the radiation patterns of the longitudinal (P, PP, PPP) and transverse (S, SS, PS) waves and can be applied to a wide range of seismograms with or without known polarities.

Two differences have to be made from how this method is applied to modern data. First, the amplitude ratio modelling has to be performed in the domain of the recorded seismogram rather than in the domain of the Green's functions (after removing the instrumental effect) as in modern studies. The reason for this is that instrumental parameters of historical data are often not precisely known and early instruments had low dynamics such that deconvolving instrumental effect can potentially introduce more errors. Furthermore, the digitisation process itself may introduce artifacts if converting from displacement/velocity to frequency domain. On the other hand, convolving the instrumental effect with the Green's functions and comparing with the recorded seismograms gives us an opportunity to identify errors in the supposed instrumental parameters and potentially correct such mistakes.

Second, in order to generate synthetic seismograms comparable to the recorded seismograms, it is also necessary to rotate the Greens' functions from the vertical-radial-transverse coordinate system into the vertical-north-east coordinate system. The reasons for rotating the Green's functions instead of rotating the recorded seismograms are twofold: (1) historical seismograms could have missing components, either due to data preservation issues or due to historical instrumental limitations, prohibiting rotation of the digitised waveforms; (2) the temporal axis of historical seismograms are often not uniform due to instability of drum rotation or due to deformation of the recording paper when it is scanned, potentially introducing errors from even minor asynchrony between components if rotated.

With the above tailoring, the amplitude ratios method has been successfully implemented to obtain focal mechanisms of a range of historical earthquakes in the early 20th century (Kulikova 2016) with the misfit η typically characterised as:

$$\eta = \frac{1}{K}\sum_{k=1}^{K}\left(\sqrt{\frac{1}{N}\sum_{i,j}^{N}\left(\left(\frac{A_i}{A_j}\right)_{obs} - \left(\frac{A_i}{A_j}\right)_{syn}\right)^2}\right)_k, \tag{2.2}$$

$$\text{where } i/j = \begin{cases} \frac{P}{S} & \text{for } \Delta < 30^\circ, \\ \frac{P}{S}, \frac{PP}{S}, \frac{P}{SS}, \frac{PP}{SS} & \text{for } 30^\circ < \Delta < 50^\circ, \\ \frac{P}{S}, \frac{PP}{S}, \frac{PPP}{S}, \frac{P}{SS}, \frac{PP}{SS}, \frac{PPP}{SS} & \text{for } 50^\circ < \Delta < 100^\circ, \\ \frac{PP}{SS}, \frac{PP}{PS}, \frac{PPP}{SS}, \frac{PPP}{PS}, \frac{PS}{SS} & \text{for } \Delta > 100^\circ \end{cases}$$

where K is the number of waveforms and N is the number of epicentral distance (Δ) dependent i/j phase combinations (Ou et al. 2020). Terms related to the PPP and SS amplitudes could be given lower weights because the 1D model results deviate from nature due to upper mantle heterogeneity.

For the 1902 Atushi (Kulikova and Krüger 2017), 1911 Chon-Kemin (Fig. 2.10) (Kulikova and Krüger 2015), good constraint on source depth was also obtained by this method.

2.3.4.2 First Motion Polarities

While the amplitude ratios method works well for historical earthquakes, it has a few shortcomings. (1) The S wave can be influenced by local structures which the synthetic waveforms will not reproduce; (2) The amplitude of the S wave can be compounded with that of the surface reflected phases, causing uncertainties which can be difficult to quantify. On the other hand, the polarities of the first motions are not contaminated by later arrivals and less affected by local site effects, thus could provide an independent result of the focal mechanism of the rupture initiation. Nevertheless, for large-magnitude earthquakes with complex rupture processes, the *first-motion* focal mechanism does not necessarily agree with the result from the P/S amplitude ratio method, which characterises the *overall* focal mechanism of the full rupture.

The first data requirement for this method is from polarities of the instruments, or the "way-up" on the seismogram. Perhaps surprisingly, many of the historical seismograms, especially those prior to 1930s, do not have this information. Amongst the almost-decadally published global station and parameter lists (Wood 1921; McComb and West 1931; Charlier and Van Gils 1953; Powell and Fries 1964), instrument polarities first appeared in McComb and West (1931). For earlier earthquakes, we have to rely on this information being labeled on the seismograms themselves (the most reliable) or in the station bulletins (not always available). Quite often, we receive, for example, a "NS" component seismogram, but it is unclear, even to the owners of the data, whether the up direction on the seismogram is pointing northward or southward.

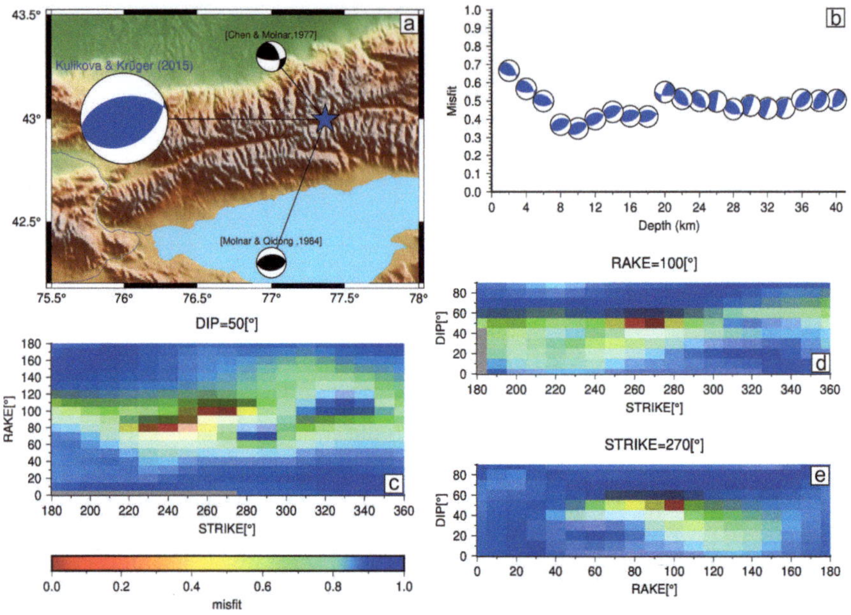

Fig. 2.10 Focal mechanism of the 1911 Chon-Kemin Earthquake determined using the P/S ratio method by Kulikova and Krüger (2015). **a** epicenter (star) and focal mechanisms (black from Chen and Molnar (1977) and Molnar and Deng (1984), and blue from Kulikova and Krüger (2015)) of the 1911 Chon-Kemin Earthquake. **b** Variation of best-fit focal mechanisms and misfits with epicentral depths, showing depth constrained between 8 and 18 km. **c–f** Misfit function dynamics for the given rake, strike, and dip accordingly

However, it is not impossible to try and infer the instrument polarity. One approach is to find a seismogram registered by the same instrument of an event of known focal mechanism, as close in time as possible from the date of the earthquake we are studying. If the polarity of the first motion of the event of known focal mechanism is clear on the seismogram, we can deduce the polarity of the instrument accordingly, as it is very unlikely for the instrument to switch its polarity once installed (Bungum et al. 2009; Kanamori et al. 2010).

Secondly, a well-constrained focal mechanism solution also requires a good azimuthal coverage by the stations (Fig. 2.11). A preview of potential data sources can be obtained by checking the International Seismological Summary (available at http://storing.ingv.it/ISS/index.html, starting in 1918) and the subsequent International Seismological Centre Bulletin which report epicentral distance, azimuth and phase time readings from global stations on an event-by-event basis. Most probably, not all of those data can be obtained. Amongst all seismograms retrieved, only a subset will have known instrument polarities and a further subset will have unequivocal first motion polarities, limiting the final azimuth coverage (Fig. 2.12).

Nevertheless, for events already with partial constraints on focal mechanism from geomorphology and tectonics, finding clear polarities from a few key stations can

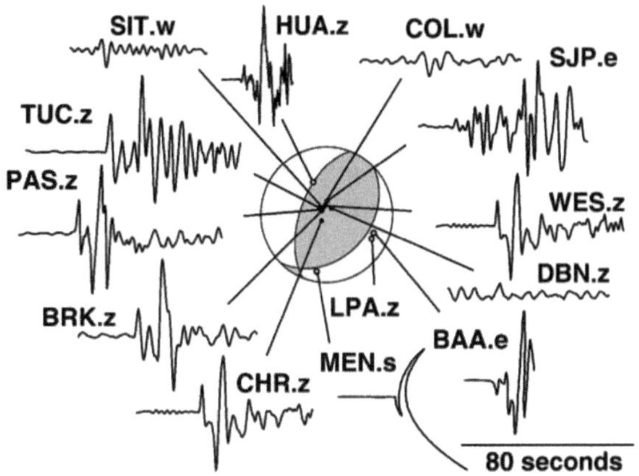

Fig. 2.11 Example of a well-constrained first-motion focal mechanism from the 1944 San Juan (Argentina) Earthquake. Solid circles indicate up-going rays and empty circles indicate down-going rays. Reproduced from Alvarado and Beck (2006)

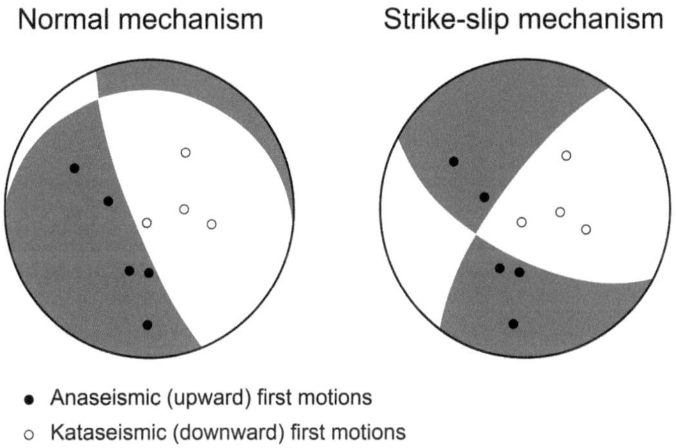

Fig. 2.12 A generic example of non-unique focal mechanism solutions from P wave first motion polarities of limited distribution

be effective in resolving ambiguities (Kanamori et al. 2010; Okal 2005). In some cases, the gain of the instrument and the magnitude of the earthquake did not allow clear first motion of the P wave, but the S wave arrivals will be more energetic. A clear first motion of the SH wave on the transverse component recorded at a strategic location can also help distinguish between competing focal mechanisms (Kanamori et al. 2010).

2.3.4.3 Waveform Modelling

In modern practice, variations of waveform modelling methods are more preferred (e.g., Global Centroid-Moment Tensor (GCMT) Ekström et al. 2012) as the full waveforms provide more data constraints than the amplitude ratios of key phases alone, and because complex ruptures can be analysed. For this method to work, on top of instrument polarities, knowledge of the exact instrumental parameters is crucial, as the natural period and damping of the instruments will affect the shape of the waveforms. Records suitable for waveform modelling should have long natural periods ($T > 5$ s) and high damping factors ($h > 0.3$). Some of the most common instruments operational in the early 20th century include Wiechert, Galitzin, Bosch-Omori and Mainka seismographs. For example, Kulikova et al. (2016) determined the focal mechanism of the 1911 Sarez-Pamir Earthquake by modelling body waveforms of good quality Wiechert and Omori records. Similarly, Alvarado and Beck (2006) modelled the P-waveforms of Benioff and Galitzin records to determine focal mechanisms of the 1944 and 1952 San Juan earthquakes. Other records could also be analysed as long as their parameters are known and suitable for earthquake studies, although it is noted that for instruments with damping factors (h) below 0.1, such as some early Milne instruments and micrometers installed in Japan, a small error in h can have a big effect on the shape of the waveform (Kanamori et al. 2010).

2.3.4.4 Spectral Amplitudes of Mantle Waves

If long- and very-long period (50–200 s) mantle waves (Fig. 2.13) are observed on seismograms from at least three stations with a good spread of azimuth sampling, a moment tensor could potentially be inverted from the spectral amplitude of mantle Rayleigh and Love waves, alone or combined (Fig. 2.14, (Reymond and Okal 2000; Okal and Reymond 2003; Okal 2005)). This method is based on the radiation patterns of the mantle Rayleigh and Love waves, which are depth- and frequency-dependent. Although it does not require knowledge of instrument polarity, it does require seismograms recorded by well-damped instruments with sufficient gains for such long-period ground motion to stand out above the noise level. The seismograms also have to be of good quality for sophisticated digital manipulation such as Fourier transform to be performed reliably. As this method has a two-fold 180° ambiguity on the strike and slip orientations, additional evidence from first motion polarity and/or geology are required to further constrain the focal mechanism (Reymond and Okal 2000; Okal and Reymond 2003; Okal 2005).

2.3.5 Kinematic Source Parameters

With the focal mechanism, known additional information of the rupture kinematics such as the source time function and rupture directivity can be extracted from the waveforms. This can be especially powerful with a seismic array with similar or known instrument responses (Kulikova and Krüger 2015), or with waveforms from

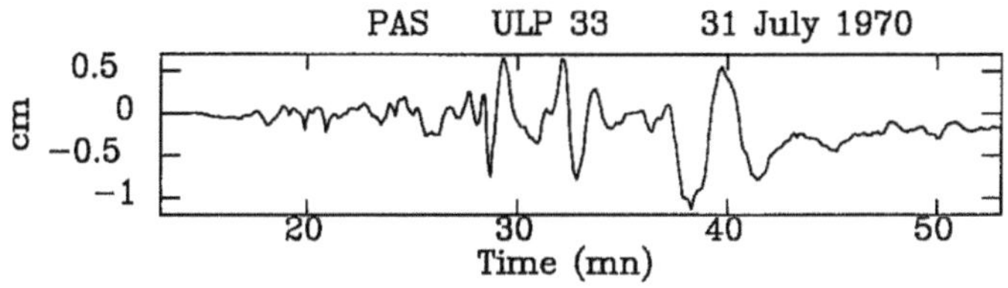

Fig. 2.13 Pasadena record of the 1970 Colombian earthquake showing a long-wavelength Rayleigh wave. Reproduced from Okal (1990)

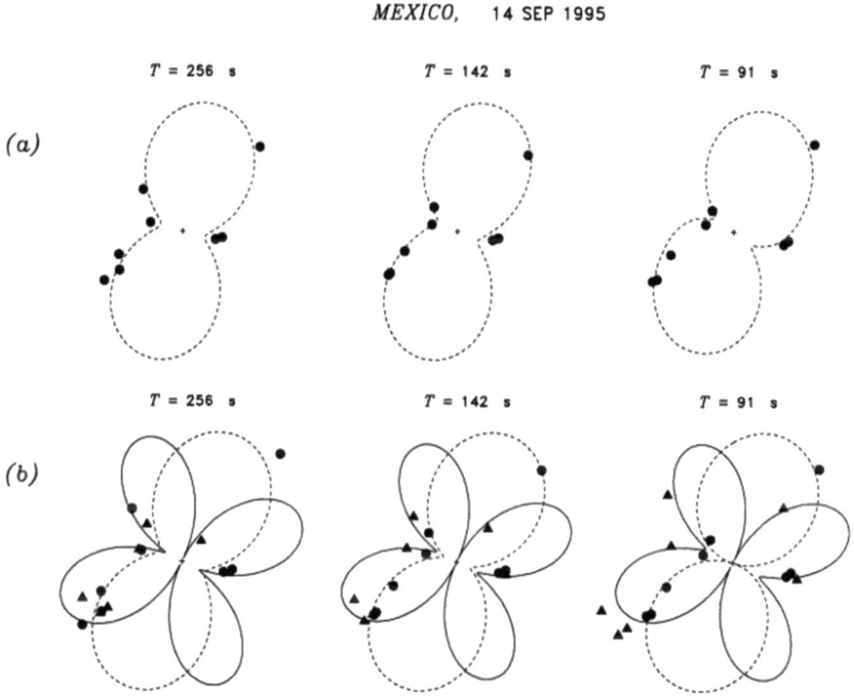

Fig. 2.14 Fits of spectral amplitudes of the 1995 Mexican earthquake: **a** inversion from Rayleigh waves only; at the three representative periods, the dashed lines are the theoretical spectral amplitudes computed from the model focal mechanism, plotted as a function of azimuth, and the solid dots the observed amplitudes at the eight stations. **b** Same as **a** after inclusion of Love waves (solid lines and triangles). Reproduced from Reymond and Okal (2000)

stations in opposite directions along strike (Rivera et al. 2002), or with waveforms of a similar modern earthquake for comparison (Kanamori et al. 2010).

Typical kinematic modelling allows evaluation of fault geometry, earthquake magnitude, position of the nucleation point, rupture directivity and velocity, as well as the sequential rupture of the fault segments. Inverting all these parameters from an often limited historical dataset can be challenging. Nevertheless, additional information such as observation of surface rupture for continental earthquakes, or tide gauge measurement and coral evidence for submarine earthquakes, can often help constrain the fault geometry and sense of slip such that the remaining unknowns can be better resolved.

2.3.5.1 Source Time Duration

Despite the common lack of low-frequency content in the waveforms from early instruments, it is still possible to estimate the source time duration from the wave trains. One way to enhance the signal-to-noise ratio is by stacking an array of same-component waveforms from similar azimuths and distances. Kulikova and Krüger (2015) stacked the European historic seismic array for both P and SH waves and observed two to three subevents with different durations for the 1911 M8.0 Chon-Kemin earthquake.

Another way of estimating the source time is through deconvolution of waveforms of modern earthquakes of similar location and focal mechanism registered by the same long-running instrument. In this case, the waveform of the smaller earthquake could be used as an empirical Green's function. The resultant pulse from the deconvolution represents the relative source time of the two earthquakes (Fig. 2.9). If the rupture duration of the more recent earthquake can be better resolved from the more abundant and higher-quality dataset, that adjusted by the relative source time will give the source time of the historical earthquake (Rivera et al. 2002).

If the more recent "repeating" earthquake was not registered by the same historical instrument, the historical waveform can still be compared with the modern waveform deconvolving its instrument response and convolving the historical instrument response. For example, Kanamori et al. (2010) extended the source time function of the modern earthquake in order to reproduce the waveform of the historical earthquake with a similar frequency content, thus estimating the rupture duration of the 1907 Sumatra earthquake.

2.3.5.2 Rupture History

When the waveforms or source time functions suggest interesting features, waveform modelling could be useful for distinguishing alternative plausible rupture histories. For example, instead of a single point source which would produce arrivals that are too short in duration and high in amplitude, the waveforms of the 1920 Haiyuan Earthquake were better fitted with a bilaterally rupturing rectangular fault (Ou et al. 2020). Segmented finite fault modelling showed that the Teregtiin fault branch also ruptured during the 1905 Bolnay earthquake, and the 1905 Tsetserleg earthquake had to start with the segment with a reverse component and rupture 60 km further northeastward compared to previously mapped surface rupture (see Fig. 2.11 of Schlupp and Cisternas (2007)).

When there are not enough waveforms to perform joint inversion, comparative studies of waveforms of historical and modern repeat earthquakes can also derive useful knowledge about the rupture history of the past event. For example, a close comparison of waveforms of the 1907 and 2002 Sumatra Earthquakes showed good agreement of the first 100 s of the S-wave, but the 1907 waveform was energetic for another 100 s, which was not observed in the 2002 waveform. Therefore, Kanamori et al. (2010) concluded that the 1907 earthquake had a similar first rupture as the 2002 earthquake, but had a more extended rupture following the initial rupture.

2.3.5.3 Directivity

If the rupture was unilateral, the directivity of rupture can best be deduced from the differences in source time durations recorded at stations along strike of the fault and opposite in azimuth. The station the fault ruptures towards will show shorter and taller source time functions whereas the station the fault ruptures away from will show longer and lower source time functions. This observation relies on instruments on both sides having sufficient gains and damping to allow clear identification of the source-time durations.

In a study of the 1935 Sumatra Earthquake, Rivera et al. (2002) deconvolved waveforms of a repeat earthquake in 1984 as empirical Green's functions from the 1935 waveforms and obtained an exceptionally short relative source time at Wellington, New Zealand. Wellington happens to be along strike of the rupturing fault hence suggesting that the 1935 earthquake ruptured towards the southeast (Fig. 2.9).

If waveforms show subevents with clear azimuthal variation of onset time delays (Fig. 2.15), it is possible to relocate the subevents relative to the epicenter, hence resolve the directivity (Cummins et al. 2002), by solving for the following equation:

$$\Delta\tau_i = t - p_i x cos(\theta_i) \tag{2.3}$$

where $\Delta\tau_i$ is the time delay of the subevent on trace i relative to the first arrival, p_i is the P-wave slowness at station i, θ_i is the angle between the fault strike and the station azimuth, x and t are the best-fit distance and time offset of the subevent along fault relative to the nucleation point (Schwartz and Ruff 1985).

If the source-time durations can be observed from both the P and S wave trains recorded at the same station which happens to be along strike, the directivity of a unidirectional rupture can be inferred with the following:

$$\Delta(T_S - T_P) = L\left(\frac{1}{\alpha} - \frac{1}{\beta}\right) cos(\theta) \tag{2.4}$$

where T_S and T_P are the source durations of the S and P wave trains, α and β are the apparent P and S-wave velocities, L is the rupture length and θ is the angle between the rupture direction and the station azimuth (Schlupp and Cisternas 2007).

If the only available waveforms with clear source time functions are from stations perpendicular fault strike, it would be difficult to distinguish the direction of rupture even if it was unilateral. However, by comparing the source time duration of the record with that of synthetic rupture models, it is still possible to discriminate between bilateral and unilateral ruptures (Schlupp and Cisternas 2007).

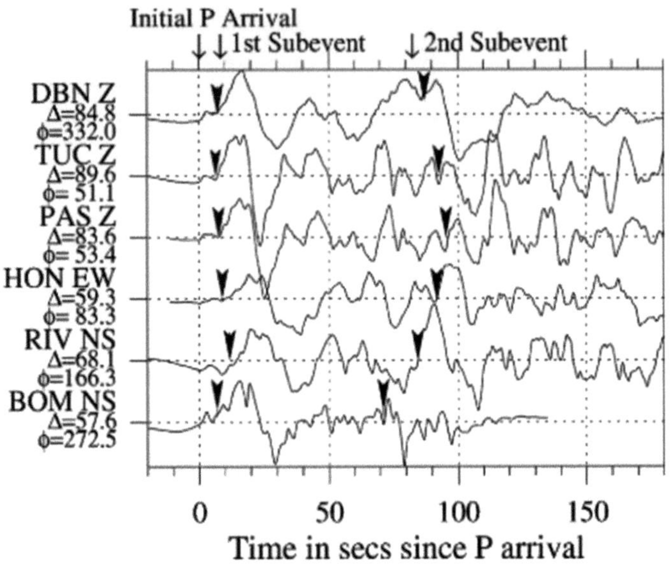

Fig. 2.15 Deconvolved displacement records of the 1946 Nankai earthquake at De Bilt, Tuscon, Pasadena, Honolulu, Riverview and Bombay. The consistent azimuthal variation of the subevent onset times suggests the energy pulses are subevents rather than depth phases (which would produce constant delays instead). Reproduced from Cummins et al. (2002)

2.4 Data Preparation

2.4.1 Data Collection

Some first stops to check whether data of interest have already been scanned include The ISC Electronic Archive of Printed Station and Network Bulletins (Di Giacomo et al. 2022), the Headquarters for Earthquake Research Promotion (HERP) database (http://www.susu.adep.or.jp/index.html, for analog seismograms preserved by Japan Meteorological Agency from 1884 to 1995); the SEISMOS website (http://storing.ingv.it/es_web/, for many European seismograms from 1896 to 2005 and station bulletins from 1899 to 1994); the inventory of the Historical Seismogram Filming Project (Glover et al. 1985, for pre-1963 seismograms from mostly US stations, but also key stations in Argentina, Bermuda, Bolivia, Chile, Japan, Philippines, Puerto Rico and the USSR, as well as station bulletins from 450 global stations); the IRIS archive (http://ds.iris.edu/seismo-archives/projects/, for post-1962 WWSSN seismograms through the USGS WWFC Pilot Scanning Project, as well as the Caltech holdings from 1928 to mid-1980's); the HRV seismogram archive (http://www.seismology.harvard.edu/HRV/archive.html, for Harvard holdings between 1933 and 1953); the Archive of Analog Seismograms from Iceland (http://seismis.hi.is/, for records from 1910 to late 1980s); the Universidad Nacional Autónoma de México (UNAM), Mexico (http://www.sismoteca.unam.mx/, for holdings from 1904 to 2015); and the website of Uppsala University http://www.snsn.se/, for scanned bulletins between 1904 and 1998. Databases of digitised seismograms are also starting to become available (Murotani et al. 2020).

On the other hand, currently the majority of the old analog seismograms are still stored in local institutional or national archives (Table 2.3) and have not been scanned let alone traces digitised due to funding limitations. Many historical seismograms also suffer from inadequate preservation conditions such that the data quality is deteriorating at an alarming rate (Okal 2015). In the worst of cases, the data may be uncatalogued and piled in unlabeled boxes in remote storage facilities. Nevertheless, archiving standards are being formulated to encourage Findable, Accessible, Interoperable, and Reusable (FAIR) data management (Hwang et al. 2020). As data users, one way to support the rescue of the historical seismic data is through actively requesting and using the dataset and citing the data in publications with Digital Object Identifiers (DOI).

On this regard, it is useful to mention that an analog holdings catalog is being created after the need being identified at a 2019 workshop entitled "Securing Legacy Data to Enable Future Discoveries" in Albuquerque, New Mexico (Hwang et al. 2019, 2020). When completed, it will be a great resource for a quick estimate of the potential data coverage for historical earthquakes of interest, and serve as a directory of contact information for future data requests. We encourage all local and national archives to join this effort and build together a supportive ecosystem in preserving, sharing and enabling the usage of this treasurable heritage from early seismologists.

Before a global inventory and online database becomes available, it is worth contacting as many archives to obtain as many original seismograms as possible. Some archives welcome users to visit in person and scan the seismograms using facilities on site. Some archives can scan and share records upon request. It is advisable to make sure the paper seismograms were scanned at high resolution (e.g., 300 ppi) to preserve the fine details, and scanned at the original paper size to preserve the true amplitudes. Given the logistical hurdles the archive managers often have to overcome to access, search for and scan the seismograms, it would be more efficient if users request data for a list of earthquakes together instead of for individual events over multiple instances. We provide a non-exhaustive list of archives with legacy seismograms in Table 2.3.

2.4.2 Digitization

Multiple tools have been developed to semi-automatically digitise historical seismograms from image scans and correct the slant and curvatures due to stylus motion (Grabrovec and Allegretti 1994; Batllo et al. 1997; Bromirski and Chuang 2003; Pintore et al. 2005a; Taylor and Yang 2011; Xu and Xu 2014; Bi et al. 2016; Wang et al. 2016; Bogiatzis and Ishii 2016b; Burkhard et al. 2019; Murotani et al. 2020). Amongst the options, Teseo (Pintore et al. 2005a), a plugin for GIMP (GNU Image Manipulation Program (Kimball and et al. 2019)), and DigitSeis (Bogiatzis and Ishii 2016b), a MATLAB software, are freely available.

Table 2.3 The contact information of selected institutions with analog seismograms

Country	Station	Institution	Access
Australia	Riverview	Australia National University, Canberra, Australia	I
Austria	Vienna	Central Institute for Meteorology and Geodynamics Vienna University of Technology	I
Bolivia	La Paz	Observatorio San Calixto, La Paz, Bolivia	I
Canada	Ottawa, Toronto	Natural Resources Canada	I
Croatia	Zagreb	Faculty of Science, University of Zagreb	I
France	Strasbourg	Ecole et Observatoire des Sciences de La Terre, University of Strasbourg	II
Germany	Göttingen, Apia (Samoa)	Institute of Geophysics, University of Göttingen	III
	Hamburg	Institute of Geophysics, University of Hamburg	II
	Helgoland, Clausthal	Institute of Geosciences, University of Kiel	II
	Leipzig	Observatory Collm, Institute of Geophysics and Geology, University of Leipzig	IV
	Hohenheim, Stuttgart, Ravensburg	Institute for Geophysics, University of Stuttgart (now at Landesarchiv Baden Württemberg)	II
	Munich	Department of Earth and Environmental Sciences, Ludwig-Maximilians-University, Munich	I
	Jena, Potsdam	Institute of Geosciences, Friedrich-Schiller-University, Jena	IV
Kazakhstan	National Collection	Seismological experience-metodology expedition, Kazakhstan National Data Center, Almaty	I
Kyrgyzstan	National Collection	Institute of Seismology, NAS, Kyrgyzstan, Bishkek	I
Iceland	National collection	University of Iceland	V
India	Bombay, Kodaikanal, Calcutta	National Seismological Data Centre (Ministry of Earth Sciences)	I
Indonesia	Jakarta	Northwestern University	I
	Jakarta	Indonesian Agency for Meteorology, Climatology and Geophysics, Jakarta, Indonesia	I
Italy	European (SISMOS)	Istituto Nazionale di Geofisica e Vulcanologia	I, VI
Japan	National collection	JMA analog seismograms	VII

(continued)

Table 2.3 (continued)

Country	Station	Institution	Access
Mexico	SISMOMex and National Seismogram Library (Pérez-Campos et al. 2020)	Universidad Nacional Autonoma de Mexico	I, VIII
Netherlands	De Bilt (Dost and Haak 2002)	Royal Netherlands Meteorological Institute in De Bilt	I
New Zealand	Apia (Samoa)	Institute of Geological and Nuclear Sciences in Lower Hutt	I
Norway	Bergen	Department of Earth Science University of Bergen	II
Portugal	Coimbra	Geophysical and Astronomical Observatory of the University of Coimbra	I
	Lisbon	Universidade de Lisboa	I
South Africa	Cape Town, Hermanus (Okal 2015)	Council for Geosciences, Pretoria	I
Spain	National collection (Batll and Bormann 2000; Batll 2004)	Instituto Geográfico Nacional	I
	Barcelona	Institut Cartogràfic i Geològic de Catalunya	I
	Ebre (Custodio et al. 2012)	Observatori de l'Ebre	I
	Fabra	El Observatorio Fabra	I
Sweden	National collection (Kulhanek and Wahlstrom 1942; Lund et al. 2021)	Svenska Nationella Seismologiska Nätet (SNSN)	I, IX
Tajikistan	National Collection	Institute of Geology, Tajikistan, Dushanbe	I
UK	National collection and WWSSN microfilm collections (Henni and Lawrie 1999; Henni et al. 2000)	British Geological Survey—National Seismological Archive (NSA)	I
US	Global microfilm collection (Meyers and Lee 1979; Glover et al. 1985)	USGS, Golden Colorado	II
	WWSSN online collection (Alejandro et al. 2019)	IRIS archive and the Albuquerque seismological Lab	X
	Harvard (Ishii et al. 2015)	Seismogram Archival Project	XI
	Berkeley	UC Berkeley Seismological Lab	I
	Cleveland	The Ohio Seismic Network	I

Access Note:

[I] Archive can scan original seismograms and transfer data online.
[II] Archive can host users to scan on site.
[III] Archive can send original records by post to users for scanning and expect return of records.
[IV] Archive can host users to scan on site with users' own scanning facilities.
[V] Online: http://seismis.hi.is/
[VI] Online: http://storing.ingv.it/es_web/
[VII] Online: http://www.susu.adep.or.jp/index.html
[VIII] Online: http://www.sismoteca.unam.mx/
[IX] Online: http://www.snsn.se/
[X] Online: http://ds.iris.edu/seismo-archives/projects/
[XI] Online: http://www.seismology.harvard.edu/HRV/archive.html

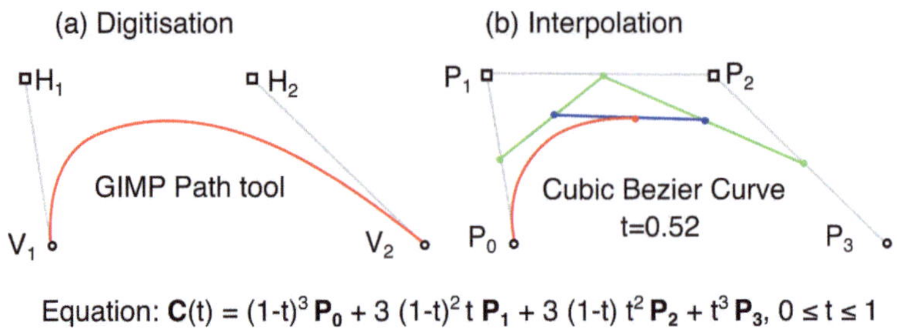

Fig. 2.16 Cartoons showing **a** digitisation of a curved segment (along waveform) with the Path tool in GIMP and **b** interpolation of the digitised curve through the cubic Bezier curve equation, C(t). The pixel positions of the Path tool's vertices, V_1 and V_2, and their associated handles, H_1 and H_2, are used as four control points, P_0 to P_3, that define the parametric curve segment. Panel (**b**) shows a snapshot of the parametric curve at t=0.52. The red (cubic Bezier) curve can be seen as a linear interpolation between linear interpolations (ends of the blue segment) between linear interpolations (ends of the green segments) between the four control points, P_0 to P_3

Due to the wide range of types of instruments installed around the world before the 1960s, seismograms collected might have different characteristics such that some bespoke software design in one tool might not be applicable for all available data. Therefore, we cover the basics of manual digitisation as an approach one can always fall back to. Our approach generally follows Pintore et al. (2005b). We first trace the waveform of interest using a Bezier curve 'path' tool in a standard vector editing graphic software (such as Adobe Illustrator or GIMP), following each turning curve using the vertices and handles (Fig. 2.16a). Then, we linearly interpolate each arc segment between every two neighbouring vertices by evaluating the cubic Bezier curve equation at 1000 intervals (Fig. 2.16b). After converting the horizontal axis of the sampled points from pixel number to time in second through the pixel-per-minute measurements from the seismogram, we then interpolate the points uniformly in time to obtain a time series with a constant sampling rate of 0.1 sec.

The slant and curvature (the bending of the otherwise symmetrical sinusoidal oscillation) of the waveform (see Figs. 2.7 and 2.17) can be corrected based on the geometry of the mechanical recording system, if known (Cadek 1987; Grabrovec and Allegretti 1994) (Fig. 2.18). Nevertheless, during our museum visits to measure the lengths of styli and radii of rotating drums (McConnell 1986), we found contrasting measurements for sometimes similarly named instruments. Therefore, we performed the curvature correction by inverting for the geometric parameters that minimises a measure of symmetry of the corrected waveform, S, which is defined as:

$$
\begin{aligned}
S &= |U| + 2B && \text{with} \\
U &= \int_c |y| \delta y \delta t && \text{for } \delta t > 0 \\
B &= \int_c \delta t && \text{for } \delta t < 0
\end{aligned}
\tag{2.5}
$$

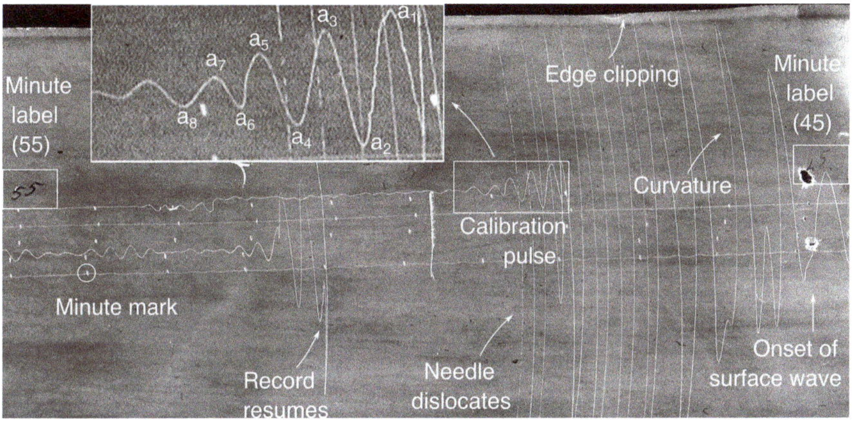

Fig. 2.17 A portion of a N-S component seismogram recorded by the Bosch-Omori instrument at the TOL station, featuring the 1920 Haiyuan Earthquake. The recording starts in the middle of the top row, with a calibration pulse blown up in the inset. It records four rows of data from right to left and spirals downwards. In the third row, the surface wave becomes clipped by the edge of the paper. The recording needle dislocated for about two minutes before recording resumed when the amplitude of oscillation decreased. Successive amplitudes of the calibration pulse, labelled a_1 to a_8, can be used to calculate the damping of the instrument. The 45th and 55th minutes of the hour are labelled by hand on the seismogram

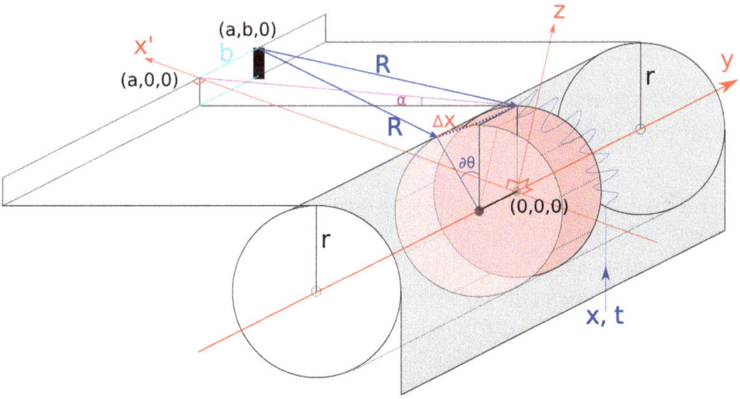

Fig. 2.18 Geometry of the recording system of typical mechanical seismographs where a stylus swings with amplified ground motion and scratches soot off the smoked paper wrapped on a rotating drum. The parameters we invert for are the length of the stylus, R, the offset of the hinge of the stylus, b, from the projection of the equilibrium line on the seismogram, the dipping angle, α, of the stylus and the radius of the drum, r. Adapted from Cadek (1987)

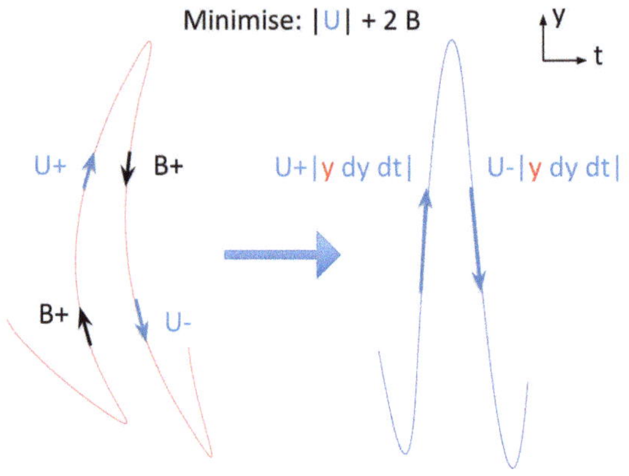

Fig. 2.19 Curvature correction by minimising the uptime, U, and the backtime, B. B strictly increases whenever a portion of the waveform moves backward in time, whereas U only changes for portions of the path moving forward in time, increasing along rising limbs and decreasing along the falling limbs. δy controls the sign. As the waveform is centred around the $y = 0$ axis, inclusion of the y term ensures the crests and the troughs are fully symmetrical

where c is the path along the waveform. In the fully symmetrical case, the uptime value U will approach zero with the rising and falling limbs of each oscillation cancelling each other out, whereas the backtime value B will be exactly zero as the time series should only progress forward in time (Fig. 2.19, Kulikova and Krüger (2015)).

Not all seismograms scanned can be digitised, however. Here we list some potential difficulties in historical seismogram processing and ways to overcome them.

- **Quality of scan.** Some scanned images have low contrast making it difficult to see the trace clearly. If unavoidable, it is better to have a darker scan which can have its contrast enhanced (e.g., in GIMP) than to have an overexposed scan. Spots, missing parts or holes can prevent further digitising. Thus, during the scanning process, it is important to softly unfold, flatten and straighten but without breaking the paper. When the seismogram is longer than the scanner, the sheet has to be scanned in parts. In order to accurately piece together the parts, it is important to have overlaps between neighbouring sections. The edges of the seismogram should also be completely scanned to ensure no time is skipped as the trace wraps around the "drum".
- **Timing.** Hour and minute marks can be recorded on seismograms as gaps in the traces (Fig. 2.7), separate dots (Fig. 2.17), short pulses or offset trace segments. The gaps can be filled during digitization by data extrapolation (Schlupp and Cisternas 2007). DigitSeis can digitise the temporal dots and automatically restores the offset trace segments (Bogiatzis and Ishii 2016b). The short pulses are tricky to remove as it may blend into the waveform. The timing of pre-1920 seismographs was recorded using local astronomical clocks and may differ from universal time. Thus, any time corrections written on the seismograms should be incorporated.

Relative time correction is also possible based on ambient noise correlation (Lee et al. 2020). If recorded by independent seismographs, two components recorded at the same station can also have temporal offsets. Therefore, it is better to analyse the recorded components instead of the radial and transverse components after rotation (Stich et al. 2005).

- **Polarity.** The polarity of the first motion can be hard to read, if the signal was weak and the gain of the historical instrument was low (e.g., 20) (Kanamori et al. 2010; Kulikova et al. 2016). In addition, the polarity of the instrument was not always marked on the paper seismogram or available from station bulletins or lists such as McComb and West (1931). In this case, some detective work would be necessary based on earthquakes of known focal mechanisms registered by the same instrument at the same station.
- **Saturation.** The mechanical pendulum instruments had limited dynamic range. Thus, for seismographs with higher gains, strong oscillations, especially of the high-amplitude surface waves, can be clipped (Fig. 2.17). Consequently, the surface waves can not be used for the determination of earthquake focal mechanism or the estimation of surface wave magnitude.
- **Interweaving waveforms.** Oscillations from large earthquakes can last for hours, leading to interweaving of waveforms from neighbouring rows (typically, seismograms register one hour or half hour per row, depending on the rate of drum rotation). Attention should be paid not to jump lines during digitising. Both Teseo (Pintore et al. 2005b) and DigitSeis (Bogiatzis and Ishii 2016b) have features to handle crossing waveforms. However, for some Galitzin records with particularly high gain and symmetrical waveforms, as they are based on electromagnetic rather than mechanical amplification, this can be challenging even for the human eye. One way of checking is to see if the equilibrium of the oscillation stays at the same level.
- **Mechanical failure.** Strong oscillations from a major earthquake can sometimes dislocate the writing needle of the instrument, resulting in interrupted (Fig. 2.17), shifted, or one-sided recording. Although such records may still contain information about the earthquake, it is easier to not use them when there are abundant other records.

2.4.3 Instrument Parameters

Essential information needed to interpret the historical waveforms are the natural period, T_o, magnification, V_o, and damping factor h (or damping ratio ϵ) of the recording instruments. Although not crucial for epicenter relocation, all three parameters are required for magnitude estimation and for focal mechanism inversion. Parameters related to the geometry of the recording system, such as length of the recording stylus and the radius of the rotation drum, are also useful information as they can help correct the curvature of the waveforms (Fig. 2.18, Sect. 2.4.2).

These parameters can be looked up from, in descending order of preference, the seismograms, station bulletins, station logbooks, and global station lists (Wood

1921; McComb and West 1931; Charlier and Van Gils 1953; Powell and Fries 1964). *Sometimes archive contacts may have more accurate knowledge about the instrument parameters than what is published in decadal summaries.*

Nevertheless, these parameters are not always available in the above mentioned sources. Sometimes, multiple instruments could be operational at a station causing ambiguities in the parameters for our records. Some early seismographs did not have damping information, which either means that the damping constant was unknown or the instrument was undamped (Okal 2015). Here we introduce some additional evidence that could be used to better ascertain or constrain the parameters.

2.4.3.1 Calibration Pulse

A calibration pulse is a decaying sequence of oscillation often appearing at the start of a seismogram (Fig. 2.17 inset). It is created when the stylus was set in motion after the paper was replaced, typically on a daily basis. Coupled with the minute mark lengths, this free oscillation gives clue to the natural period, T_o, of the instrument.

The decay of the oscillation also reveals the damping factor, h, of the instrument. According to Eqs. 4.27, 4.31 of Scherbaum (2001):

$$h = \frac{\ln(\epsilon)}{\sqrt{4\pi^2 + (\ln(\epsilon))^2}}, \quad \text{where} \quad \epsilon = \frac{a_k}{a_{k+2}} \tag{2.6}$$

where a_k and a_{k+2} are amplitudes one full wavelength apart in the calibration pulse. This first equation is also how h should be calculated from the ϵ often appearing in station bulletins and global summary books.

2.4.3.2 Paper Speed

Paper speed is listed in McComb and West (1931) for each instrument in units of mm/min. This information can be used to identify the recording instrument of the record from a list of candidate instruments operational at the station at the same time based on the dimension and the number of minute marks per row on the seismogram. In other cases, when the type of instrument is known but the true scale of the seismogram is lost, for example, because the image is scanned from a microfilm which did not contain a scale, the paper speed could be used to scale the image and restore the true size of the seismogram.

2.4.3.3 Forward Modelling

When the location and focal mechanism of the earthquake are known and only the damping of the instrument is unknown, it is possible to forward model the waveform with a range of damping factors and compare them with the recorded signal. The Green's function used to convolve with the test instrument responses could be generated from a velocity model or extracted from a waveform of a modern similar earthquake by deconvolving the modern instrument response (Kanamori et al. 2010).

Figures 2.4 and 2.5 of Kanamori et al. (2010) show such an example where a series of synthetic waveforms were created based on the 2002 Sumatra earthquake waveform to find the most likely damping factor of the Omori instrument that recorded the 1907 Sumatra earthquake.

2.4.4 Additional Data

While searching for the original seismograms can be time consuming and challenging, it has often been found that waveforms of major earthquakes have already been reproduced with time and amplitude scales in existing literature, be it station books, earthquake reports, research papers or special collections. Sometimes they present the best sources of data when the originals may have been lost or destroyed by war, fire, or other misfortunes. It is the comprehensive analysis of waveforms, seismic bulletins, catalogues, field measurements, macroseismic observations, geological and tectonic information that together constitute the best knowledge about a historical earthquake.

2.5 Earthquake Examples

Finally, we present four examples from our own studies, mainly continental earthquakes in Tienshan and around the Tibetan Plateau, which illustrate the various challenges and outcomes from analysing historical seismological data.

2.5.1 1902 Atushi Earthquake

The 1902 M$\sim$8 Atushi earthquake happened at the northwest corner of the Tarim basin with no reported surface rupture. It is a case where no original seismogram was retrieved. However, Kulikova and Krüger (2017) discovered 12 reproductions of seismic waveforms from 6 seismic stations in reports, books, and notes from Japan, Russia, Germany, Mauritius, and today's Slovakia. These records, together with additional arrival times and amplitude readings from reports and bulletins, provided nearly complete azimuth coverage for this event (Fig. 2.20a). This dataset enabled hypocenter relocation, depth constraint, focal mechanism determination and magnitude re-estimation for this significant earthquake which we had very limited knowledge about.

The search for such reproduced seismograms is an art and can depend on luck. The purposes of such historic publications might have been to introduce a new instrument, to summarise achievements at a station or to characterise seismic hazard. Waveforms of interesting large earthquakes, such as that of the Atushi earthquake, only entered as examples or case studies. Publications as such tend to be better preserved than the original seismograms and could be available at multiple libraries or even online. The

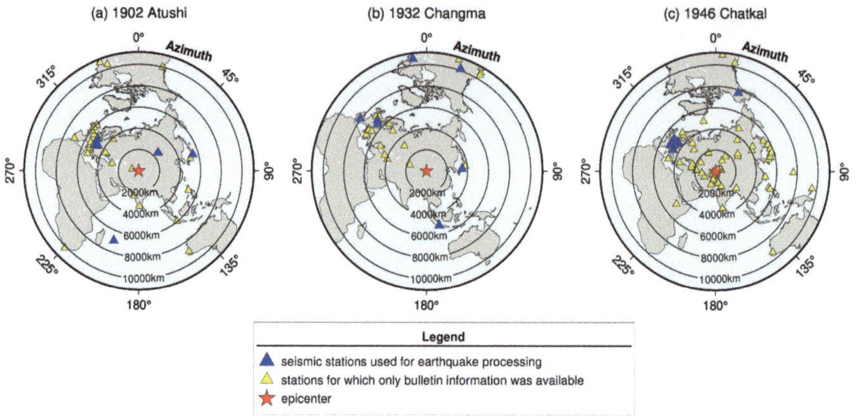

Fig. 2.20 Station distribution for waveforms (blue triangles), bulletins (yellow triangles) around the epicenter (red stars) of **a** the 1902 Atushi earthquake, **b** the 1932 Changma earthquake, and **c** the 1946 Chatkal earthquake

reproductions were usually printed as attached sheets and sometimes even reproduced at real scale allowing their reanalysis in the same way as seismogram scans.

2.5.2 1920 Haiyuan Earthquake

The Haiyuan earthquake was the second most deadly earthquake in the 20th century in the world. The amount of retrievable seismic records significantly increased by 1920, for this major strike-slip earthquake in the northeast of Tibetan Plateau. Overall, 78 stations reportedly recorded this event (ISC 2018). Ou et al. (2020) recovered 60 analog seismograms registered by 13 unique types of instruments from 27 stations in 12 countries (Fig. 2.5), out of which 2 were vertical records. However, this earthquake occurred just before the earliest archived records from Cape Town and India such that the station distribution left a gap in the southwest quarter.

From these waveforms, 123 phase amplitude and period pairs were measured for body- and surface wave magnitude calculations. Epicenter was better located using phase times picked from the waveforms and collected from bulletins. Focal mechanism derived from the amplitude ratios method agreed with geological observations and 12 waveforms were used for finite fault modelling to constrain the moment magnitude of the earthquake. The resultant moment magnitude was M_W 7.9 ± 0.2, significantly lower than the frequently quoted values M 8.7, M_S 8.5 and M_W 8.3.

2.5.3 1932 Changma Earthquake

The Changma earthquake happened at the junction between the Altyn Tagh fault and the Qilian Shan thrusts. By 1932, the quality and distribution of the collected waveforms further improved. About 90 waveforms from 40 stations were collected, out of which 14 recorded the vertical component (Fig. 2.20b). An increasing number of the waveforms were recorded by the electromagnetic Galitzin instrument with symmetrical waveforms that do not need curvature correction. Being right after the publication of McComb and West (1931), we have knowledge of the polarities of many of the instruments. The all-round azimuth distribution enabled focal mechanism determined from the first motion polarity method.

This method is better than the amplitude ratios method as it does not suffer from interferences from local velocity structures or superposition of wave arrivals. The focal mechanism was especially important for this earthquake as both left-lateral and right-lateral surface offsets were observed in the field on differently-striking faults. The good-quality mechanism determination and waveform modelling help distinguish contradictory faulting styles and unravel the slip history, shining light on the complex kinematics of this event.

2.5.4 1946 Chatkal Earthquake

The 1946 M7.5 Chatkal earthquake is the strongest known earthquake that occurred in the Fergana region, western Tien Shan, where the 600 km long dextral Talas-Fergana fault cuts through the Tien-Shan mountain belt. As all modern earthquakes happened on the neighbouring conjugate thrust faults, it was unclear how seismically active is this main strike-slip fault. Therefore, determining whether the Chatkal earthquake happened on the Talas-Fergana fault, or on the smaller adjoining faults, is crucial for understanding of the regional tectonics.

Seismograms from only 10 stations were collected for the 1946 Chatkal earthquake (Fig. 2.20c), possibly due to the disrupted operation and difficulties in data preservation during and shortly after the Second World War. Nevertheless, all 10 stations had vertical components which reflects the proliferation of vertical instruments by that era. Amplitude ratios from these waveforms show that this earthquake had a thrust mechanism, similar to those of the modern earthquakes in the region. Thus, this result argues against the speculation that this earthquake might have happened on the Talas-Fergana fault.

2.6 Conclusion and Outlook

Although historical seismograms have different characteristics from digital data recorded by standardised instruments, modern seismological methods can be applied and adapted to analyse them. Various tools have also been developed to accelerate and optimise the digitisation and processing of the historical waveforms. More and more archives are making their data available online, in scanned or even digitised forms, as well as associated metadata to help with their interpretation. Yet, many more archives are facing challenges in preserving, archiving and scanning their data.

As users, the best way we can help support the preservation of the historical seismograms is through our request, usage and proper citation of the historical data, and demonstration of what useful knowledge can be gained about past earthquakes. A niche but very supportive community of historical seismologists exists around the world. We encourage more uptake of research with this endangered data source from fellow geologists, geomorphologists, seismologists and palaeoseismologists, and we hope more researchers will find valuable information from this legacy from early seismologists for better understanding of past earthquakes and the global tectonics.

2.7 Contact Information

See Table 2.3.

References

Abe, K. (1981). Magnitudes of large shallow earthquakes from 1904 to 1980. *Physics of the Earth and Planetary Interiors, 27*, 72–92.

Ackerley, N. (2014). *Principles of broadband seismometry* (pp. 1–35). Springer.

Agnew, D. C. (2020). Time marks and clock corrections: A century of seismological timekeeping. *Seismological Research Letters, 91*(3), 1417–1429.

Alejandro, A. C., Hutt, C. R., Ringler, A. T., Moore, S. V., Anthony, R. E., & Wilson, D. C. (2019). The Albuquerque seismological Lab WWSSN Film Chip Preservation Project. *Seismological Research Letters, 90*(1), 401–408.

Alvarado, P., & Beck, S. (2006). Source characterization of the San Juan (Argentina) crustal earthquakes of 15 January 1944 (Mw 7.0) and 11 June 1952 (Mw 6.8). *Earth and Planetary Science Letters, 243*(3–4), 615–631.

Batll, J. (2004). *Catalogo Inventario De Sismografos Antiguos Espanoles.*

Batll, J., & Bormann, P. (2000). A catalog of old Spanish seismographs. *Seismological Research Letters, 71*(5), 570–582.

Batlló, J. (2014). Historical seismometer. In M. Beer, I. A. Kougioumtzoglou, E. Patelli, & I.S.-K. Au (Eds.), *Encyclopedia of earthquake engineering* (pp. 1–31). Springer.

Batllo, J., Susagna, T., & Roca, A. (1997). A processing system for old records of regional earthquakes: Analysis of the 19 November 1923 earthquake in the Pyrenees.

Bi, Y., Feng, S., Zhao, G., & Han, B. (2016). Digitalizing seismograms using a neighborhood backtracking method. In F. Lehner & N. Fteimi (Eds.), *Knowledge science, engineering and management* (pp. 391–401). Springer International Publishing.

Bogiatzis, P., & Ishii, M. (2016). DigitSeis: A new digitization software for analog seismograms. *Seismological Research Letters, 87*(3), 726–736.

Bogiatzis, P., & Ishii, M. (2016). DigitSeis: A new digitization software for analog seismograms. *Seismological Research Letters, 87*(3), 726–736.

Bondár, I., Engdahl, E. R., Villaseñor, A., Harris, J., & Storchak, D. (2015). ISC-GEM: Global instrumental earthquake catalogue (1900–2009), II. Location and seismicity patterns. *Physics of the Earth and Planetary Interiors, 239*, 2–13.

Bondár, I., & Storchak, D. (2011). Improved location procedures at the International Seismological Centre. *Geophysical Journal International, 186*(3), 1220–1244.

Bormann, P. (2012). Magnitude calibration formulas and tables, comments on their use and complementary data. Technical report.

Bormann, P., & Dewey, J. W. (2012). The new IASPEI standards for determining magnitudes from digital data and their relation to classical magnitudes. *New Manual of Seismological Observatory Practice, 2*, 1–41.

Bormann, P., Dewey, J. W., Gabsatarova, I., Gregersen, S., Gusev, A. A., Kim, W. Y., Patton, H. J., Presgrave, B. W., Ruifeng, L., Saul, J., Storchak, D., Uhrammer, R. A., Wendt, S., Firbas, P., Havskov, J., Klinge, K., & Veith, K. (2013a). Summary of magnitude working group recommendations on standard procedures for determining earthquake magnitudes from digital data. Technical report.

Bormann, P., Klinge, K., & Wendt, S. (2014). Data analysis and seismogram interpretation. In P. Bormann (Ed.), *New manual of seismological observatory practice 2 (NMSOP-2)* (pp. 1–126). Potsdam: Deutsches GeoForschungsZentrum GFZ.

Bormann, P., Wendt, S., & DiGiacomo, D. (2013). Seismic sources and source parameters. In P. Bormann (Ed.), *New manual of seismological observatory practice 2 (NMSOP2)* (pp. 1–259). Potsdam: Deutsches GeoForschungsZentrum GFZ.

Bromirski, P., & Chuang, S. (2003). SeisDig: Software to digitize scanned analog seismogram images, user's manual.

Bungum, H., Pettenati, F., Schweitzer, J., Sirovich, L., & Faleide, J. I. (2009). The 23 October 1904 MS 5.4 Oslofjord earthquake: Reanalysis based on macroseismic and instrumental data. *Bulletin of the Seismological Society of America, 99*(5), 2836–2854.

Burkhard, K., Dobrynina, A., Mackey, K. G., & Burk, D. (2019). Digitization of soviet era peaceful nuclear explosion seismograms from regional stations. *SnT2019*, (T2.5-P4), T2.5 Historical Data from Nuclear Test Monitoring.

Cadek, O. (1987). Studying earthquake ground motion in Prague from Wiechert seismograph records. *Gerlands Beitrage zur Geophysik, 96*, 438–447.

Charlier, C., & Van Gils, J. (1953). Liste des Stations Seismologiques Mondiales. Technical report.

Chen, W., & Molnar, P. (1977). Seismic moments of major earthquakes and the average rate of slip in Central Asia. *Journal of Geophysical Research, 82*(10).

Cummins, P. R., Baba, T., Kodaira, S., & Kaneda, Y. (2002). The 1946 Nankai earthquake and segmentation of the Nankai trough. *Physics of the Earth and Planetary Interiors, 132*(1–3), 75–87.

Custodio, S., Batllo, J., Martins, D., Antunes, F., Narciso, J., Carvalho, S., Lima, V., Lopes, F. C., Ribeiro, P., Sleeman, R., Alves, E. I., & Gomes, C. R. (2012). Station COI: Dusting off an old seismic station. *Seismological Research Letters, 83*(5), 863–869.

Dewey, J., & Byerly, P. (1969). The early history of seismometry (to 1900). *Bulletin of the Seismological Society of America, 59*(1), 183–227.

Di Giacomo, D., Harris, J., Villaseñor, A., Storchak, D. A., Engdahl, E. R., Lee, W. H., Verney, R., Safronova, N., Wylie, R., Baranauskaite, A., Wilson, J., & Simpson, H. (2015). ISC-GEM: Global instrumental earthquake catalogue (1900–2009), I. Data collection from early instrumental seismological bulletins. *Physics of the Earth and Planetary Interiors, 239*, 14–24.

Di Giacomo, D., Olaru, D., Armstrong, A., Harris, J., & Storchak, D. A. (2022). The ISC electronic archive of printed station and network bulletins. *Seismological Research Letters, 93*(2A), 749–752.

Dost, B., & Haak, H. W. (2002). A Comprehensive Description of the KNMI Seismological Instrumentation 60.

Ekström, G., Nettles, M., & Dziewoński, A. M. (2012). The global CMT project 2004–2010: Centroid-moment tensors for 13,017 earthquakes. *Physics of the Earth and Planetary Interiors, 200–201*, 1–9.

Engdah, E. R., van der Hilst, R., & Buland, R. (1998). Global teleseismic earthquake relocation with improved travel times and procedures for depth determination. *Bulletin of the Seismological Society of America, 88*(3), 722–743.
Ferrari, G. (1992). *Two hundred years of seismic instruments in Italy, 1731–1940.* Istituto nazionale di geofisica (Italia): SGA Storia-Geofisica-Ambiente.
Geller, R. J., & Kanamori, H. (1977). Magnitude of great shallow earthquakes from 1904 to 1952. *Bulletin of the Seismological Society of America, 67*(3), 587–598.
Glover, D. P., Meyers, H., Herrmann, R. B., & Whittington, M. (1985). Inventory of filmed historical seismograms and station bulletins at world data center A. Technical Report June.
Grabrovec, D., & Allegretti, I. (1994). On the digitizing of historical seismograms. *Geofizika, 11*, 27–31.
Gutenberg, B. (1945). Amplitudes of P, PP, and S and magnitude of shallow earthquakes*. *Bulletin of the Seismological Society of America, 35*(2), 57–69.
Gutenberg, B. (1945). Amplitudes of surface waves and magnitudes of shallow earthquakes. *Bulletin of the Seismological Society of America, 35*(1), 3–12.
Gutenberg, B. (1945). Magnitude determination for deep-focus earthquakes. *Bulletin of the Seismological Society of America, 35*, 117–130.
Gutenberg, B. (1956). Great earthquakes 1896–1903. *Eos, Transactions American Geophysical Union, 37*(5), 608–614.
Hanks, T. C., & Kanamori, H. (1979). A moment magnitude scale. *Journal of Geophysical Research B: Solid Earth, 84*(B5), 2348–2350.
Heimann, S., Kriegerowski, M., Isken, M., Cesca, S., Daout, S., Grigoli, F., Juretzek, C., Megies, T., Nooshiri, N., Steinberg, A., Sudhaus, H., Vasyura-Bathke, H., Willey, T., & Dahm, T. (2017). Pyrocko—An open-source seismology toolbox and library.
Henni, P., & Lawrie, K. (1999). British geological survey the national seismological archive WWSSN microfilm collection [Final report]. Technical report, British Geological Survey, Edinburgh.
Henni, P., Lovell, J., & Lawrie, K. (2000). *UK historical seismograms and bulletins held in the national seismological archive (NSA)*. Technical report, British Geological Survey.
Howell, B. F. (1990). An introduction to seismological research: History and development (pp. 57–72).
Hunter, R. N. (1972). Use of LPZ for magnitude. In: NOAA Technical Report ERL 236- ESL21, J. Taggart. Boulder, Colorado.
Hwang, L., Davis, U. C., Ahern, T., & Ellsworth, W. (2019). Securing legacy seismic data to enable future discoveries Albuquerque, New Mexico Organizing Committee.
Hwang, L. J., Ahern, T., Ebinger, C. J., Ellsworth, W. L., Euler, G. G., Okal, E. A., Okubo, P. G., Walter, W. R., & He (2020). Rescuing legacy seismic data FAIR'ly. *Seismological Research Letters, 91*(3), 1339–1340.
International Seismological Centre. (2022). ISC-GEM Earthquake Catalogue.
ISC, (International Seismological Centre). (2018). On-line Bulletin. https://doi.org/10.31905/D808B830
Ishii, M., Ishii, H., Bernier, B., & Bulat, E. (2015). Efforts to recover and digitize analog seismograms from Harvard-Adam Dziewoński observatory. *Seismological Research Letters, 86*(1), 255–261.
Kanamori, H. (1977). The energy release in great earthquakes. *Journal of Geophysical Research, 82*(20), 2981–2987.
Kanamori, H., Rivera, L., & Lee, W. H. K. (2010). Historical seismograms for unravelling a mysterious earthquake: The 1907 Sumatra Earthquake. *Geophysical Journal International, 183*, 358–374.
Kimball, S., et al. (2019). *GIMP—GNU image manipulation program.* GIMP is a freely distributed piece of software. http://www.gimp.org/.

Kulhanek, O., & Wahlstrom, R. (1942). **History of instrumental seismological observations in Sweden** (pp. 8–22).

Kulikova, G. (2016). *Source parameters of the major historical earthquakes in the Tien-Shan region from the late 19th to the early 20th century*. University of Potsdam.

Kulikova, G., & Krüger, F. (2015). Source process of the 1911 M8.0 Chon-Kemin earthquake: Investigation results by analogue seismic records. *Geophysical Journal International, 201*(3), 1891–1911.

Kulikova, G., & Krüger, F. (2017). Historical seismogram reproductions for the source parameters determination of the 1902, atushi (kashgar) earthquake. *Journal of Seismology, 21*(6), 1577–1597.

Kulikova, G., Krüger, F., Schurr, B., Heimann, S., & Brzoska, E. (2016). Source parameters of the Sarez-Pamir earthquake of 1911 February 18. *Geophysical Journal International, 205*(2), 1086–1098.

Lee, T., Ishii, M., & Okubo, P. (2020). Relative time corrections for historical analog seismograms using the single-day ambient noise correlation function. *Bulletin of the Seismological Society of America, 110*(6), 3185–3195.

Lund, B., Schmidt, P., Shomali, Z. H., & Roth, M. (2021). The modern Swedish national seismic network: Two decades of intraplate microseismic observation. *Seismological Research Letters, 92*(3), 1747–1758.

McComb, H. E., & West, J. C. (1931). *Bulletin of the national research council. List of seismologic stations of the world*. Published by The National Research Council of The National Academy of Sciences Washington D.C. Number 82, Second edition.

McConnell, A. (1986). Science Museusm geophysics & geomagnetism: Catalogue of the science museum collection.

Meyers, H. and Lee, W. (1979). Historical seismogram filming project: First progress report. Technical Report November.

Michelini, A., De Simoni, B., Amato, A., & Boschi, E. (2005). Collecting, digitizing, and distributing historical seismological data. *Eos, 86*(28), 2003–2005.

Molnar, P., & Deng, Q. (1984). Faulting associated with large earthquakes and the average rate of deformation in central and eastern Asia. *Journal of Geophysical Research: Solid Earth, 89*(B7), 6203–6227.

Murotani, S., Satake, K., Tsuruoka, H., Miyake, H., Sato, T., Hashimoto, T., & Kanamori, H. (2020). A database of digitized and analog seismograms of historical earthquakes in Japan. *Seismological Research Letters, 91*(3), 1459–1468.

Okal, E. A. (1990). A variable-period mantle magnitude for intermediate and deep earthquakes.

Okal, E. A. (1992a). Use of the mantle magnitude Mm for the reassessment of the moment of historical earthquakes. I: Shallow events. *Pure and Applied Geophysics PAGEOPH, 139*(1), 17–57.

Okal, E. A. (1992b). Use of the mantle magnitude mm for the reassessment of the moment of historical earthquakes. II: Intermediate and deep events. *Pure and Applied Geophysics PAGEOPH, 139*(1), 59–84.

Okal, E. A. (2005). A re-evaluation of the great Aleutian and Chilean earthquakes of 1906 August 17. *Geophys. J. Int, 161*, 268–282.

Okal, E. A. (2015). Historical seismograms: Preserving an endangered species. *GeoResJ, 6*, 53–64.

Okal, E. A., & Reymond, D. (2003). The mechanism of great Banda Sea earthquake of 1 February 1938: Applying the method of preliminary determination of focal mechanism to a historical event. *Earth and Planetary Science Letters, 216*(1–2), 1–15.

Okal, E. A., & Talandier, J. (1989). Mm: a variable-period mantle magnitude. *Journal of Geophysical Research, 94*(B4), 4169–4193.

Ou, Q., Kulikova, G., Yu, J., Elliott, A., Parsons, B., & Walker, R. (2020). Magnitude of the 1920 Haiyuan earthquake reestimated using seismological and geomorphological methods. *Journal of Geophysical Research: Solid Earth, 125*(8).

Paulescu, D., Rogozea, M., Popa, M., & Radulian, M. (2016). Digitized database of old seismograms recorder in romania. *Acta Geophysica, 64*(4), 963–977.

Pérez-Campos, X., Armendáriz-Sánchez, S., Espíndola, V. H., Castro-Escamilla, M., Perez, J., Casiano, L. M., Rasilla, I. R., Monroy, C. C., & Cárdenas, A. (2020). Preservation and reuse of historical seismic data in Mexico: SisMOmex and the online "National seismogram library". *Seismological Research Letters, 91*(3), 1482–1487.

Pintore, S., Quintiliani, M., & Franceschi, D. (2005). Teseo: A vectoriser of historical seismograms. *Computers and Geosciences, 31*(10), 1277–1285.

Pintore, S., Quintiliani, M., & Franceschi, D. (2005). TESEO: A vectoriser of historical seismograms. *Computational Geosciences, 31*(10), 1277–1285.

Powell, T., & Fries, D. (1964). *Handbook: World-wide standard seismograph network. technical report, acoustics and seismics laboratory, institute of science and technology, The University of Michigan, Ann Arbor, Michigan.*

Reid, H. F. (1912). On the choice of a seismograph. *Bulletin of the Seismological Society of America, 2*(1), 8–30.

Reymond, D., & Okal, E. A. (2000). Preliminary determination of focal mechanisms from the inversion of spectral amplitudes of mantle waves. *Physics of the Earth and Planetary Interiors, 121*(3–4), 249–271.

Richter, C. F. (1935). An instrumental earthquake magnitude scale. *Bulletin of the Seismological Society of America, 25*(1), 1–32.

Rivera, L., Sieh, K., Helmberger, D., & Natawidjaja, D. (2002). A comparative study of the Sumatran subduction-zone earthquakes of 1935 and 1984. *Bulletin of the Seismological Society of America, 92*(5), 1721–1736.

Scherbaum, F. (2001). *Of poles and zeros: Fundamentals of digital seismology*. Kluwer Academic Publishers.

Schlupp, A., & Cisternas, A. (2007). Source history of the 1905 great Mongolian earthquakes (Tsetserleg, Bolnay). *Geophysical Journal International, 169*, 1115–1131.

Schwartz, S. Y., & Ruff, L. J. (1985). The 1968 Tokachi-Oki and the 1969 Kurile islands earthquakes: variability in the rupture process. *Journal of Geophysical Research, 90*(B10), 8613–8626.

Schweitzer, J., & Lee, W. (2003). Old Seismic bulletins to 1920: A collective heritage from early seismologists. In H. K. William Lee, P. C. J. Hiroo Kanamori, & C. Kisslinger, (Eds.) *International handbook of earthquake and engineering seismology* (pp. 1665–1723). Academic Press, International Geophysics.

Stich, D., Batlló, J., Maciá, R., Teves-Costa, P., & Morales, J. (2005). Moment tensor inversion with single-component historical seismograms: The 1909 Benavente (Portugal) and Lambesc (France) earthquakes. *Geophysical Journal International, 162*, 850–858.

Storchak, D., Di Giacomo, D., Bondár, I., Harris, J., Engdahl, E., Lee, W., Villaseñor, A., Bormann, P., & Ferrari, G. (2012). ISC-GEM global instrumental earthquake catalogue (1900-2009). GEM Technical Report 2012-01, V1.0.0:128pp.

Taylor, S. R., & Yang, X. D. (2011). Development of software to digitize historic hardcopy seismograms from nuclear explosions. In *2010 monitoring research review: Ground-based nuclear explosion monitoring technologies DEVELOPMENT* (pp. 789–798).

U.S. Geological Survey. Earthquake Hazards Program (2017). Advanced National Seismic System (ANSS) Comprehensive Catalog of Earthquake Events and Products

Vanek, J., Zatopek, A., Karnik, V., Kondorskaya, N., Riznichenko, Y. V., Savarensky, E., Solov'ev, S., & Shebalin, N. (1962). Standardization of magnitude scales. *Izvestiya of the Academy of Sciences of the U.S.S.R., Geological Series, 2*, 108–111.

Villasenor, A., & Engdahl, E. R. (2005). A digital hypocenter catalog for the international seismological summary. *Seismological Research Letters, 76*(5), 554–559.

Villaseñor, A., & Engdahl, E. R. (2007). Systematic relocation of early instrumental seismicity: Earthquakes in the international seismological summary for 1960–1963. *Bulletin of the Seismological Society of America, 97*(6), 1820–1832.
Wang, L., Liu, R., & Sun, Y. (2016). Automatic arc distortion correction of seismograms using the low-rank matrix recovery method. *IEEE Geoscience and Remote Sensing Letters, 13*(2), 142–146.
Wood, H. O. (1921). *Bulletin of the national research council. A list of seismologic stations of the world.* The National Research Council of the National Academy of Sciences, Washington D.C . Part 7, No. 15.
Xu, Y., & Xu, T. (2014). An interactive program on digitizing historical seismograms. *Computers and Geosciences, 63*, 88–95.

Macroseismology

3

Susan E. Hough

> We must preserve the history, the record of what happened, and the Coast and Geodetic Survey is now doing this admirably. This, of course, must not hinder us in measuring instrumentally the earth motion in as many locations as possible. But we must remember that one set of strong motion seismographs in one basement in a town…will never give us a total picture of what happened throughout the town…
>
> What we need is more information of all kinds, not less."
>
> —Perry (Byerly, Bull. Seism. Soc. Am. 59:1735, 1969).

3.1 Intensity Scales

Macroseismic data, defined to be the documented effects of earthquake shaking on people and the natural and built environments, have been traditionally interpreted using an intensity scale that assigns a numerical value based on the severity of documented effects. Intensity scales were introduced in the technical literature in the mid-nineteenth century (e.g., Baird Smith, 1843; Mallet and Mallet, 1858). The concept, however, of ranking shaking severity with a numerical scale is so natural that it arose independently even earlier (e.g., Drake, 1815; Egen, 1828; McMurtie, 1839). The Rossi-Forel (RF) scale, introduced by de Rossi (1883), was widely used

S. E. Hough (✉)
USGS Earthquake Science Center, 350 N. Akron Road, Pasadena, CA 91106, USA
e-mail: hough@usgs.gov

A. Elliott and C. Gruetzner (eds.), *Understanding Past Earthquakes*,
https://doi.org/10.1007/978-3-031-73580-6_3

by the scientific community in the late 19th and early twentieth centuries (Musson et al., 2010). The RF scale had ten intensity levels, conventionally described by Roman numerals, ranging from I for shaking that was not felt, or perhaps felt only by an experienced observer in special circumstances, to X for shaking that caused "great disasters; overturning of rocks; fissures in the surface of the earth; mountain slides" (de Rossi, 1883). Roman numerals are still sometimes used to denote intensities, although with the advent of modern computers there has been an increasing trend towards conventional Arabic numbers. Descriptions of each RF intensity degree were brief, for example with intensity VIII defined by "Fall of chimneys; cracks in the walls of buildings." Following the introduction of the Rossi-Forel scale, a number of other scales were introduced (see Davison, 1921; Musson et al., 2010, for detailed reviews). Among the new scales was a modification of the RF scale published by Mercalli (1902). The Mercalli (1902) scale also included ten steps, but with expanded description of effects at each intensity level. Later proposed modifications expanded the scale to include 12 steps (Cancani, 1904), adding two extra degrees at the top of the scale. Sieberg (1912) expanded the scale further, providing more detailed description of effects at each intensity level, with subsequent revision by Sieberg (1923). The scale proposed by Sieberg (1923) became known as the MSC (Mercalli-Sieberg-Cancani) scale.

Much of the early work on intensity scales was published in Italian. The MSC scale was translated into English by Wood and Neumann (1931). The scale published by Wood and Neumann (1931) was very similar to the MSC scale, but included some further modifications, including shortening the descriptions of Sieberg (1923). While the scale introduced by Wood and Neumann (1931) was a slight translated modification of the MCS scale, they called it the Modified Mercalli Intensity (MMI) scale, a name that has been widely adopted, in particular within the United States. The MMI scale was used by the United States Coast and Geodetic Survey (USCGS), the government agency responsible for collection of earthquake reports in the United States from 1933 onward, as well as other organizations involved with seismological work, and went on to be widely adopted in the United States as well as elsewhere.

Following the MCS, the MMI scale formulated by Wood and Neumann (1931) included 12 steps, ranging from I, defined as not felt or rarely felt under especially favorable circumstances, to XII, defined as considerable damage to well-designed and engineered structures. While the MMI nomenclature has been used since the original formulation of the scale, experience in applying the scale led to changes in the interpretation of macroseismic effects. This led to a number of updates to the scale itself, including a minor modification by Richter (1958). These refinements were used in turn as the basis for the New Zealand MMI scale (Eiby, 1966). More recently, Stover and Coffman (1993) formally based their MMI formulation on the original Wood and Neumann (1931) scale but articulated a number of ways that their interpretations differed from the interpretations used earlier in some U.S. government agencies. For example, Stover and Coffman (1993) noted that, in their experience, subjective human responses are not reliable considerations for assigning intensities above IV. Thus, while published intensity values since 1931 have been described as MMI values, they have not necessarily been based on the same version of the scale, nor following the same conventions. As a consequence, macroseismic intensity assignments have not always been consistent over time or among different individuals.

An independent modification of the 12-step scale, based on earlier work (e.g., Sieberg, 1923; Wood & Neumann, 1931), was developed by Medvedev et al. (1964), published as the MSK (Medvedev, Sponheuer, and Karnick) scale. The MSK scale was widely used outside of the United States, notably in Europe.

In 1989 the European Seismological Commission launched a major effort to update the MSK scale, culminating in the 12-step European Macroseismic Scale (EMS, often called EMS-98), published in Grünthal (1998). Notably, the EMS scale includes greatly expanded guidelines for assigning intensity values based on the severity (grade) of damage to different types (classes) of structures, and the statistical incidence of effects. The EMS scale thus formally provides guidance that intensity assignments should not be based on the most severe observed effects, if they were few in number, but rather are assigned based on statistical incidence of effects, for which general guidelines are provided (e.g., whether levels of damage were sustained by "most," "many," or "few" buildings of different classes).

Most but not all intensity scales share a common lineage, as outlined above. While all of the modern intensity scales introduced in the United States and Europe are rooted in the original definitions of the RF scale, the Japanese Meteorological Agency (JMA) developed a 7-step JMA scale, with several levels broken into "upper" and "lower" categories (see ***Data and Resources***). Although originally introduced as a conventional intensity scale based on macroseismic observations, JMA intensities are now commonly determined from instrumental data (see Shabestari & Yamazaki, 2001). The so-called INQUA (International Union for Quaternary Research) scale, or Environmental Seismic Intensity (ESI) scale is another departure from traditional intensity scales, using a 12-step scale based only on observed environmental effects (*Michetti* et al., 2007). Using the ESI scale, for example, intensities higher than 10 are assigned based on effects such as observed surface rupture, widespread liquefaction, and significant landsliding extending to distances of 200–300 km.

Musson et al. (2010) presented a detailed comparison of modern intensity scales, including the RF, EMS-98 and JMA scales, and showed that the 10- and 12-step scales generally yield consistent intensities for a given set of macroseismic information, with some differences at the upper end of the scales. What this indicates, in effect, is that modifications of early scales did not intend to change the meaning of each intensity level, but rather focused on improving guidelines to assess intensities more accurately. This statement might appear to suggest that intensity values assigned using different scales can be combined indiscriminately. Such practice is discouraged for several reasons, many related to the approach taken by individual researchers in assigning intensities (discussed in the next section). But also, while different scales might yield consistent numerical intensities for the same macroseismic information, the different criteria used by different scales lends itself to differing tendencies towards bias. For example, EMS intensities assigned using very detailed guidelines might tend to be lower than, say, RF intensities assigned using very brief guidelines. The EMS scale includes explicit guidelines to consider statistical incidence of effects, and be mindful of sampling biases (Grünthal, 1998).

Traditionally, intensity values have been assigned with whole numbers, although intermediate values have been indicated by a range (e.g., 4–5) by some authors. Dengler and Dewey (1998) introduced the community decimal intensity (CDI) scale, reported to the nearest 0.1 unit, determined from telephone survey data using an algorithm calibrated to agree on average with MMI values assigned by the US Geological Survey. As envisioned by Dengler and Dewey (1998), the CDI scale provided a tool with which intensity estimates could be estimated rapidly using information collected on-line. As a consequence of the imposed linearity of equations used, the Dengler and Dewey (1998) algorithm produces CDI values in the range of 3.3 (not felt by anyone in the community) to 10.7 (major damage, felt by all). Dengler and Dewey (1998) noted that intensity values of XI and XII, as well as values of I, were very rarely assigned in practice. This raises an important general point about intensity data sets: although high intensity values are of greatest concern for hazard, and intensity scales have differed at the upper end of the scale (e.g., 10 *vs* 12 steps), we often have few observations of strongest shaking levels.

Drawing from the work of Dengler and Dewey (1998), *Wald* et al. (1999) defined the Community Internet Intensity (CII) scale for use by the Internet-based Community Internet Intensity Map system. (In later usage the CII term has been dropped in favor of the original CDI term. In this chapter, following *Wald* et al. (1999), I use CII to refer to intensities determined from online data.) This system, commonly known as "Did You Feel It?" (DYFI), originally determined CII values within postal ZIP codes. Musson (2006) later proposed an algorithm along the same lines to determine EMS intensities automatically. *Wald* et al. (1999) modified the Dengler and Dewey (1998) algorithm to differentiate between different levels of weakly felt shaking. The CII algorithm assigns 2.0 if any felt report is received within a given spatial footprint, and 1.0 if only "not felt" reports are received. Values between 1.0 and 2.0 are never produced by the algorithm. Boatwright and Phillips (2017) proposed an approach to correct intensity assignments for underreporting ZIP codes, which they showed would improve the resolution of the overall felt extent by defining intensities between 1.0 and 2.0 depending on the number of DYFI responses relative to the population of a postal ZIP code. To-date, this approach has not been implemented, so DYFI intensities never include decimal values between 1.0 and 2.0. At the upper end of the scale, the DYFI system combines the highest shaking levels into CII 10 +, effectively collapsing the scale back to ten levels. CII values are assigned independently for each questionnaire; where multiple responses are received within a given spatial footprint, the responses for each question are averaged to obtain an average CII. Originally, CII values were averaged within postal ZIP codes, which vary considerably in spatial size and contain an average population (today) of about 9,700. The DYFI system was later expanded to average questionnaire answers within both ZIP codes and geocoded cells, moving increasingly towards uniform generation of results within both 1- and 10-km grid cells (Quitoriano & Wald, 2020).

The DYFI system was intended to supplement but not replace assigned USGS MMI values; *Wald* et al. (1999) noted that CII values are not necessarily equivalent to traditional USGS MMI assignments. The efficiency of an internet-based approach to data collection was, however, almost immediately apparent (*Wald* et al., 2011). Internet-based macroseismic survey systems have now been implemented in a number of regions, including New Zealand (Coppola et al., 2010), Italy (Sbarra et al., 2010), and other individual countries, as well as the global system introduced by the European-Mediterranean Seismological Centre (EMSC; Amorèse,et al., 2015). The DYFI system also now collects reports from global earthquakes for reports made outside of the U.S., including both geocoded values and intensities aggregated within cities. The database of city names is, however, relatively coarse, so user-specified city names might not be recognized (Martin et al., 2015). DYFI participation can also overwhelmingly limited to relatively affluent urban areas in countries where Internet access and even basic literacy remain uneven (*Hough and Martin*, 2021). In a later section I discuss how Internet-based intensities such as DYFI compare to traditional intensities. I note here that in recent years, the usage of "intensity measure" has broadened to include scalar metrics derived from instrumental data. In this chapter, in the interests of brevity I use the term "intensity" to refer specifically to macroseismic intensity.

3.2 Macroseismic Data

The first-hand accounts of earthquake effects are themselves interpretations of macroseismic effects; in the absence of specific reasons to question reports, we cannot be sure but generally assume these interpretations are reliable. An important but sometimes forgotten note regarding numerical intensity values is that they are not data, but rather interpretations. Published intensity values are not only assigned with different scales, but also following different conventions, with different levels of conservatism. Where possible, unless an exhaustive reconsideration of a historical catalog has been undertaken (e.g., Martin & Szeliga, 2010), it is therefore desirable to consider original sources for any detailed investigation. The primary sources of macroseismic information for historic earthquakes come from extant archival accounts such as newspapers, diaries, letters, and other original sources. Exhaustive efforts have been made to compile archival sources in a number of regions, including California (Toppozada & Parke, 1982; *Salditch* et al., 2020), Italy (eg., *Guidoboni*, 1994), Japan (e.g., Ishibashi, 2004), India (Martin & Szeliga, 2010), and, recently, Canada (*Lamontagne*, oral communication, 2016), and Indonesia (*Martin* et al., 2021). For some twentieth century studies of particular earthquakes, summaries of effects were provided with published papers, for example on microfiche cards (e.g., Nuttli, 1973); however, in some focused studies, original sources were not preserved in any accessible repository. There has been a growing trend toward preserving at least a summary of documented earthquake effects (e.g., Martin & Szeliga, 2010), often in electronic supplements to published articles. But such information is not always made available.

Beginning in the nineteenth century, extensive macroseismic information was sometimes collected and published in scientific reports following notable earthquakes (e.g., Baird Smith, 1843; Taylor, 1884; *Dutton,* 1889; Oldham, 1897). In some countries, macroseismic data were collected following significant earthquakes by individual researchers or government programs, typically starting in the late 19th or early twentieth century (e.g.,.Cecić, 2018; Ulrich, 1935). In the United States, macroseismic data have been collected by a succession of government agencies starting with the U.S. Weather Service in the late nineteenth century, with collection of postal questionnaires beginning in the early twentieth century. Information from questionnaires is relatively straightforward to interpret, since questionnaires were based on indicators used to define intensities. Beginning in 1924, the USCGS left stacks of questionnaires with employees of large companies, who filled out the cards and returned them after felt earthquakes (Byerly & Dyk, 1936). Following the 1977 enactment of the National Earthquake Hazards Reduction Program in 1977 and the establishment of the U.S. Geological Survey (USGS) Earthquake Program, information was collected by the U.S.G.S. via postcards that were mailed out after a significant earthquake (e.g., Dewey et al., 1995). From 1999 onward, traditional intensity surveys were supplanted by the on-line DYFI system.

Information from USCGS postcard questionnaries was transcribed and summarized in a series of published reports, including an informal publication series, "Abstracts of Earthquake Reports from the Pacific Coast and the Western Mountain Region," (AER) that was sent to a limited distribution list by the San Francisco office of the USCGS (see ***Data and Resources***). These initial reports were superceded by the more formal "United States Earthquakes" (USE) publication series, which were published between 1928 and the mid-1980s, and were more widely distributed. USE reports included information about all US earthquakes. Notably, however, USE reports included less detailed information than the AER publications. For example, in USE reports, locations reporting weak shaking were often lumped together under "Intensity I-III," or even "Intensity I-IV," with no further details provided, whereas AER tallies included summaries of effects for locations for which low intensities were assigned. For example, for the widely felt El Centro earthquake of 18 May 1940, the AER report (see ***Data and Resources***) includes summary effects for long lists of cities where low intensities were assigned, ranging from "ratting of windows, doors, and dishes" for MMI IV to "felt by few" for lower intensities. The USE report (Neumann, 1942) includes only lists of cities where MMI IV and MMI I-III were assigned. (For some events, AER reports did not themselves include numerical intensities values for MMI I-III, but summarize effects provide information that can be used to distinguish between low intensity values.) Summary observations were also generally not included in the USE reports for intensity levels IV-V, while the AER tallies summarized effects at locations where intensities IV-V were assigned. The more obscure AER reports, of which many modern scholar are not aware, thus represent an invaluable source of information about earthquakes since 1933, although they can be difficult to find (see ***Data and Resources***).

An even more obscure but potentially more valuable data set exists for earthquakes felt in the United States: the original postal questionnaires collected by the USCGS. These postcards were scanned and the collection of digital PDF files is maintained at the U.S. Geological Survey office in Golden, Colorado. It is not currently accessible remotely due to Privacy Act restrictions that preclude the U.S. government from sharing personal data such as addresses.

3.3 The Scientific Value of "Unscientific Data"

The value of macroseismic data has long been recognized by many in the earthquake science community. From the time that instrumental data existed in seismology, however, there was tendency among some researchers to dismiss the value of intensity values as "unscientific" data. Perry Byerly, an early pioneer in network seismology, reportedly lamented that damage and other macroseismic information was neglected in the early instrumental era (James Dewey, oral communication 2019). In Byerly (1969) he summarized the unique value of macroseismic data and intensity values, including the quote used as an epigraph to this chapter. As discussed above, it is important to remember that intensities are fundamentally not basic data and have significant uncertainties. Moreover, the characterization of shaking severity by a single number clearly has limitations, as the character of ground motions will depend on duration and frequency content as well as peak velocity or peak acceleration. These limitations notwithstanding, studies have increasingly demonstrated the correspondence between consistently interpreted intensity values and instrumentally recorded ground-motion metrics. I now summarize the case for the scientific veracity and value of "unscientific" data.

Overwhelmingly, shaking intensities are controlled by frequencies between 1–8 Hz (*Trifuanc and Brady*, 1975; *Sokolev and Chernov*, 1998). Ground motions at frequencies significantly lower than 1 Hz will be potentially damaging to taller structures. For example, as a rule of thumb, 0.5 Hz (2-s) ground motions will be potentially damaging to 20-story buildings. Especially in historical times, but even in recent times, accounts of damage to tall buildings are few in number. With rare exception, intensity values do not reflect long-period (e.g., < 1 Hz) shaking effects (Hanks & Johnston, 1992; Hough, 2015). Accordingly, intensity magnitudes based on macroseismic data alone effectively provide an estimate of energy magnitude, but only limited constraint on moment magnitude (Hanks & Johnston, 1992; Hough, 2015). Without independent information to constrain rupture dimension, an intensity-based magnitude can differ from moment magnitude by a full magnitude unit (e.g., Hough, 2015).

As discussed by Hanks and Johnston (1992), the high-frequency energy that controls macroseismic intensities depends strongly on stress drop, a quantity related to the details of the rupture process of the source (e.g., Brune, 1970).

Although the definition and interpretation of stress drop are model dependent and complicated (*Atkinson and Beresnev*, 1997), so-called static stress drop generally reflects the ratio between average slip in an earthquake and rupture area (e.g. Kanamori & Anderson, 1975).

By all indications, macroseismic intensities do depend on stress drop (Hanks & Johnston, 1992), but fundamentally they provide a measure of shaking integrated over the frequencies that are damaging to most structures, and are increasingly recognized to be of tremendous value, if they are interpreted carefully, with an appreciation of the potential biases discussed in the following section. As an integrated measure, they provide a more robust indication of shaking at frequencies of engineering concern than peak ground acceleration (PGA), the parameter that engineers have long relied on to characterize shaking severity, but that fundamentally is an extremal value. Intensity values remain useful because they are directly relevant to the key issue: shaking intensities of importance to people and structures.

Intensity values are also valuable because simple considerations show that they must be generally robust measures. In addition to providing an integrated measure of shaking at frequencies of engineering concern, the range of perceptible earthquake ground motions is quantifiable, and on the order of a factor of 1000 (0.1—1 *g*). To cover this range with a 10-step intensity scale, assuming that macroseismic effects do generally scale with PGA, it follows that each unit step in intensity must correspond robustly to a factor of approximately 2 in PGA (Hough, 2000). The emergence of the DYFI and other internet-based systems has further demonstrated that consistently interpreted intensity data provide surprisingly reliable indicators of ground motions (e.g., Atkinson & Wald, 2007; Worden et al., 2012), and can provide important insights regarding ground motions (e.g., Hough, 2012, 2014a, b). In fact, some USGS-produced ShakeMaps—which depict intensity values—rely on DYFI data to flesh out shaking distributions, particularly in areas where instrumentation is sparse (*Wald* et al., 2021). A number of studies have further demonstrated that consistently interpreted intensity data can be of unique value for testing probabilistic seismic hazard (PSH) maps (e.g. Marzochhi and Jordan, 2014; Stein et al., 2015; Brooks et al., 2018).

3.4 Interpretation of Macroseismic Observations

As discussed in seminal studies by Nicholas Ambraseys (e.g., Ambraseys, 1971, 1983), in any reinterpretation of a historical earthquake, it can be important to consider original sources rather than relying on past interpretations. As summarized above, intensity scales themselves have evolved over time. In addition to issues associated with intensity scales, the assignment of traditional intensities has been affected by myriad other issues (e.g.., Ambraseys, 1983). In the following discussion, we distinguish between "traditional intensities," assigned subjectively by one or a small group of researchers, and Internet-based intensities, determined by an algorithm. Practices with traditional intensity assignment have evolved over time, driven largely by a growing appreciation for how earthquake shaking effects people, the natural environment, and the built environment:

1. Environmental effects. Some of the indicators used by early intensity scales involved environmental effects such as rock falls other non-tectonic ground deformations such as liquefaction. While these observations can potentially be valuable to constrain ground motions (Michetti et al., 2007), through the twentieth century, environmental effects were recognized increasingly as potentially unreliable indicators of overall shaking intensity (Ambraseys, 1983). For example, the fact that rock falls sometimes occur spontaneously implies that earthquake-induced rock falls can potentially occur during weak shaking. It has further been shown that even a small earthquake can generate liquefaction (e.g., Musson, 1998), and that liquefaction can have enormous spatial variability depending on the susceptibility of sediments in an area that experiences strong shaking. While the MMI scale defined by Wood and Neumann (1931) incorporates environmental effects included in earlier scales, the scale used by Stover and Coffman (1993) notes that, at MMI VIII-XII, geological effects are considered only if no other information is available. In modern practice, numerical intensities are now often not assigned at all if the only documented effects are environmental.
2. Subjective human response. Intensity questionnaires have long included questions regarding human response to shaking: would an observer describe shaking as weak, mild, strong, etc.? And how did the observer respond to felt shaking? The emotional response to earthquake shaking depends on cultural and other factors, including whether an observer has felt earthquake shaking before, and the extent to which there is a danger of catastrophic building collapse. In recognition of these issues, traditional intensity assignments have increasingly been based more on objective indicators rather than subjective human reactions. The CDI and CII scales (Dengler & Dewey, 1998; Wald et al., 1999) formalize this evolution, considering but down-weighting information pertaining to human reaction (e.g., whether an observer took any action in response to shaking, and how they would qualitatively describe shaking severity.)
3. Building vulnerability. The development of intensity scales occurred, overwhelmingly, in Europe, the United States, and Japan. As discussed by Ambraseys and Douglas (2004), and references therein, these scales were based on typical vulnerability of building stock that was different from that found in other parts of the world, including the Indian subcontinent. The catastrophic impact of the 2010 Haiti earthquake was largely attributed to the prevalence of highly vulnerable structures in that country (e.g., DesRoches et al., 2011). While traditional construction in some parts of the world is sometimes surprisingly resilient (Ambraseys, 1983), vernacular construction in many regions is more vulnerable than what is found commonly in Europe, Japan, and the United States. Highly vulnerable masonry buildings, including those of adobe and rubble stone masonry, can sustain catastrophic damage and collapse at intensities of VII-VIII, and in extreme cases at even lower levels (Fig. 3.1). In effect, if all

local building stock is highly vulnerable, most intensity scales saturate at VII-VIII. The EMS-98 scale does not formally saturate until intensity X, although it may tend to saturate in practice if information is not available to estimate the statistical incidence of effects. I.e., while shaking might have been commensurate with a higher intensity, it is impossible to determine a higher intensity in the absence of observed damage to less vulnerable structures. If building stock includes even a small number of well-built structures, for example the well-built commercial building shown in Fig. 3.1, they can provide especially valuable information to constrain shaking levels. Failure to consider vulnerability carefully can result in inflation of assigned intensity values, but can also result in intensity assignments being effectively providing a lower bound on shaking intensity and ground-motion amplitude.

In contrast to the situation in countries where building stock is especially vulnerable, as noted by Maxwell Allen, simple, ordinary wood-frame structures in some places, including parts of the United States where wood-frame construction is common, can be surprisingly resilient: "In the frame house with loose fitting windows and terra cotta pipe chimneys guyed to the roof, any shock from intensity IV to VII or even VIII will produce about the same results." (*Maxwell*, 1934). Another illustrative example, ironically, is Haiti during historical times: A visitor to that country wrote in 1830 that buildings in Port-au-Prince were constructed chiefly of wood, seldom exceeding two stories in height, with a "paltry appearance." He added that, "this style of house was adopted by the French in consequence of

Fig. 3.1 During the 2010 Haiti earthquake the well-built commercial building on the left sustained no damage, and non-structural effects commensurate with MMI V. The building on the right collapsed catastrophically

frequent earthquakes, which were found to overthrow more substantial edifices (McKenzie, 1830, page 7).

The growing appreciation of the above issues through the twentieth and twenty-first centuries, and beyond, not only fueled refinements of intensity scales themselves, but has also led to an evolution in how intensity values have been assigned over time. Increasingly, care has been taken to avoid assigning intensity values based on isolated, unrepresentative effects.

3.5 Representative Versus Dramatic Effects

Any researcher who visits an area impacted by a major earthquake is invariably struck by not only the large-scale variability of damage in a region (often due to site response), but also the small-scale variability of damage, with undamaged buildings standing next door to collapsed structures. Even in Port-au-Prince following the 2010 Haiti earthquake, nearly half of all buildings sustained significant damage or collapse, but many structures survived with limited or no damage. In this case, undamaged structures included both especially well-built structures (Fig. 3.1) but also many that were clearly poorly built. Differences in vulnerability usually explain most but not all of the variability in damage, which must result in part from local and complex wave propagation effects.

As defined by modern scales, intensities should be assigned within a defined spatial footprint, based on the statistical incidence of effects. Postal questionnaire programs in the United States often relied on postmasters, whose responses provided an overall indication of observed effects within communities. Postal ZIP codes were introduced in 1963, providing a straightforward although imperfect way to define a spatial footprint. On average, today about 9,700 people live in each individual ZIP code, but with considerable variability. The spatial extent of individual ZIP codes also varies enormously. Although conventions were never laid out in detail, a perusal of reports (US Earthquakes, AER series) reveals that, in the interpretation of postcard questions, USCGS and later USGS practice was generally to assign an intensity based on the most severe effect documented with a ZIP code. For example, a MMI value of VI or VI-VII was typically assigned to an account described "rattling of windows and dishes" and no displacement of small objects, but "some chimneys cracked." Accounts of this type are not uncommon among both postcard questionnaires and traditional archival sources such as newspaper articles and diary entries (Fig. 3.2). (While one can always speculate that a given account might have been colored by an agenda, for example to downplay or overdramatize the severity of effects, one usually has no basis to discount the veracity of an account like that shown in Fig. 3.2.) In effect, a map based on this practice will tend to reflect the most severe damage observed within a given spatial footprint, and be biased high relative to a map based on representative effects. Hough (2013) showed that, for the 2011 Mineral, Virginia, earthquake, the distribution of intensities in individual ZIP codes within a large city follow a quasi-normal distribution (Fig. 3.3). (A strict normal distribution would include longer tails than are possible with intensity values.) A traditional intensity map may thus be controlled by the tail of the distribution, if not the extremal value, then certainly closer to the mean plus one standard deviation than to the mean, or representative, value.

Over time, practices have evolved increasingly towards assignment of representative intensity values within a given spatial footprint, based on the statistical

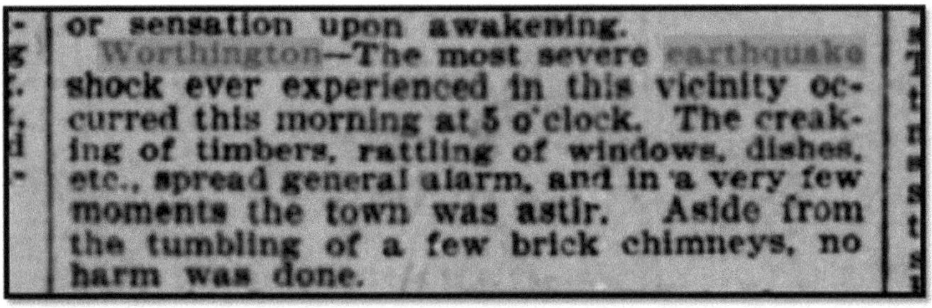
or sensation upon awakening.
Worthington—The most severe earthquake shock ever experienced in this vicinity occurred this morning at 5 o'clock. The creaking of timbers, rattling of windows, dishes, etc., spread general alarm, and in a very few moments the town was astir. Aside from the tumbling of a few brick chimneys, no harm was done.

Fig. 3.2 Newspaper account of the Charleston, Missouri, earthquake of 1895. Conventionally, an intensity of VI-VII—the MMI level at which "weak chimneys are broken at the roof line"—would have been assigned for an account like this based on the reported chimney damage. However, apart from the "tumbling" of a few chimneys, the overall effects are consistent with MMI 4–5

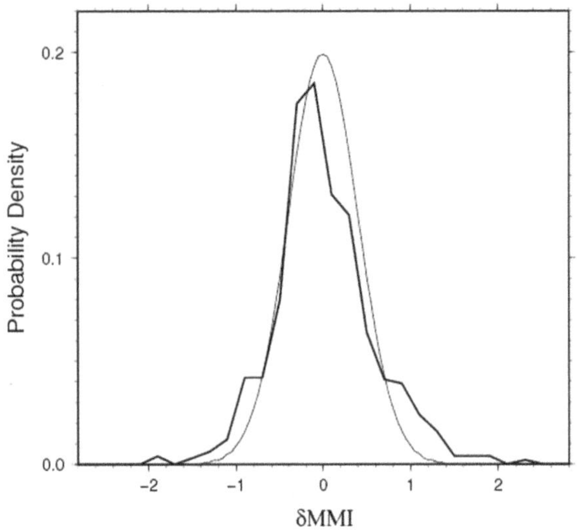

Fig. 3.3 Reprinted from Hough (2013), shows the distribution of intensity residuals within 20 cities for which CII values for the 2011 Mineral, Virginia, earthquake, were determined from at least 20 individual ZIP codes. Residuals are calculated as the difference between the average CII in each ZIP code and the overall average for the city. The distribution is well fit by a Gaussian, with high-intensity tails including residuals of + 1 and even + 2 intensity units, although a strict Gaussian distribution implies longer tails than would be possible given intensity values

incidence of effects. Algorithms such as that used by the DYFI system, which assess intensities for individual questionnaires but calculates average intensity within a ZIP code or geocoded cell by averaging responses to individual questions, effectively represents the codifiation of this evolution. It is important to note that an intensity map based on this practice will not indicate the most severe effects that occurred in an area.

Apart from the evolution of practices among researchers, earthquake reports themselves often reflect a fundamental reporting bias whereby the media and other eyewitnesses often focus on the most dramatic instances of damage rather than providing a good indication of the overall level of damage. This effect has been striking to many generations of earthquake professionals, as well as others, who have observed the disconnect between the media depiction of damage and the overall situation (e.g., Golden Era, 1865; Richter, 1958; Hough, 2015; Fig. 3.4). With some important historical earthquakes, there have been notable cases where some media accounts deliberately downplayed observed damage (e.g., *Hanson* et al., 1989); in general, however, media articles and photographs are not inaccurate per se, but overwhelmingly focus on instances of dramatic damage (Hough, 2014a, 2014b). For example, the day after the 29 June 1925 M6.5 Santa Barbara earthquake, the story was headline news in the New York Times (30 June, 1925): "SEVERE EARTHQUAKE WRECKS SANTA BARBARA; HOTELS COLLAPSE; BUSINESS AREA IN RUINS; 12 PERSONS KILLED; PROPERTY LOSS OF MISSIONS." Several weeks later the New York Times (24 July 1925) published an article downplaying the severity of damage, noting that it was overwhelmingly due to shoddy construction; this single column article appeared on page 25. The media bias tends to be strongest when written reports are short; more extensive accounts are likely to provide a better indication of overall effects. Thus, the bias tends to especially plague investigations of historical earthquakes, for which newspaper and other accounts are often especially fragmentary, although even recent news reports of shaking can certainly often be fragmentary as well.

3.6 Every Historical Earthquake is Historical in Its Own Way: A Case Study

To illustrate the issues discussed above, I now discuss a case study provided by a relatively recent earthquake: the 9 November 1968 Southern Illinois earthquake. This event is illuminating to consider because it is large and recent enough to have a well-constrained instrumental magnitude (5.3; Hermann, 1973) and a rich supply of media accounts, and early enough to have a generated a large volume of USCGS postal questionnaires. An independent postcard survey was also conducted by St. Louis University (SLU; Robert Hermann, written communication, 2018). Intensities were assigned to the questionnaires by Martin and Hough (2019). In most of the states where the earthquake was felt, postal questionnaires were widely distributed, providing a good characterization of the limit of the felt extent. In several

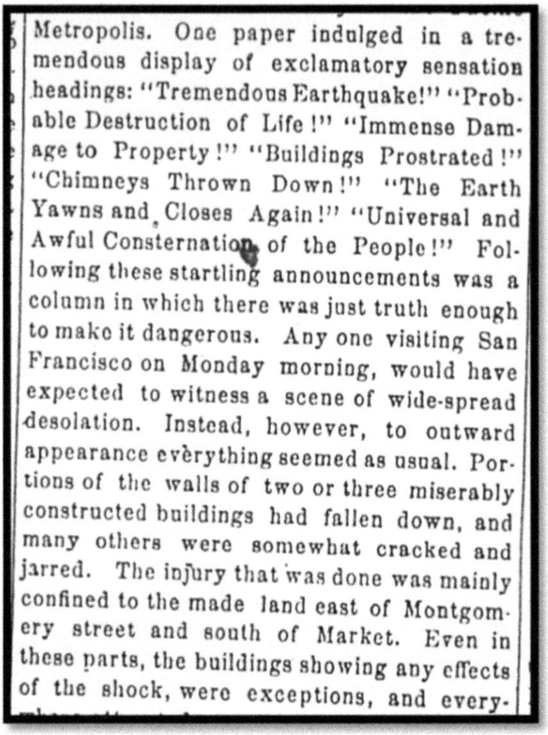
Metropolis. One paper indulged in a tremendous display of exclamatory sensation headings: "Tremendous Earthquake!" "Probable Destruction of Life!" "Immense Damage to Property!" "Buildings Prostrated!" "Chimneys Thrown Down!" "The Earth Yawns and Closes Again!" "Universal and Awful Consternation of the People!" Following these startling announcements was a column in which there was just truth enough to make it dangerous. Any one visiting San Francisco on Monday morning, would have expected to witness a scene of wide-spread desolation. Instead, however, to outward appearance everything seemed as usual. Portions of the walls of two or three miserably constructed buildings had fallen down, and many others were somewhat cracked and jarred. The injury that was done was mainly confined to the made land east of Montgomery street and south of Market. Even in these parts, the buildings showing any effects of the shock, were exceptions, and every-

Fig. 3.4 Account from "The Golden Era" newspaper, 15 October, 1865, following the "great San Francisco earthquake" of 1865

other states, postcards were not as widely distributed; in these states, responses appear to have been received preferentially from locations where shaking was felt. Nonetheless, the dataset provides a spatially rich view of the shaking distribution. In theory, the postcard-based distribution should be akin to the map that would have been produced by a DYFI system, the algorithm which was designed to reproduce subjective intensity assignments.

A summary of questionnaire responses was published by Coffman and Cloud (1984), and are available from the NOAA online data base (see ***Data and Resources***). It is thus possible to compare original assignments from Coffman and Cloud (1984) with values assigned by Martin and Hough (2019) from the original USCGS and SLU postcards. One can also compare intensities estimated from postcard questionnaires with values that would be estimated from media sources alone.

The NOAA database includes intensity values for the 1968 earthquake from 806 Zip codes. Of these, all are whole-number assignments, and 436 are listed as MMI 4. The locations assigned MMI 1–4 are combined by Coffman and Cloud (1984) as "intensity 1–4"; in the database, all of these locations are listed as MMI 4 The practice of lumping together values in this way, with no details provided by the USE series, makes it impossible to use USE/NOAA values to characterize the low-intensity field in detail. Even at regional distances, combined values will affect the averages at distances at which a range of low intensities were reported. The original postal questionnaires, on the other hand, generally do provide information sufficient to distinguish between MMI 2 (felt by very few) and MMI 4 (felt by many; windows, dishes, and doors rattled, etc.). Apart from the low intensity field, the intensities interpreted by Martin and Hough (2019) tend to be slightly lower than original values, on average by less than 0.5 units. This difference is partly due to the assignment of half intensity units; in most cases, these assignments are a half unit lower than original values. Using the SLU survey, Martin and Hough (2019) also assigned intensities to more locations: a total of 1270.

Following conventional practice, Martin and Hough (2019) assigned MMI values of 6–7 for locations where any damage to chimneys or brick buildings is described. It is possible that at some locations, the severity of shaking did not exceed MMI 4–5, and that a small percentage of chimneys was damaged either because they were especially vulnerable or because they were damaged by weak but prolonged shaking. It is possible that chimney damage is not a reliable indicator of overall shaking intensity in regions were prolonged *Lg* wave trains develop, including central and eastern North America. A summary of all accounts and intensity values is provided in an electronic supplement to Martin and Hough (2019).

For locations within $\approx$100 km epicentral distance, intensities reinterpreted by Martin and Hough (2019) are generally consistent with the original USCGS assignments (Coffman & Cloud, 1984; Fig. 3.5). At distances above 100 km, the reinterpreted values are progressively lower than USCGS values, reflecting the more detailed consideration of accounts describing weak shaking.

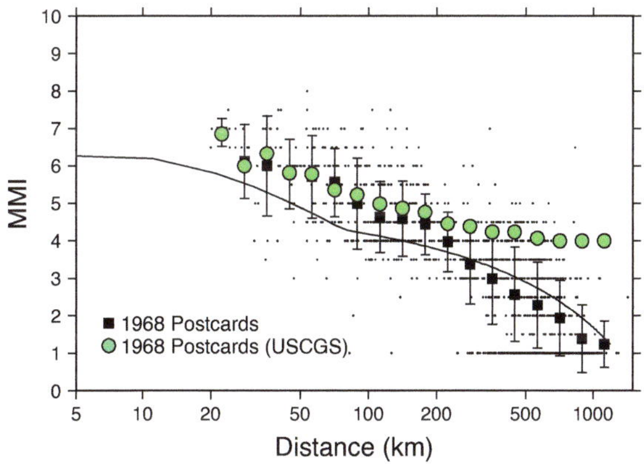

Fig. 3.5 Bin-averaged intensity values based on original postcard questionnaires, including combined values listed in the NOAA database as MMI IV (green dots; *Stover and Coffman*, 1984), and reinterpreted postal questionnaires (Martin & Hough, 2019), including both individual values (black dots) and bin-averaged values (black squares, mean $\pm$ 1). Solid line indicates predicted MMI versus distance curve for M5.3 (Atkinson & Wald, 2007)

I now consider how the postcard-based intensity distribution compares to a media-based intensity distribution for the same earthquake. To develop a media-based map, I searched a subscription-based online archive of historical newspapers for articles mentioning each of the towns for which Martin and Hough (2019) include an intensity value. While not an exhaustive archival search, the online database is extensive. This exercise reveals that the number of locations for which a media account is available and sufficient to assign a numerical intensity is much lower than the number for which a postcard was returned: 181 versus 1270, respectively. Overwhelmingly, locations mentioned in media accounts fall in one of three categories: (1) Cities large enough to have local newspapers, where articles about the earthquake mentioned effects in that location, (2) Smaller cities that experienced especially severe damage, and (3) Locations on the periphery of the felt extent, mentioned in articles that commented specifically on the extent of felt shaking. As an example of the last category, an article from the Arkansas Gazette that noted that "dishes and windows rattled over a section of northeast Arkansas from the borders of the state to Salem, Batesville, Stuttgart, and Helena." In the second and third cases, reports are clearly available preferentially from locations where shaking was relatively strong compared to the surrounding area. I.e., articles commenting on where shaking was felt are more likely to mention places where effects were documented than to specifically note that shaking was not felt in other places. And while accounts from larger cities do not suffer this inherent reporting bias, they do still tend to mention relatively dramatic effects within a given city. For this earthquake, media-based intensity values are within the range of intensities assigned from postcards, but systematically at the upper edge of the distribution at a given distance (Fig. 3.6).

The intensity prediction equation (IPE) determined by Atkinson and Wald (2007) using DYFI data for earthquakes in central and eastern North America is consistent within the $\pm 1\sigma$ limits with the questionnaire-based intensities for the 1968 earthquake (Fig. 3.6). Observed intensities are relatively higher than the IPE at distances within about 200 km, and slightly lower at lower distances.

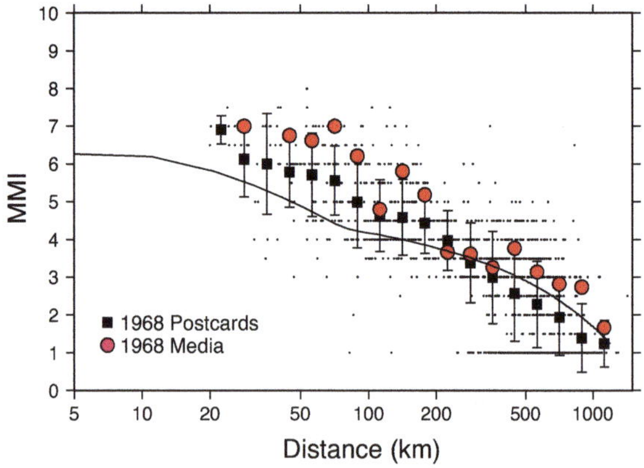

Fig. 3.6 Bin-averaged intensity values based on media accounts (red dots), and reinterpreted postcard questionnaires (this study), including both individual values (black dots) and bin-averaged values (black squares, mean $\pm 1\sigma$). Solid line indicates predicted MMI versus distance curve for M5.3 (Atkinson & Wald, 2007)

These systematics could be explained by the relatively deep focal depth (estimated 21 km; see ***Data and Resources***), and possibly a higher than average stress drop, of the earthquake. Even so, the reinterpreted questionnaire-based intensities are consistent with the Atkinson and Wald (2007) IPE. In contrast, the media-based intensities, even interpreted following modern conservative practices, tend to track the upper range of the distribution of intensity values estimated from individual postcards, with a less steep decay out to distances of ~ 100 km. The extensive macroseismic data sets for this earthquake thus confirm earlier results (Hough, 2013, 2014a, b) suggesting that, primarily due to fundamental reporting biases, intensity values estimated subjectively from extant archival accounts are not generally consistent with intensities estimated from questionnaires designed to measure representative effects in a given spatial footprint.

I note that fundamental questions arise regarding intensity maps: What information do they convey? And what information should they convey? From the earliest intensity scales through EMS-98, intensity values are determined based on the statistical incidence of effects: e.g., whether shaking was felt by "most," "many," or "few" people. By increasingly well-established convention, intensity can be estimated but not robustly constrained from point measurements. The question then arises, over what spatial footprint should macroseismic information be aggregated? The USGS DYFI system implicitly adopts 1-km as the standard. A further question is, what information does an intensity map shown? A map based explicitly or implicitly on representative effects within a given spatial footprint (city, ZIP-code, geocoded cell) will not characterize the most severe effects observed in a given place, while a map based on the most severe effects will not provide a good characterization of representative effects. The evolution of intensity maps over time has moved away from the latter and towards the former. While either choice is arguably defensible, with any intensity map, past or present, it is critical to understand what information is presented.

3.7 Modern Analysis Methods

As discussed above, the spatially rich nature of intensity values makes such information valuable for understanding detailed ground motion distributions from even recent earthquakes (e.g., Atkinson & Wald, 2007; Hauksson et al., 2008; *Hough*, 2011; *Adhikari* et al., 2018). In keeping with the scope of this volume, I now focus on the question, how can intensity values be used to better characterize historical earthquakes? I expand the question slightly to include early instrumental earthquakes for which only limited instrumental data typically exist.

Fundamentally, intensity distributions are of critical value to constrain the magnitude of earthquakes for which no instrumental data are available. They supplement other available types of information, including surface rupture, tsunami observations, and other geological effects, but for many earthquakes, intensity observations are the only source of information available to constrain magnitude and location.

For any earthquake, the distribution of ground motions, and therefore intensities, depends on three things: earthquake source properties, so-called path effects including attenuation and geometrical spreading, and local site effects. In general, the strongest shaking occurs at the closest near-fault distances; an idealized intensity map is thus expected to look like a bulls-eye around a small source or an elongated set of ovals around an extended rupture. In practice, however, intensity distributions are complex, reflecting detailed source properties as well as site and path effects.

These complications notwithstanding, methods to analyze intensity distributions conventionally rely on comparisons to intensity values from instrumentally recorded calibration events in the same or perhaps tectonically similar regions. It is well established that propagation effects differ significantly in different regions, underscoring the need to use appropriate calibration relationships (i.e., intensity prediction equations). Different approaches have been taken to develop calibration relationships, including some that are based on isoseismal area (e.g., Johnston, 1996) and some that are based on intensity decay as a function of distance (e.g., Bakun & Wentworth, 1997; Gasperini et al., 2010).

It is well established that shaking can be strongly influenced by local site effects, including sediment-induced amplification (e.g., Borcherdt, 1970), topographic amplification (e.g., *Sanchez-Sesma*, 1985), deep-basin response (Joyner, 2000), and non-linear deamplification of high frequencies (e.g., Trifunac, 2003). In extreme cases, the spatial distribution of damage can be potentially grossly misleading without the consideration of such effects, as illustrated by the 1985 Michoacan, Mexico, earthquake, which caused catastrophic damage 320 km away in Mexico City. More commonly, the distribution of intensities can be potentially skewed by site response, such that an intensity centroid does not necessarily reflect source location. The 31 October 1895 Charleston, Missouri, earthquake is an example: based on the overall distribution of intensities, Bakun et al. (2003) estimated a location over 100 km north of the conventionally estimated location near the town of Charleston, Missouri. Reviewing original sources in detail, Martin and Hough (2019) concluded that overall intensity distribution was controlled to some extent by amplification along major river valleys, and that the earthquake was indeed centered near Charleston. An intensity distribution can also be biased towards high values if extant accounts are preferentially from locations where ground motions were locally amplified, as Hough et al. (2000) concluded was the case for the 1811–1812 New Madrid sequence.

Site response affects some subset of intensity values in any intensity distribution. One does not want to remove site response per se; rather one needs to understand how it contributes to the intensity distribution. Although it might be tempting to correct estimated intensities for site response, this approach is fraught with peril for several reasons. An immediate practical difficulty is that it is often not possible to characterize the local site geology of any given site. For macroseismic data, it is moreover often impossible to know the exact location of a historical site, which might not be the same as the modern location. I.e., while a town might have been clustered in immediate proximity to a river during its earliest history, the later population distribution can be quite different.

Non-linear site response poses a further challenge. Within the seismological community, it has long been assumed that low-impedance near-surface sediments will universally amplify earthquake shaking. While this is generally the case, during strong shaking, sediments can respond non-linearly, such that the sediments no longer respond as an elastic solid. In extreme cases, gross non-linear behavior such as liquefaction or ground slumping occurs. More commonly, but less obviously, pervasive non-linear response can deamplify high frequency shaking (Trifunac, 2003; *Rajaure* et al., 2017.) In light of these issues, development of calibration equations generally proceeds without explicit consideration of site response, which is assumed to be responsible for much of the scatter in the data, but not to generate systematic bias.

Limitations notwithstanding, recently developed calibration-relationship methods have been used with success to determine magnitude estimates and locations for historical earthquakes around the world. The method developed by Bakun and Wentworth (1997) has been widely used in the United States as well as elsewhere. The BOXER method has been widely used in Europe (e.g., Gasperini et al., 2010; Stucchi et al., 2013).

Magnitudes estimated from intensity values are conventionally denoted M_I, to distinguish them from magnitude types based on instrumental observations. If magnitudes are estimated using set of calibration events for which moment magnitude, *Mw*, values are used, then M_I should correspond on average to *Mw*. However, the aforementioned strong dependence of intensities on stress drop introduces a significant uncertainty. Using results from random vibration theory, Boore (1983) relates peak acceleration and velocity, respectively PGA and PGV to magnitude and stress drop, σ in bars.

$$\log \text{PGA} \approx 0.31 \text{ Mw} + 0.80 \log(s) \tag{3.1}$$

$$\log \text{PGV} \approx 0.55\text{Mw} + 0.64 \log(s) \tag{3.2}$$

This result is illustrated in Fig. 3.7, which shows theoretical (omega-squared; Brune, 1970) velocity spectra for a range of moment magnitude values and a given stress drop versus spectra for a range of stress drop values for a given magnitude, assuming $\sigma = Mo(f_c/0.42\beta)^3$, where is β is the shear-wave velocity near the source and *fc* is the corner frequency (Madariaga, 1976).

Since estimated intensities correlate well with PGA and/or PGV (e.g., Atkinson & Wald, 2007; Worden et al., 2012, it is clear from Eqs. (3.1) and (3.2) (also Fig. 3.7) that magnitude estimated from intensity values will depend critically on stress drop. If intensities are controlled by peak velocity, for example, a stress drop difference of factor of 10 is roughly equivalent to a magnitude difference of 1.2 units. If intensities are controlled by peak acceleration, the stronger control of stress drop on PGA (relative to PGV) implies that an order of magnitude difference in stress drop will correspond to an even bigger different in magnitude..

Fig. 3.7 (**a**, bottom) Theoretical (omega-squared) amplitude-normalized source (velocity) spectra for a given magnitude and a range of stress drop values. (**b**, top) Theoretical spectra for a given stress drop value and a range of magnitudes

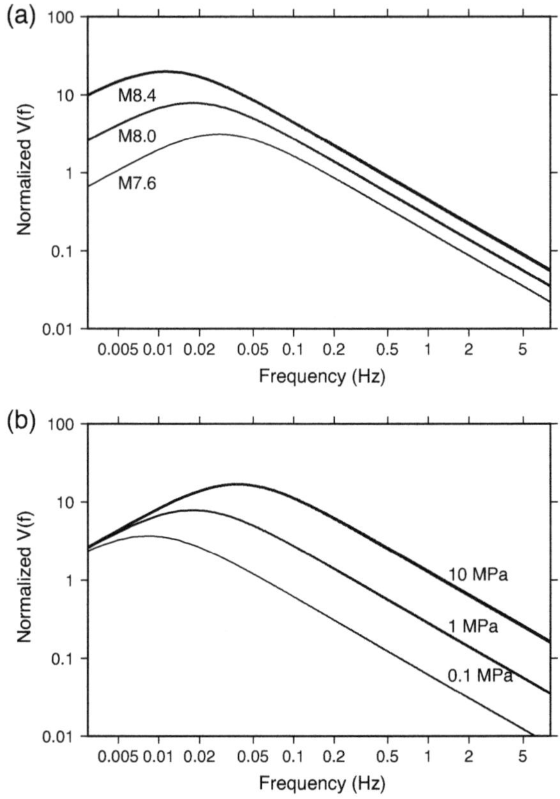

It remains unclear whether intensities are more controlled by peak acceleration or peak velocity; Eqs. (3.1) and (3.2) are moreover only approximations based on random vibration theory. Nonetheless, these simple calculations show that, for example, if magnitude is estimated from intensity values, the uncertainty associated with the stress drop of a historical earthquake will give rise to magnitude uncertainties of a full magnitude unit, if not more. Put another way, the shaking distribution from an earthquake of a given magnitude and a relatively low stress drop value, say 2 MPa, will be effectively indistinguishable from that of an earthquake ~ 1 unit smaller and a 20 MPa stress drop. Stress drop is an elusive parameter in seismology, believed to vary over at least two orders of magnitude (e.g., Abercrombie, 1995). For a historical earthquake, stress drop is unknowable for all but some large events for which rupture area and average slip can be estimated from geological observations. Thus, under the best circumstances, apart from uncertainties associated with site response, calibration relationships, etc., a moment magnitude estimate based on intensities alone will be inherently uncertain by 0.5 magnitude units, if not more. One notes, however, that this statement could be recrafted to say that moment magnitude itself does not provide a strong control on shaking at frequencies of engineering concern for most structures. This issue has been raised in recent years in studies focused on characterization of ground motions (e.g., Bindi et al., 2018).

3.8 Discussion and Conclusions

Intensity values assigned based on reported macroseismic effects have always and likely will always provide an invaluable source of data to investigate earthquake ground motions and earthquake sources. Even as the scientific community increasingly exploits data from ubiquitous commercial devices (e.g., Cochran, 2018), "large-N" deployments (*Lin* et al., 2013), fiber-optic cables for distributed monitoring (e.g., Lindsey et al., 2017), in many areas, instrumental data cannot match the spatial density provided by macroseismic information. Macroseismic data are also, by definitely, concentrated in populated areas that are of most concern for risk. As summarized in this chapter, care and consistency are needed in the interpretation of macroseismic information, including an understanding of its limitations. The value of carefully and consistently interpreted intensity distributions has been demonstrated by many recent studies. To cite a few examples, Trifunac (2003) used damage observations to investigate non-linear site response during the 1933 Long Beach, California, earthquake. *Hough* et al. (2005) used the intensity distribution of the 1905 Kangra, India, earthquake to identify a likely significant triggered earthquake that occurred within minutes of the Kangra mainshock. Lozos (2016) used historical observations and dynamic modeling to conclude that the 1812 southern California earthquake likely involved rupture on both the San Andreas and San Jacinto faults. *Stein* et al. (2017) used historically observed intensities as an independent test of the PSHA map for Japan, work that was followed up by testing of the one-year induced earthquake hazard map produced by the U.S. Geological Survey (Brooks et al., 2018). *Adhikari* et al. (2017) present a synthesis of intensity values determined by different groups for the 2015 Gorkha, Nepal, earthquake, to explore the effects of pervasive nonlinear site response in the Kathmandu Valley. Martin et al. (2019) used detailed macroseismic information to conclude that the 1907 Sumatra "tsunami earthquake" was in fact two separate earthquakes. In recent years, intensity data contributed from online systems such as DYFI have also been exploited increasingly to improve earthquake response products (e.g., *Wald* et al., 2021).

The above and other studies demonstrate the enormous value of macroseismic data, even a half-century after Byerly (1969) made the same point. It is important, however, that intensity values not be interpreted without an appreciation of the issues discussed in this chapter. In the absence of previous work documenting the consistency of intensity assignments and the basis on which they are assigned, it remains critical to not blindly analyze or combine published intensity values, but rather to consider the original sources on which they are based. It is also critical to preserve those sources, at a minimum including detailed summaries and source citations. In addition to inherent subjectivity of traditional intensity assignments, one should be aware of the fundamental reporting bias, whereby dramatic effects are more likely to be reported. Traditional intensities differ in this regard from Internet-based intensities such as Did You Feel It?, which determines an intensity value within a given spatial footprint by averaging all available responses. Because of the overall evolution in practices, intensity maps have evolved generally over time towards depiction of representative effects, and therefore do not reflect the most severe effects in a given spatial footprint.

3.9 Data and Resources

The JMA intensity scale is described at http://www.jma.go.jp/jma/en/Activities/inttable.html (last accessed 7 October 2021).

The NOAA intensity database, including intensity assignments for U.S. earthquakes from 1638—1985, can be accessed at https://www.ngdc.noaa.gov/hazard/eq-intensity.shtml (last accessed 7 October 2021).

The "Abstracts of Earthquake Reports from the Pacific Coast and the Western Mountain Region" series were informally published and distributed. Digital scans of some reports can be accessed at https://babel.hathitrust.org (last accessed 7 October 2021).

Did You Feel It? data and earthquake parameters can be accessed from USGS events pages, which can be accessed via http://earthquake.usgs.gov/earthquakes/search/ (last accessed 8 October 2021).

Acknowledgements The author's appreciation of issues discussed in this chapter benefited over the years from discussions with many colleagues, including Nicolas Ambraseys, Jim Dewey, Roger Bilham, Paola Albini, and Stacey Martin. I thank Eric Thompson and Jessica Saunders for constructive reviews that significantly improved this manuscript, and the editors for their guidance and patience.

References

Abercrombie, R. E. (1995). Earthquake source scaling relationships from− 1 to 5 ML using seismograms recorded at 2.5-km depth. *Journal of Geophysical Research: Solid Earth, 100*(B2), 24015–24036.

Adhikari, S.R., G. Baysal, A. Dixit, S.S. Martin, M. Landes, R. Bossu, and S.E. Hough, Towards a unified near-field intensity map of the, 2015 Adhikari, S. R., Baysal, G., Dixit, A., Martin, S. S., Landes, M., Bossu, R., & Hough, S. E. Towards a unified near-field intensity map of the 2015 Gorkha, Nepal, earthquake. *Earthquake Spectra, 33* (S1), S21-S34.

Allen, M. W. (1934) Unpublished letter to Harry Oscar Wood, 1934, papers of Harry Oscar Wood, Box 1, folder 1, California Institute of Technology Archives.

Ambraseys, N. N. (1971). Value of historical records of earthquakes. *Nature, 232*(5310), 375–379.

Ambraseys, N. (1983). Notes on historical seismicity. *Bull. Seism. Soc. Am., 73*(6A), 1917–1920.

Ambraseys, N., & Douglas, J. J. (2004). Magnitude calibration of north Indian earthquakes. *Geophysical Journal International, 159*, 165–206. https://doi.org/10.1111/j.1365-246X.2004.02323.x.

Amorèse, D., Bossu, R., & Mazet-Roux, G. (2015). Automatic clustering of macroseismic intensity data points from internet questionnaires: Efficiency of the partitioning around medoids (PAM). *Seism. Res. Lett, 86*(4), 1171–1177.
Atkinson, G. M., & Wald, D. J. (2007). "Did You Feel It?" Intensity data: A surprisingly good measure of earthquake ground motion. *Seism. Res. Lett., 78*(3), 362–368.
Beresnev, I. A., & Atkinson, G. M. (1997). Modeling finite-fault radiation from the ω n spectrum. *Bull. Seism. Soc. Am., 87*(1), 67–84.
Bindi, D., Spallarossa, D., Picozzi, M., Scafidi, D., & Cotton, F. (2018). Impact of magnitude selection on aleatory variability associated with ground-motion prediction equations: Part 1—Local, Energy, and Moment magnitude calibration and stress-drop variability in central Italy. *Bull. Seism. Soc. AM., 108*(3A), 1427–1442.
Baird Smith, R. (1843). Memoir of Indian earthquakes—part II. *J. Asiatic Soc. Bengal, 12*(144), 1029–1056.
Bakun, W. H. (2006). Estimating locations and magnitudes of earthquakes in southern California from modified Mercalli intensities. *Bull. Seism. Soc. Am., 96*(4A), 1278–1295. https://doi.org/10.1785/0120050205.
Bakun, W. H., & Wentworth, C. M. (1997). Estimating earthquake location and magnitude from seismic intensity data. *Bull. Seism. Soc. Am., 87*(6), 1502–1521.
Bakun, W. H., Johnston, A. C., & Hopper, M. G. (2003). Estimating locations and magnitudes of earthquakes in eastern north America from modified Mercalli intensities. *Bulletin of the Seismological Society of America, 93*(1), 190–202.
Boatwright, J., & Phillips, E. (2017). Exploiting the demographics of "Did You Feel It?" responses to estimate the felt area of moderate earthquakes in California. *Seismological Research Letters, 88*(2A), 335–341.
Boore, D. M. (1983). Stochastic simulation of high-frequency ground motions based on seismological models of radiated spectra. *Bull. Seism. Soc. Am., 73*(6A), 1865–1894.
Borcherdt, R. D. (1970). Effects of local geology on ground motion near San Francisco Bay. *Bull. Seism. Soc. Am., 60*(1), 29–61.
Brooks, E. M., Stein, S., Spencer, B. D., Salditch, L., Petersen, M. D., & McNamara, D. E. (2018). Assessing earthquake hazard map performance for natural and induced seismicity in the central and eastern United States. *Seismological Research Letters, 89*(1), 118–126.
Brune, J. N. (1970). Tectonic stress and the spectra of seismic shear waves from earthquakes. *Journal of Geophysical Research, 75*(26), 4997–5009.
Byerly, P. (1969). Seismic intensity scales, letter to the editor. *Bull. Seism. Soc. Am., 59*(4), 1735.
Byerly, P., & Dyk, H. S. (1936). Questionnaire program for collecting earthquake data (in earthquake investigations in California, 1934–35]. (In Coast and Geodetic Survey, Special Publication 201, 43–48 and map.
Cancani, A. (1904). Sur l'emploi d'une double echelle sismique des intensites, empirique et absolue. *Gerlands Beitrage Zur Geophysik, 2*, 281–283.
Cecić, Ina. (2018). Using Damage Evaluation Forms for a Historical Earthquake-A Case of 29 January 1917 Brežice, Slovenia, Earthquake.
Cochran, Elizabeth S. (2018). To catch a quake. *Nature Communications, 9*(1), 2508.
Coffman, J. L., & Cloud, W. K. (1984). United States earthquakes, 1968, U.S. Geol. Surv. Prof. Pap. 1527, 418 pp, Washington D.C.
Coppola, J. M., Cowan, L. X., Downes, G. L., Fenaughty, K. F., Grimwood, P. D., Leach, P., & de J. Robertson, E. (2010). Felt earthquake reporting via the Internet in New Zealand. *Seism. Res. Lett., 81* (6), 984-991.
Davison, C. (1921). On scales of seismic intensity and on the construction of isoseismal lines. *Bull. Seism. Soc. Am., 11*(2), 95–129.
Dengler, L. A., & Dewey, J. W. (1998). An intensity survey of households affected by the Northridge, California, earthquake of 17 January, 1994. *Bull. Seism. Soc. Am., 88*(3), 441–462.
de Rossi, M. S. (1883). Programma dell'osservatorio ed archivio centrale geodinamico presso il R. Comitato Geologico d'Italia. *Bull. Vulcanismo Ital, 10*, 3–128.

DesRoches, R., Comerio, M., Eberhard, M., Mooney, W., & Rix, G. J. (2011). Overview of the 2010 Haiti earthquake. *Earthquake Spectra.* https://doi.org/10.1193/1.3630129.

Dewey, J. W., Reagor, B. G., Dengler, L. A., & Moley, K. (1995). Intensity distribution and isoseismal maps for the Northridge, California, earthquake of January 17, 1994. U.S. Geol. Survey Open File Rep. 95–92, 35, Washington D.C.

Drake, D. (1815). Natural and statistical view, or picture of Cincinnati and the Miami County, illustrated by maps, Looker and Wallace}, Cincinnati.

Dutton, C. (1889).

Egen, P. N. C. (1928). Uber das erdbeben in den Rhein und Niederlanden vom 23. Feb.1928, *Ann Phys., 13,* 153–163.

Eiby, G. A. (1966). The modified Mercalli scale of earthquake intensity and its use in New Zealand. *New Zealand J. Geol. and Geophys, 9*(1–2), 122–129.

Gasperini, P., Vannuccii, G., Tripone, D., & Boschi, E. (2010). The location and sizing of historical earthquakes using the attenuation of macroseismic intensity with distance. *Bull. Seism. Soc. Am., 100*(5A), 2035–2066.

Era, G. (1865). *The earthquake, Golden Era, 15 October 1865, pg 4.*

Gomez-Capera, A. A., Rovida, A., Gasperini, P., Stucchi, M., & Viganò, D. (2015). The determination of earthquake location and magnitude from macroseismic data in Europe. *Bulletin of Earthquake Engineering, 13*(5), 1249–1280.

Grünthal, G. (*Editor*). (1998). The European Macroseismic Scale EMS-98, Vol. 15, Conseil de l'Europe, Cahiers du Centre Europeén de Geodynamique et de Seismoligie, Luxembourg, 101.

Guidoboni, E., Comastri, A., Traina, G., & Rom Istituto Nazionale di Geofisica. (1994). *Catalogue of Ancient Earthquakes in the Mediterranean Area up to the 10th Century,* 504 pp. Rome: Istituto nazionale di geofisica.

Hanks, T. C., & Johnston, A. (1992). Common features of the excitation and propagation of strong ground motion for North American earthquakes. *Bull. Seism. Soc. Am., 82*(1), 1–23.

Hansen, G. et al. (1989). Denial of disaster.

Hansen, S. M., & Schmandt, B. (2015). Automated detection and location of microseismicity at Mount St. Helens with a large-N geophone array. *Geophys. Res. Lett., 42*(18), 7390–7397.

Hauksson, E., Felzer, K., Given, D., Giveon, M., Hough, S., Hutton, K., Kanamori, H., Sevilgen, V., Wei, S., & Yong, A. (2008). Preliminary report on the 29 July 2008 mw 5.4 chino Hills, eastern Los Angeles basin, California, earthquake sequence. *Seism. Res. Lett., 79*(6), 855–866.

Hermann, R. B. (1973). Surface-wave generation by the south-central Illinois earthquake of November 9, 1968. *Bulletin of the Seismological Society of America, 63*(6), 2121–2134.

Hough, S. E. (2000). On the scientific value of unscientific data. *Seismological Research Letters, 71,* 483–485.

Hough, S. E., Armbruster, J. G., Seeber, L., & Hough, J. F. (2000). On the modified mercalli intensities and magnitudes of the 1811–1812 new Madrid. *Central United States, Earthquakes, Journal of Geophysical Research, 105*(B10), 23839–23864.

Hough, S. E., Bilham, R., Ambraseys, N., & Feldl, N. (2004). Revisiting the 1897 Assam and 1905 Kangra earthquakes in northern India: Site response. *Moho Reflections, and a Triggered Earthquake, Current Science, 88,* 1632–1638.

Hough, S. E. (2012). Initial assessment of the intensity distribution of the Mw5.8 Mineral, Virginia, earthquake. *Seism. Res. Lett., 83*(4), 649–657.

Hough, S. E., & Page, M. (2011). Towards a consistent model for strain accrual and release for the New Madrid seismic zone, central United States. *J. Geophys. Res., 116*(B3), 17. https://doi.org/10.0129/2010JB007783.

Hough, S. E. (2013). Spatial variability of "Did You Feel It?" intensity data: Insights into sampling biases in historical earthquake intensity distributions. *Bull. Seism. Soc. Am., 103*(5), 2767–2781. https://doi.org/10.1785/0120120285.

Hough, S. E. (2014a). Quantifying and correcting for reporting biases in historical intensity assignments. *Bulletin of Earthquake Engineering, 12,* 135–155. https://doi.org/10.1007/s10518-013-9573-x.

Hough, S. E. (2014b). Shaking from injection-induced earthquakes in the central and eastern United States. *Bull. Seism. Soc. Am.*, *104*(5), 2619–2626. https://doi.org/10.1785/0120140099.
Hough, S. E. (2015). Ground motions from induced earthquakes. The Leading Edge, June, 690–697.
Ishibashi, K. (2004). Status of historical seismology in Japan. *Annals of Geophysics, 47*(2–3), 339–368.
Johnston, A. C. (1996). Seismic moment assessment of earthquakes in stable continental regions—III. New Madrid 1811–1812, Charleston 1886 and Lisbon 1755. *Geophysical J. Int., 126*(2), 314–344.
Joyner, W. B. (2000). Strong motion from surface waves in deep sedimentary basins. *Bull. Seism. Soc. Am., 90*(6B), S95–S112.
Kanamori, H., & Anderson, D. L. (1975). Theoretical basis of some empirical relations in seismology. *Bull. Seism. Soc. Am, 65*(5), 1073–1095.
Madariaga, R. (1976). High-frequency radiation from crack (stress drop) models of earthquake faulting, Geophys. *J. Royal Astron. Soc., 51*, 625–651.
Lindsey, Nathaniel J., Eileen R. Martin, Douglas S. Dreger, Barry Freifeld, Stephen Cole, Stephanie R. James, Biondo L. Biondi, & Jonathan B. Ajo-Franklin. (2017). Fiber-optic network observations of earthquake wavefields. *Geophy. Res. Lett. 44, 3,* 11.792.
Lozos, J. C. (2016). A case for historic joint rupture of the San Andreas and San Jacinto faults. *Science Advances, 2*(3), e1500621.
Mallet, R. and J.W. Mallet (1858). The earthquake catalogue of the British Association with the discussion, curves, and maps, etc., Transactions of the British Association for the advancement of Science, 1852Mallet, R., & Mallet, J. W. (1858). The earthquake catalogue of the British Association with the discussion, curves, and maps, etc., Transactions of the British Association for the advancement of Science, 1852 to 1858, Taylor and Francis, London.
Martin, S., & Szeliga, W. (2010). A catalog of felt intensity data for 589 earthquakes in India, 1636–2009. *Bull. Seism. Soc. Am., 100*(2), 536–569. https://doi.org/10.1785/0120080328.
Martin, S. S., Hough, S. E., & Hung, C. (2015). Ground motions from the 2015 M7.8 Gorkha, Nepal, earthquake constrained by a detailed assessment of macroseismic data. *Seism. Res. Lett. 86, 6,* 1524–1532.
Martin, S., & Hough, S. E. (2019). Where was the 31 October 1895 Charleston, Missouri, earthquake? *Bull. Seism. Soc. Am., 109*, 4. https://doi.org/10.1785/0120180328.
Martin, S. S., Li, L., Okal, E. A., Morin, J., Tetteroo, A. E., Switzer, A. D., & Sieh, K. E. (2019). Reassessment of the 1907 Sumatra "Tsunami Earthquake" based on Macroseismic, seismological, and tsunami observations, and Modeling. *Pure and Applied Geophysics, 176–7*, 1–38.
Marzocchi, W., & Jordan, T. H. (2014). Testing for ontological errors in probabilistic forecasting models of natural systems. *Proceedings of the National Academy of Science of the United States of America, 111*(33), 11973–11978.
McKenzie, C. (1830). Notes on Haiti made during a residence in that republic, p. 335, A.J. Valpy, London.
McMurtrie, H. (1819). Sketches of Louisville and its environs; including, among a great miscellaneous matter, a Florula Louisvillensis; or, a catalogue of nearly 400 Genera and 600 Species of plants, that grow in the vicinty of the town, exhibiting their generic, specific, and vulgar English names, S. Penn, Jun. Main-street, Louisville.
Medvedev, S., Sponheuer, W., & Karnik, V. (1964). Neue seismische Skala Intensity scale of earthquakes, 7. Tagung der Europaischen Seismologischen Kommission vom 24.9 bis 30.9.1962. In: Jena, Veroff, Institut fur Bodendynamik und Erdebenforschung in Jena, 77, Deutsche Akademie der Wissenschaften zu Berlin, pp. 69–76.
Mercalli, G. (1902). Sulle modificazioni proposte alla scala sismica De Rossi-Forel. *Boll. Socc. Seismol. Ital, 8*, 184–191.
Michetti, A. M. (2004). The INQUA scale: An innovative approach for assessing earthquake intensities based on seismically-induced ground effects in natural environment, **67** SystemCart
Musson, R. M. (1998). Intensity assignments from historical earthquake data: Issues of certainty and quality. *Annals of Geophysics, 41*, 1. https://doi.org/10.4401/ag-3795

Musson, R. M. W. (2000). Intensity-based seismic risk assessment (2000). *Soil Dyn. and Earthq. Eng., 20*, 353–360.
Musson, RMW (2006). Automatic assessment of EMS-98 intensities, British Geological Survey, Internal Report IR/06/048.
Musson, R. M., Grünthal, G., & Stucchi, M. (2010). The comparison of macroseismic intensity scales. *Journal of Seismology, 14*(2), 413–428.
Neumann, F. (1942). United States Earthquakes, 1940, p. 75 , U.S. Government Printing Office, Wash. D.C.
Nuttli, O. W. (1973). The Mississippi Valley earthquakes of 1811 and 1812: Intesities, ground motion and magnitudes. *Bulletin of the Seismological Society of America, 63*(1), 227–248.
Oldham, R. D. (1897). *Report of the great earthquake of 12th June, 1897* (p. 29). Memoirs of Geological Surv.
Quitoriano, V., & Wal, D. J. (2020). https://doi.org/10.3389/feart.2020.00120.
Rajaure, S., Asimaki, D., Thompson, E., Hough, S., Ampuero, P., Martin, S. S., Inbal, A., & Dhital, M. (2016). Strong motion observations of the Kathmandu valley response during the M7.8 Gorkha earthquake sequence, Tectonophysics, https://doi.org/10.1016/j.tecto.2016.09.030.
Richter, C. (1958). *Elementary seismology*. Freeman.
Sanchez-Sesma, F. J. Diffraction of elastic SH waves by wedges, Bull. Seism. Soc. Am. *75*, 5, 1435–1446.
Sbarra, P., Tosi, P., & De Rubeis, V. (2010). Web-based macroseismic survey in Italy: Method validation and results. *Natural Hazards, 54*(2), 563–581.
Shabestari, K. T., & Yamazaki, F. (2001). A proposal of instrumental seismic intensity scale compatible with MMI evaluated from three-component acceleration records. *Earthquake Spectra, 17*(4), 711–723.
Sieberg, A. (1912). Uber die makroseismische Bestimmung der Erdebebenstarke. *Gerlands Beitrage Zur Geophysik, 11*, 227–239.
Sieberg, A. (1923). Geologische, physikalische und angewandte Erdebenkunde, G. Fischer, Jena.
Sokolov, V. Y., & Cherov, Y. K. (1998). On the correlation of seismic intensity with Fourier amplitude spectra. *Earthquake Spectra, 14*(4), 679–694. https://doi.org/10.1193/1.1586022
Stein, S., Spencer, B. D., & Brooks, E. M. (2015). Metrics for assessing earthquake-hazard map performance. *Bulletin of the Seismological Society of America, 105*(4), 2160–2173.
Stover, C. W., & Coffman, J. L. (1993). Seismicity of the United States, 1568–1989, *U.S. Geol. Surv. Prof. Paper* 1527, p. 418
Stucchi, M., Rovida, A., Capera, A. G., Alexandre, P., Camelbeeck, T., Demircioglu, M. B., Gasperini, P., Kouskouna, V., Musson, R. M. W., Radulian, M., & Sesetyan, K. (2013). The SHARE European earthquake catalogue (SHEEC) 1000–1899. *J. Seismology, 17*(2), 523–544.
Taylor, J. E. (1884). An earthquake in England. *Nature, 29*, 602–603.
Toppozada, T. R., & Parke, D. L. (1982). Areas damaged by California earthquakes, Open-File Rep. 82–17 SAC, *Calif. Div. of Mines and Geol.*, Sacramento.
Trifunac, M. D., & Brady, A. G. (1975). A study on the duration of strong earthquake ground motion. *Bull. Seism. Soc. Am., 65*(3), 581–626.
Trifunac, M. D. (2003). Nonlinear soil response as a natural passive isolation mechanism. The 1933 Long Beach, California, earthquake. *Soil Dynamics and Earthq. Eng., 23*,*7*, 549–562.
Ulrich, F. P. (1935). The California strong-motion program of the United States coast and geodetic survey. *Bull. Seism. Soc. Am., 25*(1), 81–95.
Wood, H. O., & Neumann, F. (1931). Modified Mercalli intensity scale of 1931. *Bull. Seism. Soc. Am., 21*(4), 277–283.
Worden, C. B., Gerstenberger, M. C., Rhoades, D. A., & Wald, D. J. (2012). Probabilistic relationships between ground-motion parameters and Modified Mercalli intensity in California. *Bull. Seism. Soc. Am., 102*, 204–221. https://doi.org/10.1785/0120110156.

Archaeoseismology: Identifying Earthquake Effects in Ancient Sites

4

Manuel Sintubin

4.1 Archaeseismology

Earthquake-prone regions around the world often show a long cultural history, resulting in a rich archaeological heritage (Force & McFadgen, 2010). In this respect, chances are that archaeological sites have recorded earthquakes that have occurred during or after the occupancy of the site. The archaeoseismological potential of an archaeological site indeed depends on the time span of occupancy (Sintubin & Stewart, 2008). Such pre-instrumental earthquakes, that can only be identified through indirect evidence in the archaeological record, we define as **ancient earthquakes**. Most of ancient earthquakes we identify, are prehistorical, not being recorded in written—historical—records (e.g. Similox-Tohon et al. 2006). Historical earthquakes can, however, have left their marks in the archaeological record. Archaeological evidence can then complement the knowledge with respect to the historical earthquake (e.g. Stiros, 2010). Jusseret (2014) moreover advocates to apply archaeoseismological practices also on recent earthquakes in efforts to help addressing specific social problems associated with disaster recovery and so to play a central role in prevention and mitigation efforts. Archaeoseismological practices, applied after recent earthquakes can furthermore contribute to the development of methods to protect cultural heritage in the face of future earthquakes (e.g. Davis et al., 2020).

Archaeoseismology—a term first introduced by Karcz and Kafri (1978)—is the scientific discipline studying ancient earthquakes using evidence in the archaeological record. As the other disciplines of earthquake science (palaeoseismology, historical seismology, instrumental seismology, neotectonics), archaeoseismology

M. Sintubin (✉)
KU Leuven, Celestijnenlaan 200e-box 2408, 3001 Leuven, Belgium
e-mail: Manuel.Sintubin@kuleuven.be

A. Elliott and C. Gruetzner (eds.), *Understanding Past Earthquakes*,
https://doi.org/10.1007/978-3-031-73580-6_4

aims at contributing to the reconstruction of the earthquake history in a region (Caputo & Helly, 2008).

In the beginning of the twentieth Century, Sir Arthur Evans, the excavator of the Bronze Age Minoan site of Knossos on the island of Crete (Greece) in the Eastern Mediterranean Sea, was the first to consider earthquakes as a destructive agent leaving evidence in the archaeological stratigraphy (Jusseret & Sintubin, 2013) (Fig. 1). Modern archaeoseismolgy developed by the end of the twentieth Century in the Mediterranean and the Middle East, primarily relying a structural damage features to monumental buildings at archaeological sites (e.g. Stiros, 1996). Since the publication of the book *Archaeoseismology* (Stiros & Jones, 1996), modern archaeoseismology progressively evolved towards an ever increasingly interdisciplinary discipline (cf. McGuire et al., 2000; Galadini et al., 2006; Caputo and Pavlides, 2008; Sintubin et al., 2010; Silva et al., 2011; Jusseret and Sintubin, 2017), responding to the skepticism about the early studies (Ambraseys, 2005).

4.2 Earthquake Archaeological Effects

Individual evidence for an earthquake in an archaeological context, is named an **earthquake archaeological effect** (EAE) (Rodríguez-Pascua et al., 2011), analogous to the macroseismic earthquake environmental effect (EEE), that forms the base for the Environmental Seismic Intensity Scale (ESI2017) (Michetti et al., 2007). Jusseret et al. (2013) rather defines a **potential earthquake archaeological effect** (PEAE), to emphasize the uncertainty associated with the use of an individual archeological effect as evidence of an ancient earthquake.

Two groups of PEAEs can be distinguished: structural effects on manmade constructions, and stratigraphical effects in the archaeological stratigraphy. Furthermore, PEAEs can be classified into three categories. Primary PEAEs are directly related to coseismic surface rupturing or ground failure, related to morphogenic earthquakes. Secondary PEAEs are related to coseismic ground shaking and ground motion. Tertiairy PEAEs are postseismic, delayed-response effects of an earthquake, such as fire damage, clearing of earthquake debris, repairs, recycling of building material, complete or partial abandonment, as well as architectural and/or cultural changes.

Primary structural PEAEs are manmade structures in an archaeological context that are displaced due to the coseismic surface rupturing on an active fault (Fig. 2). The study of such faulted archaeological relics is very comparable to paleoseismological work (McCalpin, 2009). A faulted archaeological relic not only enables the identification of the trace of an active fault in the landscape, it moreover allows the identification of the type of faulting (normal, reverse, strike-slip), the amount of coseismic slip related to individual earthquakes, as well as the cumulative fault slip caused by multiple surface-rupturing earthquakes during the lifespan of the faulted relic. Knowing the age of the manmade construction, a time-averaged fault-slip rate, possibly spanning centuries to millennia, can be derived. Examples of such faulted relics, exposing a sequence of fault-rupturing

Fig. 4.1 The *House of the Fallen Blocks* at Knossos (Crete), a PEAE *'avant la lettre'*, inspiring Sir Arthur Evans to eventually consider earthquakes as the primary destructive agent responsible not only for clear markers in the archaeological stratigraphy, but also leading to major cultural changes as evidenced in ceramic style and architecture. From Sintubin (2011)

earthquakes, are the Crusaders Fortress of Vadum Iacob (Ateret) along the Jordan river in Israel (Marco et al., 1997) (Fig. 2B), the Roman and Byzantine relics at Tiberias along the Sea of Galilee in Israel (Ferrario et al., 2020), faulted relics at the Roman city of Hierapolis in Turkey (Hancock & Altunel, 1997), the faulted Temple of Aphrodite at the Hellenistic city of Cnidus in Turkey (Altunel et al. 2003) (Fig. 2A), the Al Harif Roman Aqueduct in Syria (Sbeinati et al., 2010), and the qanats in Iran (Hessami et al., 2003; Jackson, 2006).

Secondary structural PEAEs comprise typical strain structures in the building fabric. These earthquake-related damage features are most obvious on monumental masonry constructions, such as temples, bath houses, and aqueducts (Figs. 3

Fig. 4.2 Examples of primary structural PEAEs or faulted relics. (A) The *Temple of Aphrodite Euploia* at Cnidus (Datça peninsula, SW Turkey) displaced by a normal fault. From Sintubin (2011) (B) The northern wall of the Crusaders fortress of Vadum Iacob (Ateret, Israel), displaced by a left-lateral strike-slip fault (i.e. Dead Sea transform fault)

and 4). Coseismic ground shaking can generate a wide range of damage features, such as penetrative fractures in masonry walls (Fig. 3C), and monolithic columns (Fig. 3A), rotated and displaced drums in columns (Fig. 4E), dropped keystones in arches (Fig. 4C), collapsed stairways, domino-type collapsed walls (Fig. 4A) and columns (Fig. 4D), U-shaped gaps in walls (Fig. 4B), folded and fractured pavements (Fig. 3B), and so many more (e.g. Sintubin et al., 2003; Silva et al., 2009). Also off-fault, coseismic grond-failure features, such as landslides and rockfalls, subsidence and uplift, and liquefaction, can be recorded in archaeological contexts (e.g. Silva et al., 2009). The main challenge is, though, to unambiguously relate these damage features to a single earthquake, or to multiple earthquakes, excluding any other, natural or human, destructive agent, such as natural failure, storm, vandalism, looting, or warfare. This uncertainty forces us to be cautious in this process, thus playing safe by using PEAEs for each individual effect observed. It is furthermore challenging to attribute damage features to an individual earthquake. In earthquake-prone regions archaeological sites have most probably experienced ground shacking from multiple earthquakes, both nearby and farther away, not only during occupancy but also after final abandonment. Due to the poor temporal resolution, it is very difficult to determine whether PEAEs are the result of a single earthquake, the sequence of main shock and its aftershocks, or even a clustered series of earthquakes. Poor temporal resolution may lead, by attributing damage features that may have been caused by multiple earthquakes to a single earthquake, to an erroneous interpretation of an oversized—geologically unrealistic—earthquake catastrophe (Guidoboni, 2002; Ambraseys et al., 2002). Therefore, Jusseret et al. (2013) introduced the concept of the **'same earthquake'** as the 'collection of earthquakes consisting of the mainshock, its immediate aftershocks, as well as possibly triggered earthquakes on the same of neighbouring fault segments during weeks to months after the mainshock that initiated the PEAEs'. By using this concept of the 'same earthquake', one at least recognizes the temporal challenges facing archaeoseismological work.

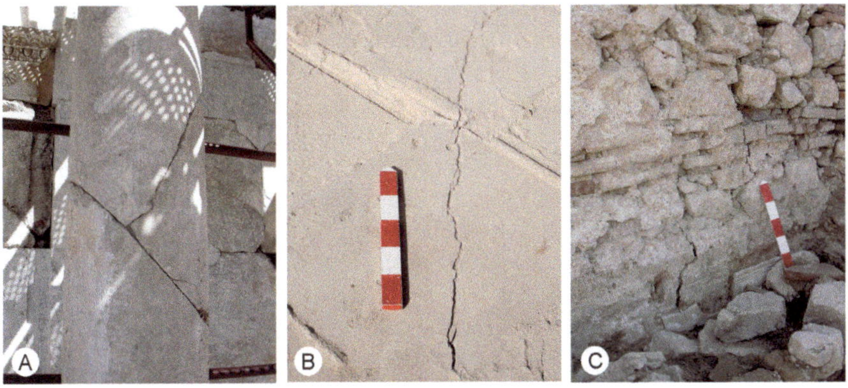

Fig. 4.3 Examples of secondary structural PEAEs. (A) Typical fracture pattern in monolithic columns caused by axial loading in the *Nymphaeum* at Sagalassos (SW Turkey) (cf. Sintubin et al. 2003). (B) Fractured pavement in the *Roman Baths* at Sagalassos (SW Turkey). (C) Fractured wall in the *Roman Baths* at Sagalassos (SW Turkey)

Fig. 4.4 Examples of secondary structural PEAEs. (A) Domino-type collapsed wall near the *Byzantine Gate* at Hierapolis (Pamukkale, SW Turkey). (B) U-shaped gap in wall of *Roman Theatre* at Sagalassos (SW Turkey) (cf. Sintubin and Stewart 2008). (C) Subsided keystone in arch of the *Roman Baths* at Sagalassos (SW Turkey) (cf. Sintubin et al. 2003). (D) Domino-type collapsed column at Petra (Jordan). (E) Shifted masonry column of the *Marcellum* at Baelo Claudia (SW Spain) (cf. Silva et al. 2009). From Sintubin (2011). (F) Rotated masonry block within a wall of the *Doric Temple* at Sagalassos (SW Turkey) (modified from Sintubin and Stewart 2008)

Stratigraphical PEAEs are reflected in specific archaeological destruction layers or deposits (Fig. 5), showing evidence of sudden collapse caused by human and/or natural agents, commonly covering a living surface, evidenced by e.g. in situ broken vessels, buried valuable objects, or skeletons of victims. Archaeological destruction layers may also contain burned material, charcoal, collapsed architectural debris, and crushed and/or toppled objects. In archaeological contexts dominated by rubble architecture, such as in the case of Bronze Age civilizations around the Mediterranean (Jusseret 2017) (Fig. 6) or the Harappan civilization in the Indus Valley (Kovach et al., 2010), destruction layers may provide the only evidence for ancient earthquakes. As is the case with structural PEAEs, also the interpretation of stratigraphical PEAEs is hindered by a poor temporal resolution, although it is more robust than the age constraints of structural PEAEs. Relating a destruction layer to an individual earthquake remains a challenge for the archaeoseismologist. Another major issue with respect to the interpretation of destruction layers is their preservation and/or disturbance. Earthquake debris may have been cleared from buildings and streets and disposed of in dumpsites (e.g. Similox-Tohon et al., 2006) (Fig. 5B), leaving no obvious traces of a destructive earthquake in the archaeological stratigraphy. Valuable objects and/or victims may have been recovered from the debris, disturbing the stratigraphy. Material may have been recycled in rebuilding efforts, again disturbing stratigraphical earthquake evidence. Dumpsites, recycled building material, or disturbed stratigraphy are all examples of **tertiary PEAEs**. The preservation of pristine destruction layers may depend strongly on the degree to which the affected community was able to cope with the aftermath of an earthquake, much more than on the physical parameters of the earthquake itself (e.g. Forlin & Gerrard, 2017). The same preservation issue applies to structural earthquake evidence. One can thus expect a rather incomplete archaeological record of ancient earthquakes, which is moreover not evenly distributed through time. Archaeoseismic evidence may suffer from an observation bias, focusing on periods of social upheaval, total destruction, complete and sudden abandonment, and so on. During periods of cultural, social and political stability and flourishing economies, most of the indications of an earthquake are most probably covered up, making it difficult for archaeoseismologists to identify ancient earthquakes. Contrary, during periods of upheaval and economic crisis, there may not be any impetus or funds to fully recover from an earthquake, thus leaving clear earthquake evidence for archaeoseismologists to discover (Jusseret & Sintubin, 2012).

4.3 Methods

Initially archaeoseismological investigations, primarily grafted on archaeological work in the Mediterranean region and the Middle East, depended largely/solely on identifying structural PEAEs to monumental buildings (Stiros & Jones, 1996). Elsewhere, such as in Japan (Barnes, 2010) or the US (e.g. Tuttle & Schweig, 1995), stratigraphical PEAEs received from the beginning more attention to identify earthquakes in archaeological contexts (e.g. liquefaction phenomena in burial mounds; Barnes, 2010). Through the years, different attempts were made to develop, primarily qualitative, schemes of structural PEAEs to be used during excavations works at an archaeological site. The most recent attempt by Rodríguez-Pascua et al. (2011) resulted in a comprehensive classification of structural PEAEs

Fig. 4.5 Examples of stratigraphical PEAEs. (A) Archaeological destruction layer underneath a series of toppled columns at Cnidus (Datça peninsula, SW Turkey), allowing an age constraint of this PEAE. From Sintubin (2011). (B) Remarkable Roman dumpsite of earthquake debris at Sagalassos (SW Turkey) in the direct hanging wall of the 'active' Sagalassos fault (cf. Similox-Tohon et al. 2006). (C) Potential tsunami or debris flow deposit at the Minoan site of Palaikastro (Crete, Greece) (cf. Bruins et al. 2008)

Fig. 4.6 Examples of secondary structural PEAEs in archaeological contexts dominated by rubble architecture. (A) Tilted base of wall at Amnisos (Crete, Greece). (B) Domino-type collapsed wall segment at Malia (Crete, Greece). (C) Rotated block in a wall at Zakros (Crete, Greece). Compared to equivalent secondary structural PEAEs in monumental masonry constructions (Fig. 4), proving an earthquake hypothesis in these rubble contexts remains very challenging

that is comparable to the classification of earthquake environmental effects, or EEs, in the framework of the macroseismic Environmental Seismic Intensity Scale (ESI2007; Michetti et al., 2007). Since then, a lot of studies use the classification of Rodríguez-Pascua et al. (2011) to derive parameters of ancient earthquakes based on structural damage features in archaeological sites, or even of instrumental earthquakes based on earthquake-induced structural damage features in recent buildings (e.g. Lorca, Rodríguez-Pascua et al., 2012; Kathmandu, Davis et al., 2020).

Many published archaeoseismological studies solely focus on PEAEs. Danger of overinterpreting potential earthquake-related evidence looms. As already mentioned, there is the preservation issue. More importantly, there is inevitable ambiguity with respect to an earthquake hypothesis of PEAEs. One must prove that particular damage features, attributed to earthquakes, cannot be caused by other destructive agents, which in the majority of cases is nearly impossible. Furthermore, distinguishing far-field from near-field PEAEs makes any valid interpretation of PEAEs very difficult. To cope with the uncertainties, Hinzen (2005) has introduced a feasibility matrix to evaluate the probability of occurrence of a proposed ancient earthquake at an archaeological site. Sintubin and Stewart (2008) developed a logic-tree formalism to assess the degree of certainty to which archaeological site could have recorded the proposed ancient earthquake (e.g. Yerli et al., 2011).

Probably the greatest challenge in archaeoseismology is constraining the timing of the ancient earthquake based solely on PEAEs. Classically, a chronological control of PEAEs is achieved by bracketing the PEAEs with a minimum (*terminus ante quem*) and maximum (*terminus post quem*) age. Furthermore, archaeological dating is based on e.g. architectural styles, ceramic style changes, coins in destruction layers, or inscriptions. In the most favourable conditions, radiometric dating methods (e.g. charcoal in destruction layers) can be called upon to constrain the timing of the earthquake evidence. The accuracy and precision of the age control largely depends on the type of dating method and the number of complementary dating methods applied. The temporal resolution is commonly rather poor, ranging from decennia to centuries. Once an ancient earthquake has been dated, it is often equated to an historical earthquake, present in an inherently incomplete earthquake catalogue. Here looms the danger of a circular argument, when subsequently the archaeological earthquake evidence is used as supplementary proof for the historical earthquake (Rucker & Niemi, 2010). Poor temporal resolution may also lead to amalgamate PEAEs as being caused by a single—oversized—earthquake (cf. Guidoboni 2002; Ambraseys et al., 2002), while in reality the PEAEs may result from a sequence of a main shock and its aftershocks, or a cluster of earthquakes spanning decades to centuries (e.g. Bindi et al., 2014; Mosca et al., 2019).

Poor age control furthermore hampers spatial correlations of PEAEs both within a single archaeological site and between different archaeological sites, again giving rise to the danger of amalgamating archaeological earthquake evidence at neighbouring archaeological sites, that may effectively be the result of regionally distinct ancient earthquakes. Such a regional amalgamation of earthquake evidence in Late Bronze Age sites around the Eastern Mediterranean by Schaeffer (1948) eventually lead to the myth of the Late Bronze Age earthquake catastrophe c. 1200 BC, ending the Late Bronze Age civilization in the Eastern Mediterranean (Nur & Cline, 2000). Reconstructing the spatial distribution of variable ground motion intensities of a pre-instrumental earthquake is crucial to be able to quantitatively estimate the magnitude of the pre-instrumental earthquake. Archaeological evidence for ground shaking could be used for this purpose, whether or not in combination with EEEs. Besides the poor temporal resolution, the very sparse spatial distribution of archaeological sites over a region, remains a limiting factor to use PEAEs in any effort to determine the magnitude of an ancient earthquake, in particular in regions which lack a long and rich archaeological heritage.

To fully make advantage of archaeological earthquake evidence in our pursuit to reconstruct the earthquake history of a region, archaeoseismological work should be integrated into a more interdisciplinary and quatitative approach. Such an approach also requires shared protocols and standardized methodologies (Sintubin & Stewart, 2008 and references therein). This tendency towards a more quantitative approach of PEAEs in archaeological sites is fully exemplified in the rapid advances in **quantitative archaeoseismology**. Firstly, quantitative archaeoseismology calls upon modern visualization techniques, such as 3D laser scanning (e.g. ground-based LIDAR), to obtain a quantitative, high-resolution, three-dimensional structural model of the damaged archaeological structure, allowing the construction of a very precise structural damage inventory (e.g. Yerli et al., 2010; Hinzen et al., 2016). Using earthquake engineering models, the dynamic behavior of the ancient manmade structure can be evaluated (e.g. Hinzen, 2005; Hinzen et al., 2016). Scenario-based earthquake ground motion simulations subsequently test earthquake hypotheses to explain the PEAEs in the ancient building. Such quantitative approach allows not only to reduce the subjective influence during damage interpretation, but also to quantify ground motion parameters and eventually magnitude estimates of the ancient earthquakes (Hinzen et al., 2011). In this respect, archaeological structures become **seismoscopes**, defined by Richter (1958) as *"devices which indicate the occurrence of an earthquake but don't write a record"* (Schweppe et al., 2017). In Hinzen et al. (2011) a list can be found of archaeoseismological studies using quantitative modeling and advanced measuring techniques.

Emphasis is thus no longer put on trying to identify physical parameters of pre-instrumental, ancient earthquakes, such as the exact date of the earthquake, its magnitude, its seismic source, or its epicenter, so that the ancient earthquake identified can be entered in an earthquake catalogue. As ancient seismoscopes, archaeological sites become testing grounds to quantitatively assess site-specific ground effects (e.g. Schweppe et al., 2017). Archaeological sites may hold the eventuality to have narrowed down macroseismic parameters associated with the maximum credible earthquake, irrespective of the time of occurrence and the physical parameters of ancient earthquakes that have affected the archaeological site during and after its occupancy.

Another type of seismoscopes, providing crucial information with respect to the long-term seismicity in a region, are precariously balanced rocks (PBRs), not being toppled by earthquake-induced ground motions (e.g. Ludwig et al., 2015). Archaeological remains, such as free-standing columns (e.g. Ambraseys & Psycharis, 2012), may act as such precariously balanced rocks, becoming an instrument to determine the maximum upper ground motion bounds at a site. Schweppe et al. (2017) refer to these ancient manmade structures as **precariously balanced archaeological structures** (PBASs) (Fig. 7). These PBASs may furthermore provide valuable parameters with respect to preservation efforts of the cultural heritage. Remarkably, structural engineering analysis of such structures suggests that ancient manmade construction are quite resistant to earthquakes (Stiros, 2020).

But what may turn out to be even more important in assessing PEAEs from a structural engineering perspective, is that such a quantitative approach allows to discriminate between damage caused by earthquakes and damage caused by other natural or anthropogenic agents. This engineering approach reveals that often single-event collapses are rather exceptional. The gradual demise of a manmade construction seems mostly the result of a sequence of natural and

Fig. 4.7 Example of a precariously balanced archaeological structure (PBAS). Free-standing column in the *Temple of Aphrodite* at Afrodisias (SW Turkey). The displaced drums in the column may serve as a PEAE, indicative for a potential earthquake that did not though topple the column

anthropogenic destructive events (e.g. Stiros, 2020). In a number of cases, this approach demonstrates that certain PEAEs, that were conventionally been considered earthquake-related, were most probably caused by alternative destructive agents (e.g. Sarcophagus at Lycian Pinara, Turkey; Hinzen et al. 2010). Moreover, classical damage typologies attributed to earthquake-related ground motions, such as perfectly aligned toppled columns (e.g. toppled columns at Byzantine Sussita, Israel; Hinzen, 2009a & 2009b) (Fig. 8), could not be validated by modelling, thus

Fig. 4.8 The perfectly aligned toppled columns at Susita (Golan Heights), commonly seen as a classical damage typology attributed to earthquake-related ground motion, but not validated by ground-motion modelling (cf. Hinzen 2009a & b). From Sintubin (2011)

seriously questioning the classical archaeoseismological practice solely based on the interpretation of PEAEs.

While this novel quantitative archaeoseismological practice is very effective on sites with monumental architecture, a more semiquantitative, **integrated territorial approach** is more appropriate in archaeological contexts dominated by rubble architecture and associated destruction layers (e.g. Jusseret et al., 2013). This approach starts from the seismotectonic context of the archaeological site, identifying different types of potential seismic sources, as well as their ground-motion relationships. In the search for ancient earthquakes well-documented, high-visibility archaeological contexts are selected. These high-visibility archaeological context are e.g. characterized by distinctive and rapid ceramic changes, narrowing down the time window substantially and thus improving the temporal constraints of the PEAEs. Subsequently, different earthquake scenarios are tested in an iterative process, eventually leading to an earthquake scenario that provides the best explanations for the PEAEs observed in the archaeological context studied.

4.4 Perspectives

The question remains whether or not archaeoseismology can really contribute in seismic hazard assessments. A seismic hazard assessment requires an accurate catalogue of earthquakes and their physical parameters, spanning a time range as wide as possible. Besides the instrumental database (instrumental seismology), we call upon the incomplete records of historical and pre-historical earthquakes, the former derived from historical records (historical seismology), the latter from archaeological (archaeoseismology) and geological records (paleoseismology; neotectonics) (Caputo & Helly, 2008). Considering the incompleteness of the archaeological record, its limited spatial and temporal resolution, and all the uncertainties inherent to PEAEs, skepticism towards the applicability of archaeoseismology in a seismic hazard assessment remains legitimate (Sintubin, 2011). Only in particular cases, such as the use of faulted relics in an attempt to better constrain earthquake slip rates on a millennial scale, or the use of PEAEs in efforts to better characterize historical earthquakes, archaeoseismology may provide complementary information in support of a seismic hazard assessment.

A promising development in archaeoseismology resides in archaeological sites becoming testing grounds to quantitatively assess site-specific ground motion effects, eventually to help calibrating specific earthquake scenarios in a region. Archaeological sites, especially with a long and lasting history, have the potential to have recorded the effects of the characteristic earthquake on one of the nearby active faults, irrespective of the time of occurrence.

Finally, the strength of archaeoseismology may just lie in a better understanding of how our ancestors coped with the threat of earthquakes (Fig. 9), improving local earthquake preparedness in earthquake-prone regions. In the end, 'telling a good story' (Richter, 1958) may turn out to have more effect than trying to derive crude parameters for ancient earthquakes (Sintubin, 2008).

Fig. 4.9 The world-famous *Lion Gate* at Mycenea (Greece). The fortification wall was built on top of an active, Aegean-type, limestone fault scarp (see Stewart and Hancock 1988 for further reading), eliciting the question whether or not Myceneans were aware of the earthquake hazard threatening their city

References

Ambraseys, N. N. (2005). Archaeoseismology and Neocatastrophism. *Seismological Research Letters, 76*, 560–564.

Ambraseys, N. N., & Psycharis, I. N. (2012). Assessment of the long-term seismicity of Athens from two classical columns. *Bulletin of Earthquake Engineering, 10*(6), 1635–1666.

Ambraseys, N. N., Jackson, J. A. & Melville, C. P. (2002). Historical seismicity and tectonics: the case of the eastern Mediterranean and the middle East. In W. H. K. Lee, H. Kanamori, P. C. Jennings, & C. Kisslinger, *International handbook of earthquake and engineering seismology*. International Geophysics Series 81A, pp. 747–763

Barnes, G. L. (2010). Earthquake archaeology in Japan: an overview, in Ancient Earthquakes, edited by Sintubin, M., Stewart, I. S., Niemi, T. and Altunel, E. *Geological Society of America, Special Papers, 471*, 81–96.

Bindi, D., Parolai, S., Gómez-Capera, A., Locati, M., Kalmetyeva, Z., & Mikhailova, N. (2014). Locations and magnitudes of earthquakes in Central Asia from seismic intensity data. *Journal of Seismology, 18*(1), 1–21.

Bruins, H. J., MacGillivray, A., Synolakis, C. E., Benjamini, C., Keller, J., Kisch, H. J., Klügel, A. and van der Plicht, J. (2008), Geoarchaeological tsunami deposits at Palaikastro (Crete) and the Late Minoan IA eruption of Santorini, *Journal of Archaeological Sciences, 35*, 191–212.

Caputo, R., & Helly, B. (2008). The use of distinct disciplines to investigate past earthquakes. *Tectonophysics, 453*, 7–19.

Caputo, R., & Pavlides, S. B. (2008). Earthquake geology: methods and applications. *Tectonophysics, 453*, 1–6.

Davis, C., Coningham, R., Acharya, K. P., Kunwar, R. B., Forlin, P., Weise, K., Maskey, P. M., Joshi, A., Simpson, I., Toll, D., Wilkinson, S., Hughes, P., Sarhosis, V., Kumar, A., & Schmidt, A. (2020). Identifying archaeological evidence of past earthquakes in a contemporary disaster scenario: case studies of damage, resilience and risk reduction from the 2015 Gorkha Earthquake and past seismic events within the Kathmandy Valley UNESCO World Heritage Property (Nepal). *Journal of Seismology, 24*, 729–751.

Ferrario, M. F., Katz, O., Hillman, A., Livio, F., Amit, R. and Michetti, A. M. (2020), The Mid-eighth Century CE Surface Faulting Along the Dead Sea Fault at Tiberias (Sea of Galilee, Israel), *Tectonics*, 39, e2020TC006186.

Force, E. R., & McFadgen, B. G. (2010). Tectonic environments of ancient civilizations: opportunities for archaeoseismological and anthropological studies, in Ancient Earthquakes, edited by Sintubin, M., Stewart, I. S., Niemi, T. and Altunel, E. *Geological Society of America, Special Papers, 471*, 21–28.

Forlin, P., & Gerrard, C. M. (2017). The archaeology of earthquakes: The application of adaptive cycles to seismically-affected communities in late medieval Europe. *Quaternary International, 446*, 95–108.

Galadini, F., Hinzen, K.-G., & Stiros, S. (2006). Archaeoseismology: methodological issues and procedure. *Journal of Seismology, 10*, 395–414.

Guidoboni, E. (2002). Historical Seismology: the long memory of the inhabited world, in International Handbook of Earthquake and Engineering Seismology, edited by Lee, W. H. K., Kanamori, H., Jennings, P. C. and Kisslinger, C. *International Geophysics Series, 81A*, 775–790.

Hancock, P. L., & Altunel, E. (1997). Faulted archaeological relics at Hierapolis (Pamukkale), Turkey. *Journal of Geodynamics, 24*(1–4), 21–36.

Hessami, K., Pantosti, D., Tabassi, H., Shabanian, E., Abbassi, M. R., Feghhi, K., & Solaymani, S. (2003). Paleoearthquakes and slip rates of the North Tabriz Fault, NW Iran: preliminary results. *Annals of Geophysics, 46*(5), 903–915.

Hinzen, K.-G. (2005). The use of engineering seismological models to interpret archaeoseismological findings in Tolbiacum, Germany: A case study. *Bulleting of the Seismological Society of America, 95*, 521–539.

Hinzen, K.-G. (2009). Sensitivity of earthquake-toppled columns to small changes in ground motion and geometry. *Israel Journal of Earth Sciences, 58*(3), 309–326.

Hinzen, K.-G. (2009). Simulation of toppling columns in archaeoseismology. *Bulletin of the Seismological Society of America, 99*(5), 2855–2875.

Hinzen, K.-G., Fleischer, C., Reamer, S. K., Schreiber, S., Schütte, S., & Yerli, B. (2011). Quantitative methods in archaeoseismology. *Quarternary International, 242*, 31–41.

Hinzen, K.-G., Schwellenbach, I., Schweppe, G., & Marco, S. (2016). Quantifying earthquake effects on ancient archers, Example: The Kalat Nimrod Fortress, Dead Sea Fault Zone. *Seismological Research Letters*, 87(3), 751–764.

Jackson, J. (2006). Fatal attraction: living with earthquakes, the growth of villages into megacities, and earthquake vulnerability in the modern world. *Philosophical Transactions of the Royal Society A: Mathematical, Physical and Engineering Sciences, 364*(1845), 1911–1925.

Jusseret, S. (2014). Earthquake archaeology: A future in ruins? *Journal of Contemporary Archaeology, 1*, 277–296.

Jusseret, S., & Sintubin, M. (2012). All that rubble leads to trouble: Reassessing the seismological value of archaeological destruction layers in minoan crete and beyond. *Seismological Research Letters, 83*(4), 736–742.

Jusseret, S., & Sintubin, M. (2013). The Origins of an Old Myth: Sir Arthur Evans, Claude Schaeffer and the Seismic Destruction of Late Bronze Age Easter Mediterranean Civilizations. *Seismological Research Letters, 84*(1), 94–100.

Jusseret, S., & Sintubin, M. (2017). *Minoan Eartquakes* (p. 408). Leuven University Press, Leuven, Belgium, Studies in Archaeological Sciences.
Jusseret, S., Langohr, C. & Sintubin, M. (2013). Tracking earthquake archaeological evidence in Late Minoan IIIB (~1300–1200 B.C.) Crete (Greece): A proof of concept. *Bulletin of the Seismological Society of America*, 103(6), 3026-3043.
Jusseret, S. (2017), Archaeoseismological research on Minoan Crete: past and present, in *Minoan Eartquakes. Breaking the Myth through Interdisciplinarity*, edited by Jusseret, S. and Sintubin, M., Leuven University Press, Leuven, Belgium, Studies in Archaeological Sciences, 223-248.
Karcz, I., & Kafri, U. (1978). Evaluation of supposed archaeoseismic damage in Israel. *Journal of Archaeological Sciences, 5*, 237–253.
Kovach, R. L., Grijalva, K. & Nur, A. (2010). Earthquakes and civilizations of the Indus valley: A challenge for archaeoseismology. In M. Sintubin, I. S. Stewart, T. Niemi, & E. Altunel, *Ancient earthquakes*. Geological Society of America, Special Papers, 471, pp. 000–000.
Ludwig, L. G., Brune, J. N., Anooshehpoor, A., Purvance, M. D., Brune, R. J., & Lozos, J. C. (2015). Reconciling Precariously Balanced Rocks (PBRs) with Large Earthquakes on the San Andreas Fault System. *Seismological Research Letters, 86*(5), 1345–1353.
Marco, S., Agnon, A., Ellenblum, R., Eidelman, A., Basson, U., & Boas, A. (1997). 817-year old walls offset sinistrally 2.1 m by the Dead Sea transform, Israel. *Journal of Geodynamics, 24*, 11–20.
McCalpin, J. (2009). *Paleoseismology*, vol. 95. Academic Press, XXX, International Geophysics Series, p. 629.
McGuire, W. J., Griffith, D. R., Hancock, P. L., & Stewart, I. S. (Eds.). (2000). *The archaeology of geological catastrophes* (p. 417). Geological Society.
Michetti, A. M., Esposito, E., Guerrieri, L., Porfido, S., Serva, L., Tatevossian, R., Vittori, E., Audemard, F., Azuma, T., Clague, J., Comerci, V., Gürpinar, A., McCalpin, J., Mohammadioun, B., Mörner, N. A., Ota, Y., & Roghozin, E. (2007). Intensity scale ESI 2007, APAT, Rome. *Memorie Descrittive della Carta Geologica d'Italia, 74*, 42p.
Mosca, I., Baptie, B., Sargeant, S., & Walker, R. T. (2019). Integrating outcomes from probabilistic and deterministic seismic hazard analysis in the Tien Shan. *Bulletin of the Seismological Society of America, 109*(2), 688–715.
Nur, A., & Cline, E. H. (2000). Poseidon's horses: Plate tectonics and earthquake storms in the late Bronze age Aegean and Eastern Mediterranean. *Journal of Archaeological Science, 27*, 43–63.
Richter, C. F. (1958). *Elementary Seismology* (p. 578p). Freeman.
Rodríguez-Pascua, M. A., Pérez-López, R., Giner-Robles, J. L., Silva, P. G., Garduño-Monroy, V. H., & Reicherter, K. (2011). A comprehensive classification of earthquake archaeological effects (EAE) in archaeoseismology: Application to ancient remains of Roman and Mesoamerican cultures. *Quaternary International, 242*, 20–30.
Rodríguez-Pascua, M. A., Pérez-López, R., Martín-González, F., Giner-Robles, J. L., & Silva, P. G. (2012). Efectos arquitectónicos del terremoto de Lorca del 11 de mayo de 2011. *Neoformación y reactivación de efectos en el Patrimonio Cultural, Boletin Geológico y Minero, 123*(4), 487–502.
Rucker, J. D., & Niemi, T. M. (2010). Historical earthquake catalogues and archaeological data: achieving synthesis without circular reasoning, in Ancient Earthquakes, edited by Sintubin, M., Stewart, I. S., Niemi, T. and Altunel, E. *Geological Society of America, Special Papers, 471*, 97–106.
Sbeinati, M. R., Meghraoui, M., Suleyman, G., Gomez, F., Grootes, P., Nadeau, M.-J., Al Najjar, H., & Al-Ghazzi, R. (2010). Timing of earthquake ruptures at the Al Harif Roman aqueduct (Dead Sea fault, Syria) from archaeoseismology and paleoseismology. In M. Sintubin, I. S. Stewart, T. Niemi, & E. Altunel, *Ancient earthquakes*. Geological Society of America, Special Papers, 471, pp. 243–267.
Schaeffer, C. F. A. (1948). *Stratigraphie compare et chronologie de l'Asie occidentale (IIIe en IIe millénaires)*. Oxford University Press.

Schweppe, G., Hinzen, K.-G., Reamer, S. K., Fishcer, M., & Marco, S. (2017). The ruin of the Roman Temple of Kedesh, Israel; example of a precariously balanced archaeological structure used as a seismoscope. *Annals of Geophysics, 60*(4), S0444.

Silva, P. G., Sintubin, M., & Reicherter, K. (2011). New advances in studies of earthquake archaeology and palaeoseismology. *Quaternary International, 242*, 1–3.

Silva, P. G., Reicherter, K., Grützner, C., Bardaji, T., Lario, J., Goy, J. L., Zazo, C., & Becker-Heidmann, P. (2009). Surface and subsurface palaeoseismic records at the ancient Roman city of *Baelo Claudia* and the Bolonia Bay area, Cadiz (south Spain). In K. Reicherter, A. M. Michetti, & P. G. Silva (eds), *Palaeoseismology: historical and prehistorical records of earthquake ground effects for seismic hazard assessment*. The Geological Society, London, Special Publications, 316, pp. 93–121.

Similox-Tohon, D., Sintubin, M., Muchez, P., Verhaert, G., Vanneste, K., Fernandez, M., Vandycke, S., Vanhaverbeke, H., & Waelkens, M. (2006). The identification of an active fault by a multidisciplinary study at the archaeological site of Sagalassos (SW Turkey). *Tectonophysics, 420*, 371–387.

Sintubin, M. (2008). Earthquake archaeology—Just a good story? *Seismological Research Letters, 79*(6), 767–768.

Sintubin, M., & Stewart, I. S. (2008). A logical methodology for archaeoseismology: a proof of concept at the archaeological site of Sagalassos, southwest Turkey. *Bulletin of the Seismological Society of America, 98*, 2209–2230.

Sintubin, M. (2011). Archaeoseismology: Past, present and future, *Quaternary International 242*(1) 4-10.

Sintubin, M., Muchez, P., Similox-Tohon, D., Verhaert, G., Paulissen, E., & Waelkens, M. (2003). Seismic catastrophes at the ancient city of Sagalassos (SW Turkey) and their implications for seismotectonics in the Burdur-Isparta area. *Geological Journal, 38*, 359–374.

Sintubin, M., Stewart, I., & S., Niemi, T. M. & Altunel, E. (Eds.). (2010). *Ancient earthquakes* (p. 471). Geological Society of America.

Stewart, I. S. and Hancock, P. L. (1988). Normal fault zone evolution and fault scarp degradation in the Aegean region, *Basin Research, 1*(3), 139–163.

Stiros, S. C. (1996). Identification of earthquakes from archaeological data: methodology, criteria and limitations. In S. C. Stiros & R. E. Jones (Eds.), *Archaeoseismology* (pp. 129–152). British School at Athens.

Stiros, S. C. (2020). Monumental articulated ancient Greek and Roman columns and temples and earthquakes: archaeological, historical and engineering approaches. *Journal of Seismology, 24*, 853–881.

Stiros, S. C. (2010). The 8.5+ magnitude, AD 365 earthquake in Crete: Coastal uplift, topography changes, archaeological and historical signature. *Quaternary International, 216(1-2)*, 54–63.

Stiros, S., & Jones, R. E. (eds). (1996). *Archaeoseismology*. British School of Athens, Oxford, British School of Athens Fitch Laboratory Occasional Paper 7.

Tuttle, M. P., & Schweig, E. (1995). Archeological and pedological evidence for large prehistoric earthquakes in the new madrid seismic zone. *Central United States, Geology, 23*(3), 253–256.

Yerli, B., ten Veen, J., & Sintubin, M. (2011). Testing a logic tree approach for archaeoseismology to the ancient city of Pinara (SW Turkey). *Quaternary International, 242*, 52–64.

Yerli, B., ten Veen, J. H., Sintubin, M., Karabacak, V., Yalciner, C. & Altunel, E. (2010). Assessment of seismically induced damage using LIDAR: The ancient City of Pinara (SW Turkey) as a case study. In M. Sintubin, I. S. Stewart, T. Niemi, E. Altunel (eds) *Ancient earthquakes*, Geological Society of America, Special Papers, 471, pp. 157–170.

Past Earthquakes in Continental Settings—A Geomorphologic Perspective

5

O. Zielke and Y. Klinger

This chapter addresses the measurement of displaced geomorphic landforms to constrain the along-fault distribution of surface displacement from individual earthquakes and their recurrence. This line of work is motivated by the desire to constrain the size and other characteristics of past earthquakes to permit estimates of a fault's likely future behavior to further improve seismic hazard assessment. For a number of reasons, discussed in this chapter, taking such measurements is not trivial. With this chapter, we want to outline why such measurements are important, what data sets are used to take them, how measurements can be taken and what assumptions are adopted to take those measurements. We also provide our recommendations for taking such offset measurements as well as a conceptual model for earthquake recurrence, based on findings from numerous researchers over the last few decades.

5.1 Introduction/Motivation

Earthquakes occur to release previously accumulated elastic strain, primarily along pre-existing faults. The alternation of interseismic strain accumulation and coseismic strain release is commonly referred to as the "earthquake cycle" or "seismic

O. Zielke (✉)
Physical Sciences and Engineering, King Abdullah University of Science and Technology (KAUST), Thuwal, Saudi Arabia
e-mail: olaf.zielke@kaust.edu.sa

Y. Klinger
Institut de Physique du Globe de Paris, Université de Paris, CNRS, 75005 Paris, France
e-mail: klinger@ipgp.fr

A. Elliott and C. Gruetzner (eds.), *Understanding Past Earthquakes*,
https://doi.org/10.1007/978-3-031-73580-6_5

cycle". Motivated by the severe devastation that may accompany large earthquakes, it is of interest to better understand (a) where active and potentially hazardous faults are located, (b) what sizes of earthquakes these faults may generate, and (c) how the temporal occurrence of large earthquakes is characterized. Addressing these questions becomes more and more relevant as an ever-increasing fraction of the global population is living in dense population centers at, or very near, to seismically active faults.

One approach to gain this understanding lies in the careful documentation of active fault traces, past earthquake ruptures, and their characteristics (e.g., surface rupture trace, spatial distribution and temporal evolution of slip along a rupture surface, and variation from earthquake to earthquake). This approach starts generally with the detailed analysis of the most recent rupture along the fault of interest. Carefully analyzing past earthquakes follows the assumption that the past behavior and rupture characteristics give insights into the underlying physical processes as well as a fault's likely future behavior (e.g., Schwartz & Coppersmith, 1984; Shimazaki & Nakata, 1980): If we are able to properly document the past behavior of a fault, then we might be able to determine its likely future behavior. The characterization of the most recent earthquake (MRE) plays a special role in that regard, as it is often used as a template to interpret measurements that are associated with older earthquakes (along the same fault). If the MRE occurred recently enough (i.e., within the last few decades), then instrumentally recorded seismograms, high-resolution geodetic measurements, or imagery correlation approaches may be used to solve for the coseismic distribution and temporal evolution of slip along the fault surface (e.g., Ide, 2007; Kurtz et al., 2018), enabling us to constrain the surface rupture length, the complexity of the rupture geometry, or the ratio between localized and distributed deformation, and eventually the earthquake's size. In many instances, however, the MRE did not occur recently enough to be recorded instrumentally, macroseismically, nor with pre- and geodetic benchmarks. Furthermore, even when the MRE is recorded instrumentally, this seismological observation does not provide information on the penultimate and prior earthquakes nor the time interval between them. In order to improve seismic hazard assessment by better understanding how earthquakes along a fault recur and on average how fast faults move at the Holocene time scale, we need to identify how ruptures along a given fault differ from one earthquake to the next (e.g., in terms of slip), how long the time interval between them is, and whether this time interval is correlated with other rupture characteristics. For example, was the slip in the penultimate earthquake similar to the slip observed for the MRE? Is there spatial and/or temporal correlation between earthquake timing and the distribution of slip along the fault? Addressing these questions requires documentation of not only the MRE but also the sequence of earthquakes that occurred before, extending as far back in time as possible. Instrumental recordings are currently and for a foreseeable future too short to document the repeated succession of large earthquakes along the same portion of a fault except in a few rare instances (Cite: Imperial Fault, SAF Parkfield, Papua New Guinea, Italy 2016). Other records must be used instead and they need to fulfill some requirements:

- Their temporal resolution must be sufficiently high to distinguish the effects of individual earthquakes that occurred along a fault.
- They must be spatially sensitive enough to resolve potentially small surficial disturbances caused by earthquake rupture.
- They must store that evidence of faulting over a sufficiently long time period (i.e., they must have a high preservation potential), and
- They must be reasonably accessible and abundant in order to constitute robust and reliable documentation of earthquake effects.

Geomorphic and stratigraphic records may fulfill these requirements: exemplified by many recent ruptures, we know that earthquakes may disrupt the ground surface and displace landscape and stratigraphic units along its surface rupture trace. The succession of surface-rupturing earthquakes along a fault will—step by step—disrupt landscape and stratigraphic units, forming distinctive features that may be used to identify the along-fault accumulation of slip, the size of the causative earthquakes, as well as the long-term slip-rate of a fault (if the displaced units have also been dated—an endeavor beyond the scope of this chapter). An advantage of stratigraphic and geomorphic records is that they are—depending on the specific geologic and topographic setting—able to rejuvenate themselves. That is, an active and dynamically evolving landscape forms new or modifies existing geomorphic features and deposits new stratigraphic units over time. Under favorable site conditions this modification of the landscape may occur at a frequency that is higher than the occurrence frequency of earthquakes that disrupt these (geomorphic and stratigraphic) records. Thus, geomorphic and stratigraphic data can provide the "temporal" resolution to distinguish between different earthquakes, and are used to constrain rupture characteristics of earthquakes that occurred during the pre-instrumental period. Stratigraphic and geomorphic records are therefore used to constrain some of the aforementioned rupture characteristics, including: along-fault distribution of slip, slip-accumulation through time, number of fault sections involved, fault slip-rate. Another important reason to use geomorphic and stratigraphic data is that they are comparably abundant and—for continental settings—relatively easy to access. As a result, much of the previous work on earthquake rupture physics and the recurrence of earthquakes has concentrated on continental faults, strike-slip faults in particular where data are comparably abundant and acquisition is reasonably straightforward.

In this chapter we will address the measurement and interpretation of on-fault displacement that is caused by continental earthquakes and stored in the tectono-geomorphic record, with a brief excursion to shallow sub-surface stratigraphic evidence of faulting (paleoseismic investigations). We will first discuss the features found in geomorphic records that are commonly associated with coseismic rupture (an earthquake) and that can be used to constrain an earthquake's displacement. Following, we will describe the data types that were previously and are currently used to document geomorphic and stratigraphic records, also highlighting some of the recent contributions that these data types have made to improve our understanding of past earthquakes. These highlights are separated based on whether they were collected only after an earthquake or also prior to it. We will continue to summarize these contributions to formulate the current understanding of slip accumulation by continental earthquakes as it is resolved in stratigraphic and geomorphic records—with an emphasis on the variety of uncertainties associated with those records and their interpretation. Lastly, we provide some guidelines that should be considered when using geomorphic data in active tectonics studies, and give an outlook on future developments.

5.2 Geomorphic Markers for Offset Measurement, Methodology

When an earthquake occurs, displacement takes place along the fault surface to release the previously accumulated elastic strain. The strain release occurs through rebounding motion of the crust on either side of a fault. For recent earthquakes, the deformation that is associated with this rebound motion and slip along the fault may be measured with a range of different approaches, employing seismological records, Global Navigation Satellite Systems (GNSS) positioning, Interferometric Synthetic Aperture Radar (InSAR), optical image correlation and pixel-tracking methods, and field observations. Utilization of these data sets may resolve an earthquake's on-fault displacement as well as off-fault deformation, and they show the distribution of slip along the fault surface itself (through kinematic source inversion; e.g., Mai et al., 2016).

Large-enough earthquakes rupture the ground surface along the trace of the slipping fault, juxtaposing stratigraphic units and geomorphic features. Detailed mapping campaigns along the traces of active faults and recent surface-rupturing earthquakes have revealed a high level of geometric complexity of these surface rupture traces. Fault surfaces are not planar but instead rough, sometimes anastomosing surfaces that may be described statistically by self-similar or self-affine fractal distributions (e.g., Candela et al., 2012; Power & Tullis, 1991; Sagy et al., 2007). A fault trace, being a cross-sectional profile across a fault surface, shares these geometric characteristics. As we will discuss in a later section, acknowledging this complexity (e.g., in form of fault bends, steps, jogs, branches, etc.) is important to ensure meaningful offset measurement and data interpretation.

Measuring the amount of surface displacement that took place in a given earthquake at a given location along an earthquake's rupture trace requires the presence of preexisting landscape features (i.e., landforms) that were subsequently displaced (Burbank & Anderson, 2012) and whose pre-earthquake morphology can be estimated with some level of confidence. Landforms that enable us to perform this task (i.e., geomorphic lineaments or surfaces that provide a reference frame against which to assess relative or absolute deformation; Burbank & Anderson, 2012) are commonly referred to geomorphic markers. Measuring the displacement of a geomorphic marker (i.e., reconstructing its pre-earthquake morphology) is not a trivial endeavor. It works best if (a) the marker (i.e., piercing point, line, or plane) can be uniquely identified across the fault zone; (b) we have a good understanding of the marker's pre-earthquake(s) morphology, (c) the marker's morphology is simple (e.g., sub-linear, or approximately planar, discrete/sharp, and with similar morphologic expression on either side of the fault zone); (d) the displacement is sufficiently large with respect to the marker's spatial scale to unambiguously disrupt it, and (e) the fault zone itself is well localized, preferably onto a single strand. Naturally, not all of these conditions will be met for each marker and some of them are more likely to be met for recent earthquakes, where geomorphic processes have not been able to extensively alter the morphology of the fault scarp and displaced marker. Deviations from this ideal scenario will diminish our confidence in the validity of an offset reconstruction and will therefore be expressed in the assigned measurement uncertainties and quality ratings.

Of particular interest are geomorphic markers that form or get modified frequently and exhibit length-scales similar to those of the disturbance caused by the earthquake. This prior condition (the frequent formation or rejuvenation of geomorphic markers) must be present in order to discern the contribution of successive earthquakes—measurements of individual coseismic slip events require geomorphic markers that formed in-between earthquakes or become distinctly modified by them in order to separate the effects of individual earthquakes (Fig. 5.1). Consequently, markers that form in fluvial and alluvial settings, where deposition and erosion are frequent and ongoing, are most commonly used to investigate the accumulation of slip over multiple earthquake cycles within continental fault zones. In these settings the complex interaction of tectonic and climatic (i.e., hydrologic) forcing may provide the conditions to frequently form new markers. From the perspective of a tectonic geomorphologist it is fortunate that large earthquakes—those we are primarily interested in because they dominate seismic strain release and pose the largest hazard—have the potential to significantly disrupt and reorganize a drainage system. For example, the emplacement of coseismically induced landslides might change a river's drainage system significantly, forcing it to adjust to new conditions of sediment supply, gradient, and drainage area, etc. As a result, it is not uncommon for the occurrence of large earthquakes to be followed by an increased rate of marker formation or modification (Burbank & Anderson, 2012). Hence, fluvial and alluvial geomorphic markers are most commonly used to study the accumulation of slip along a fault of interest. Figure 5.2 illustrates geomorphic markers that are commonly used when investigating the accumulation of slip along a fault of interest i.e., its longer-term fault slip-rate.

For faults with predominantly lateral motion (strike-slip faults) we may observe geomorphic markers such as offset channels, beheaded channels, offset landslides, debris flows, moraines, or alluvial fans (where the focus may be on their respective edges or on their overall shape), or offset channel risers and terraces edges (Fig. 2A). The relative abundance of these markers depends on the local site conditions (geology, topography, hydrology) and may vary distinctly from one fault system to the next as well as along individual faults. Despite variability in their presence, offset or beheaded channels are frequently observed and enable us to

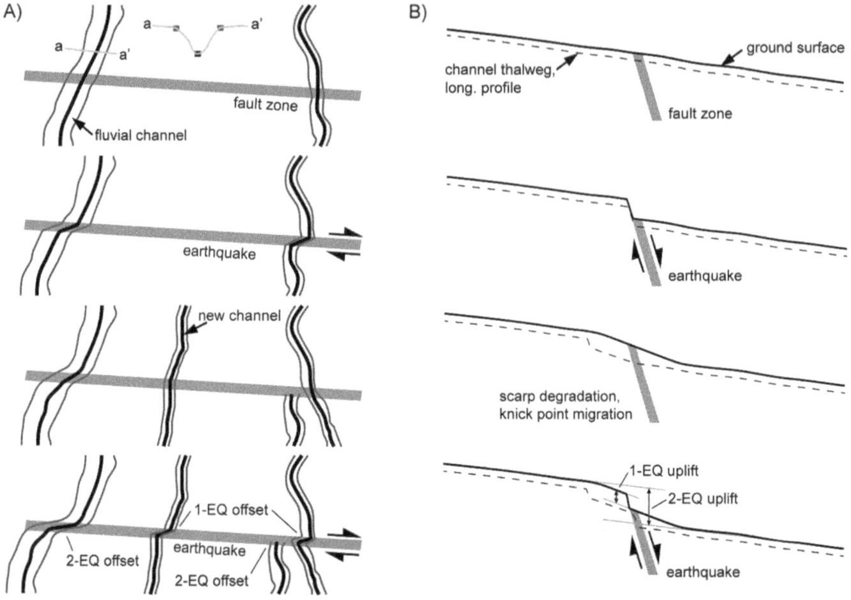

Fig. 5.1 Schematic representation of marker formation and modification, necessary to discern the amount of surface slip that is caused by a succession of earthquakes for faults with predominantly strike-slip (**A**, map view) or normal (**B**, cross-section view) fault motion. In between earthquakes the marker morphology adjusts to the modified conditions e.g., abandoning channel tail, degrading channel bend at fault, or degradation and backward-tilting of fault scarp and channel knick points. During the occurrence of an earthquake all existing markers are displaced. If markers were formed in-between earthquakes then the offset contributions of individual events may be extracted. The cross-sectional channel view (**A**, top) indicates piercing lines (a channel thalweg and both riser edges) that may be used to discern offset amounts

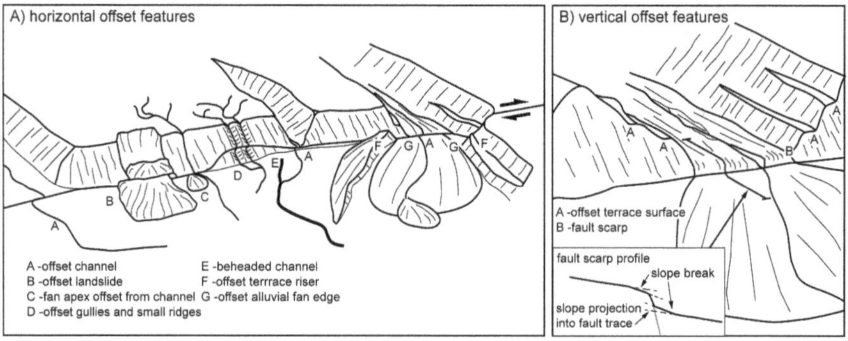

Fig. 5.2 Common expressions of surface faulting, focusing on geomorphic markers that are frequently used to constrain a fault's surface slip due to one or multiple earthquakes for faults with predominantly horizontal offsets (**A**) or vertical offsets (**B**). See text for further detail

resolve very small as well as very large offsets (ranging from dm- to km-scale offsets; depending on the channel's own length scale and time since initial disruption; Wallace, 1968). Given that channels form in an erosional environment (as opposed to a depositional one) it is not trivial to assign an age to these features. Instead, it requires extensive paleoseismic excavations to expose the youngest stratigraphic units the channel was incised into as well as the oldest channel fill deposits to bracket the age of the incision. Similarly, the edges of alluvial fans, fluvial terraces, and other narrow, sublinear piercing lines that cross the fault trace (preferably at a high angle) provide good means to measure lateral displacement. Given that the time of terrace formation (e.g., due to formation of a new flood plain) or fan abandonment (e.g., when the active fan shifts its location) pre-dates the observable accumulation of slip, we may employ geochronological methods that date the duration of surface exposure, or the latest date of burial. Alluvial fans in which the apex has been displaced laterally from the parent channel also provide the means to measure lateral slip but their presence is distinctly less common. A range of other geomorphic features may also form in response to earthquake faulting, such as scarps, benches, linear valleys and ridges, sag ponds, pressure- or shutter ridges and pull-apart basins. While these features provide important information, helping to map a fault trace or provide a framework that needs to be considered when interpreting observed marker displacements or deflections, they are themselves not used to constrain earthquake slip, slip accumulation, or fault slip-rates, for which distinct piercing lines are essential.

Faults with predominantly dip-slip motion (normal and reverse faults) also provide and store evidence of past earthquakes in the geomorphic record—often in the form of cumulative fault scarps or laterally extensive surfaces that were displaced vertically by a succession of earthquakes (e.g., the top of an alluvial fan surface or fluvial terraces, Fig. 2B). For example, abandoned fluvial terraces may be used to determine the vertical separation or offset to the current channel bed (i.e., the active flood plain) and may provide an age via exposure or shallow burial dating. Fault scarps may also be used to infer cumulative as well as individual offset contributions via detailed analysis of scarp profiles and fault scarp diffusion modeling that relates geomorphic form to age (e.g., Zielke & Strecker, 2009; Arrowsmith et al., 1998; Hanks & Andrews, 1989). Two limitations are still worth mentioning here: The generation of geomorphic features such as river terraces could originate from processes fully independent of tectonic processes and therefore any tectonic interpretation of vertical separation for landforms should be examined with extra care. In addition, especially in the case of thrust motion, each earthquake rupture tends to go, at least partially, through and in the process destroying older deformation evidence, making the interpretation of paleoearthquake deformation more difficult. Thus it is most of time unlikely to resolve sequences with >3 earthquakes as is more frequently reported for strike-slip faults. As a result, the focus to study earthquake recurrence and along-fault slip accumulation via earthquake rupture lies on strike-slip faults.

The next step, after identifying a displaced geomorphic marker along the previously mapped fault trace, is the actual offset measurement. As briefly mentioned before, doing so requires projecting the marker's correlative sections on either side of the fault into piercing points on the fault (Fig. 3A). These piercing points define a distance by which we may back-slip and therefore reconstruct the marker's pre-earthquake(s) morphology. If a marker consists of more than one piercing line (e.g., an offset channel consists of two channel edges and a channel thalweg that may be distinct enough to provide a datum for offset measurement) then we would aim to align all of them within the same reconstruction to define the preferred offset estimate as well as an offset range (Fig. 3B). This offset range—in essence a probability density function (PDF) for offset values—is considered to incorporate the (full) range of plausible offset values that are able to sufficiently well reconstruct the assumed morphology (i.e., in analogy to a normal distribution being equivalent to ± 2 standard deviations, $\pm 2\sigma$). Currently, different authors use different shapes for these PDFs such as uniform, triangular, or normal distributions depending on the displaced feature in question and the underlying assumption to take respective measurement (Fig. 3E; Salisbury et al., 2015; Scharer et al., 2014). Other approaches use a goodness-of-fit metric that is derived via cross-correlation of up-fault and down-fault cross-sectional profiles of the displaced geomorphic marker (e.g., Zielke & Arrowsmith, 2012).

Complex marker morphology, a wide fault zone, and distinct geomorphic alteration can make it difficult to assess a marker's pre-deformation morphology with confidence. Hence, in addition to the assigned offset range, a rating scheme of a marker's "quality" is often used to distinguish between markers for which the pre-earthquake morphology can be assessed with high confidence (high quality rating) versus those for which the morphology is difficult to assess (low quality rating; Table 5.1, after Brothers et al., 2020; Salisbury et al., 2014).

If not only one but multiple measurements were taken along a fault trace, then we can plot these measurements as a function of distance along fault (Fig. 3C), potentially enabling us to approximate the along-fault surface slip distribution of the most recent and prior earthquakes. A convenient way to identify slip contributions from individual earthquakes is the superposition of offset PDFs along an investigated section of a fault, producing cumulative offset probability densities (COPD; Fig. 3D; e.g., Klinger et al., 2011; McGill & Sieh, 1991; Zielke et al., 2010, 2015). Such a COPD plot helps to identify what the commonly observed offset measurements were and therefore help to identify potentially existing patterns in the accumulation of slip increments (McGill & Sieh, 1991). This approach follows the assumption that individual peaks are generally formed by individual large earthquakes. We will elaborate on this assumption in a later section.

The concept of using geomorphic observations to assess earthquake recurrence and along-fault slip accumulation was introduced in the late 1960s (Wallace, 1968) and led in combination with stratigraphic observations to the formulation of a number of end member scenarios to describe the accumulation of slip along a fault i.e., the size distribution and timing (Schwartz & Coppersmith, 1984). These models are widely applied and form the templates against which recent observations are compared, allowing researchers to state that a fault's behavior is following either one of these models. Naturally, because these are end member models, no fault

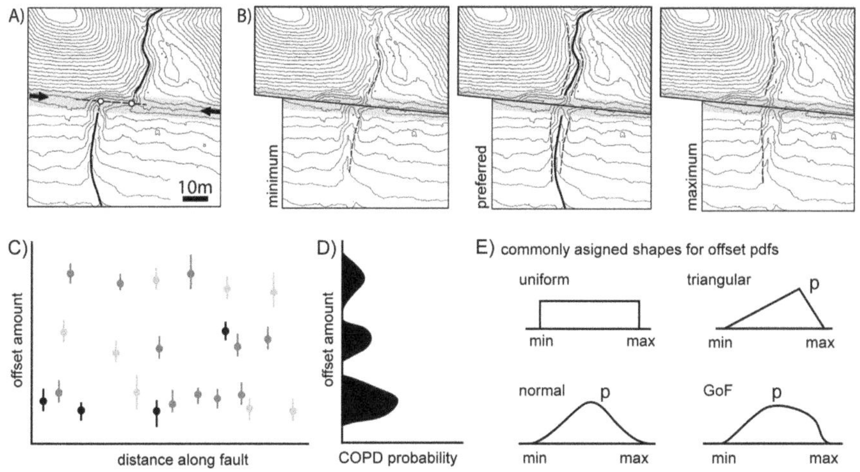

Fig. 5.3 **A**, **B** Example of measuring a geomorphic marker's horizontal displacement by projecting the assumed pre-earthquake morphology into the fault zone. The underlying elevation model is from Elkhorn Hills, Carrizo Plain, California along the San Andreas Fault (Bevis et al., 2005). Contour lines have a 1-m spacing. After identifying the fault trace (between arrows) and displaced marker (right-laterally offset channel) the markers morphology can be reconstructed via back-slipping, enabling to identify the minimum, preferred, and maximum offset amounts that are able to produce the assumed pre-earthquake(s) marker morphology. **C** Individual offset measurements are commonly plotted with respect to their relative distance along the fault, enabling to identify patterns in slip accumulation. The different shading indicates the assumed reliability (i.e., quality rating, Table 5.1) of the observed marker while the vertical bars indicate the assigned offset measurement uncertainty. **D** Summation of the individual PDFs provides a cumulative offset probability density (COPD). Individual, well separated peaks are thought to represent the subsequent offset contributions of (large) surface rupturing earthquakes. **E** Different kinds of shapes are commonly assigned for the identified offset ranges, where p indicates the preferred offset value. In general, they are meant to represent a $\pm 2\sigma$ range, meaning that they are meant to cover the majority of plausible offset values. GoF refers to Goodness of Fit where cross-correlation metrics are used when comparing two cross-sectional profiles

will follow them directly and over the 40+ years since these models were formulated, numerous studies have provided additional information and insight. We want to combine these insights later on to formulate an alternative model that is, in essence, able to include the main concepts of all three end member cases.

Thus, in order to constrain the behavior of past earthquake we first identify active fault traces and map them carefully. Then we locate displaced geomorphic markers along the fault trace and measure their separation across the fault zone. Lastly, we can plot these offset observations as function of distance along fault to identify potentially existing patterns in slip accumulation. Performing these tasks requires detailed representations of the earth's surface in form of high-resolution imagery and, importantly, topographic data.

Table 5.1 List of quality ratings, indicating the assumed reliability of a displaced geomorphic marker for measuring single-event or accumulated slip

Quality rating	Quality criteria
High	Markers have sharply defined morphology with correlative features on either side of the fault that do not have an observable alternative due to the morphological uniqueness
Moderate	Markers are less unique in their reconstruction, but their offset matches that of surrounding features as part of a clustered set, which increases the uniqueness of the correlation based on the reconstructed offset for the entire cluster
Low	Markers are either standalone and lacking in sharpness, or part of a cluster of features that are poorly expressed individually, but collectively provide a consistent offset reconstruction
<Low	Markers for which those criteria are not met should not be considered i.e., used in offset measurements and slip reconstruction

5.3 Data Sets and Analysis Techniques

High-resolution representations of the Earth's surface are very important for the identification of displaced landforms and measurement of corresponding slip i.e., reconstruction of its pre-earthquake(s) shape. These data sets, often remotely acquired, typically from space- or airborne sensors, provide a detailed view of a study area while maintaining its reference to the regional geologic and geomorphic conditions. What "high-resolution" actually means depends on the scale of the research question at hand but generally refers to decimetric-scale ground resolutions (e.g., revealing single-event slip contributions which are on the scale of 0.1–10 m). A range of data sets is currently available to perform these tasks. Either they rely on passive remote sensing, meaning optical imagery, or on active sensors such as InSAR (Burgmann et al., 2000; Massonnet & Feigl, 1998; Wang et al., 2015). In both cases, beyond specifications of each method, these data can be used to build digital elevation models and to measure crustal deformation due to earthquakes. Thus, both can greatly inform our understanding of current and past earthquake ruptures and the associated on- and off-fault deformation features.

Aside from those remotely sensed representations of the Earth's surface, field-based approaches, including field mapping and offset measurements provide additional important detail as well as insight and control on the interpretation of elevation and optical imagery data. Field investigations are important in "ground-truthing" crucial findings that were based on remotely sensed data. Following, we will briefly discuss some of these data sets, providing examples of their recent applications in earthquake geology. We do this by zooming into the field site, starting with spaceborne data sets and moving through airborne data sets to field-based measurements.

5.3.1 Spaceborne Data Sets

5.3.1.1 Optical Imagery

The acquisition of optical imagery by space-borne sensors for scientific purposes started in the 1970s with the *Landsat* program. This program was the first civilian space mission dedicated to imaging the surface of the Earth in a systematic manner. Early on, the imagery had ground resolutions ranging between 40 and 80 m. Since then, numerous additional sensors were sent into space and ground resolution of space-borne optical imagery has improved by several orders of magnitude to currently reach a ground resolution of less than 0.5 m for panchromatic images. As time goes on, resolution continues to improve, including for multi-spectral images. Similarly, the time interval between the acquisition of two images of the same location has also been reduced significantly. Currently, it has become possible—thanks to large fleets of satellites—to acquire several images *per day* of the same place. Furthermore, substantial improvements in computational power and visualization schemes have made it comparably easy to display and manipulate ever larger image files—which continue to increase in size with greater ground resolution and spatial coverage. Web-based applications such as Google Earth or Bing Maps provide access to a subset of these high-resolution images for free and for nearly any place on Earth to a large audience, including in Earth sciences.

These remote sensing optical data have been key to several paradigm shifts in Earth sciences. First, they allowed the appraisal of active continental faulting in the late 1970s (Tapponnier & Molnar, 1977), which helped secure huge advances in our understanding of tectonic deformation and earthquake hazard within the continents. With the emergence of the sub-metric ground resolution images, remote sensing optical data now allow detailed mapping of ground rupture for large continental ruptures, including the documentation of fine scale features that reveal the mechanics and dynamics of fault rupture propagation and behavior. One of the seminal studies about extensive mapping of coseismic ruptures using optical high-resolution imagery was the mapping of about 100 km of surface rupture at the meter scale for the Mw7.8, 2001, Kunlun earthquake in China (Klinger et al., 2005), which occurred in a remote and hostile environment, at an average elevation above 4000 m (Fig. 5.4). Such mapping allowed recognizing a specific rupture behavior, the co-seismic slip partitioning (King et al., 2005), which might have been missed without such extensive mapping effort that was only possible due to the high-resolution imagery. Similarly, taking advantage of arid climate and remoteness of the environment, such images proved to be key to document preserved ruptures associated to past earthquakes, enabling us to describe rupture geometry as well as slip distribution for earthquakes more than a century old. Other examples of such documentation of past earthquakes are the M8, 1905, Bulnay-Tsetserleg earthquake in Mongolia (Choi et al., 2018), or the M7.8, 1931, Fuyun earthquake in China (Klinger et al., 2011).

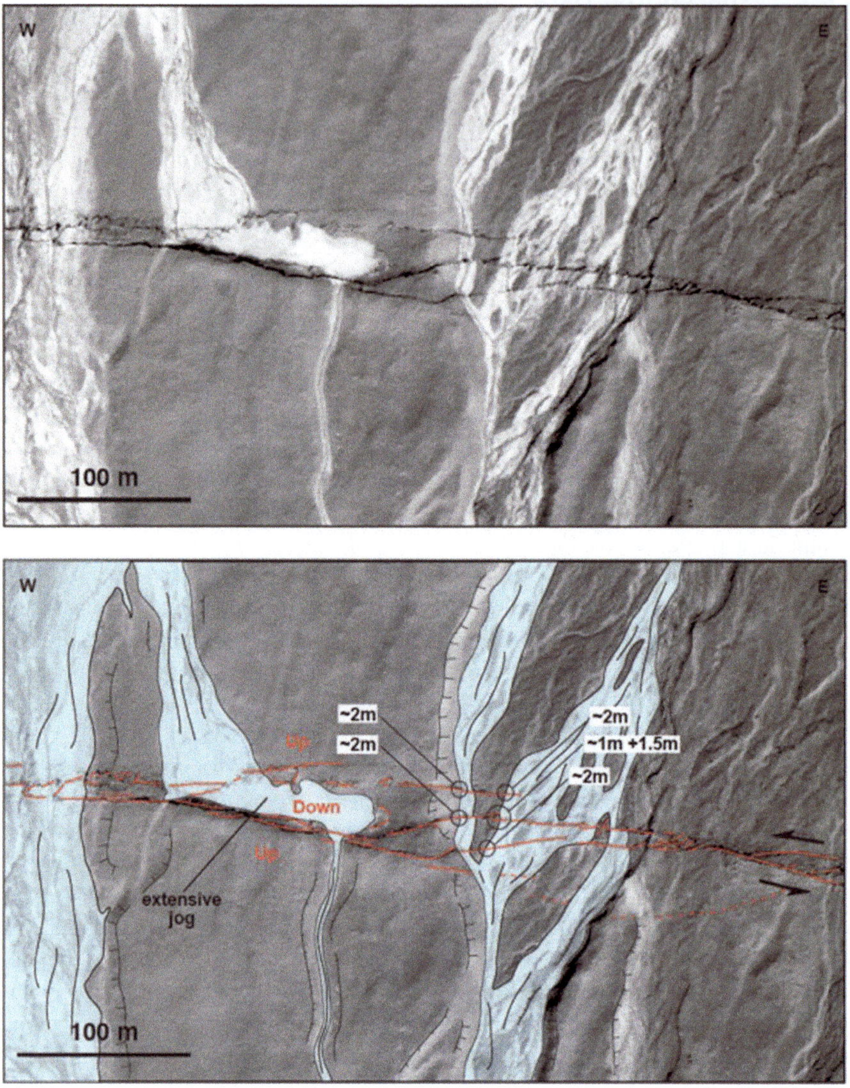

Fig. 5.4 After Klinger et al. (2005). The top panel shows some satellite image acquired in the months after the Mw7.8 Kunlun earthquake, China. Thanks to the arid environment, the surface rupture can be seen without any ambiguities, allowing for detail mapping (lower panel) of the full ground rupture, including secondary faulting down to a metric scale

5.3.1.2 Elevation Data

In addition to being able to image zones of active faulting, a key observation to understand earthquake deformation at the time scale of several earthquake cycles is the topography, storing the geomorphologic record. Since the early 2000s, the scientific community has seen a full change of paradigm with the progressive availability of high-resolution topography for the Earth. In a first place, this topography resulted from dedicated space missions, the most important one being the Shuttle Topography Radar Mission (SRTM) flown in 2000, which acquired topography for 80% of the planet with a ground resolution of 30 m at the equator. These data have been progressively made publicly available, first at a degraded resolution of 90 m, and now at the full 30 m resolution. Several other space missions, based on optical image acquisitions were operated in the mid-2000's that also provided topography coverage to complement SRTM and are still at the best of what is available at global scale (Fig. 5.5). The US-Nippon ASTER program, the HRS Spot5 program from Airbus, and later the TerraSAR—Tandem-X SAR missions, all provided topographic data of the surface of the Earth at decametric scale. The development of fleets of high-resolution optical satellites, often equipped with agile sensors, during the first decade of 2000 has also revived classical photogrammetry methods, which have been adapted to remote-sensing data. The combination of such methodologies with high-resolution imagery allows producing local topography with ground posting at metric scale.

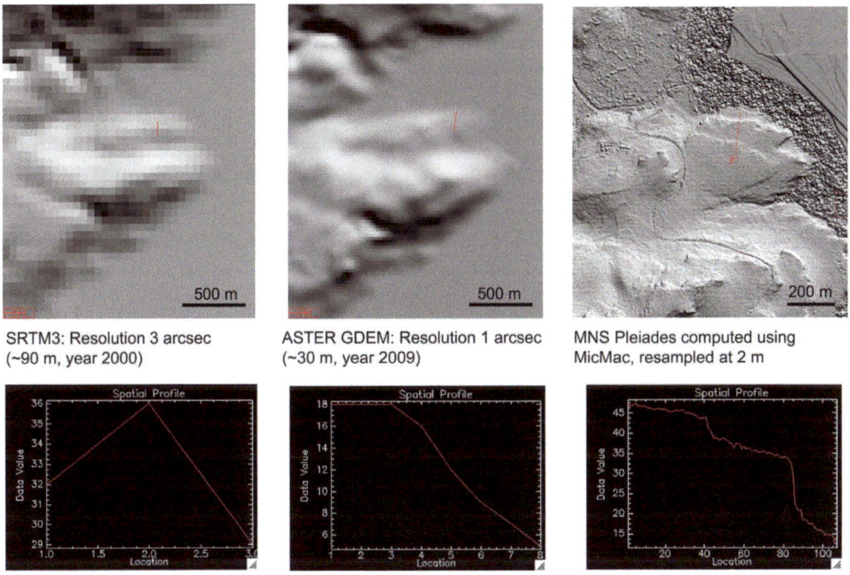

Fig. 5.5 Example of different topographies of the same place using different data and processing. In about 20 years resolution was improved by almost 2 order of magnitudes

5.3.1.3 Synthetic Aperture Radar (Sentinel)

In parallel with optical imaging sensors, satellites based on active radar technology have been developed in the last 30 years that are now routinely used to measure topography and crustal deformation associated with contemporary earthquakes. However, it is not of direct use to address issues related to past earthquake deformation.

5.3.2 Airborne Data Sets

5.3.2.1 Optical Imagery

Aerial photos have been acquired for more than 100 years—since the advent of aviation. Since the 1950s broad coverage of aerial photos exists for many continental areas around the world. These data sets have increasingly become available to the public after military declassification. They provide sub-meter to meter-scale resolution photography of the ground. Careful processing of aerial photos allows one to generate orthorectified images and digital elevation models, which in turn can be used to map fault traces and measure displaced landforms. In many cases, however, for old aerial photo collections, critical acquisition parameters have been lost (such as elevation of the sensor or detailed description of the optical parameters of the sensor), making orthorectification impossible. Although it might be possible in some cases to retrieve a-posteriori these parameters, it is often difficult; where it is *not* possible to obtain these original aerial photo parameters, the imagery can only be used for qualitative studies, as the optical distortions included in each photo cannot be correctly accounted for, leading to biases in measurements. Orthorectification is specifically critical when the area of interest includes rugged topography, which leads to the greatest internal distortions of the imagery.

Aerial photos are used directly to map and measure geologic features and surface expression of earthquake rupture. Their use is becoming less common with the increasing availability of regular, high-resolution imagery provided by space-borne instruments. However, 20th century aerial photos may still prove critical in studies of older earthquakes, especially when photos from before and after the earthquake exist. Recent developments in computational algorithms and power have permitted researchers to revisit past earthquakes using differencing approaches (image correlation) on these old datasets. In that sense, air photos still provide valuable (and unique) information—particularly for earthquakes that occurred prior to the advent of space-borne image generation (image correlation for earthquakes between e.g., 1950s and 2000s).

Rockwell and Klinger (2013) used air photos to map in detail the surface slip along the Imperial Valley earthquake of 1940 using crop lines that were coseismically displaced. The close spacing of those crop lines and the resolution of the air photos allowed very high-resolution offset measurements, with spacing between individual measurements of less than a meter, to document slip variability over short spatial scale. The case of the Goby Altay, M8, in 1957 in Mongolia is

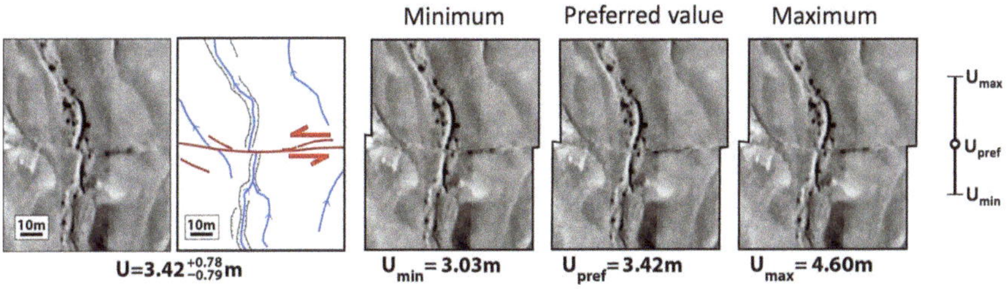

Fig. 5.6 Offset landform associated to the 1957 Gobi Altay earthquake surface rupture derived from air-photos shot right after the earthquake (see Kurtz et al., 2018 for details)

also interesting in that respect as the rupture was flown and shot by a team of the former USSR in 1958.

Luckily, these air photos were kept in good shape, including acquisition parameters, allowing for later orthorectification. Thus, they have been recently used for determination of an accurate slip distribution associated with the 1957 Goby Altai earthquake, 60 years after the earthquake occurred (Fig. 5.6; Kurtz et al., 2018). Similar datasets are likely still buried in archives that need to be discovered and exploited. Milliner et al. (2015, 2016) performed image correlation based on optical imagery acquired from airborne cameras to reevaluate the on-fault distribution of coseismic slip and off-fault deformation caused by the 1992 Landers and the 1999 Hector Mine earthquakes. They found that a significant but also very variable amount of coseismic deformation occurs not on the main fault trace but rather off-fault in form of distributed deformation. Their study further documented the self-affine character of single-event earthquake surface slip along a fault trace.

5.3.2.2 Elevation Data

As mentioned in the previous section, elevation data can be generated from optical imagery acquired with an airborne camera. This approach has been used for more than half a century and provided many important high-resolution topographic data sets available for analysis in Earth sciences and other scientific disciplines. Over the last 20 years, another acquisition method for elevation data with an airborne instrument has been available, namely Lidar—which stands (in reference to radar) for light detection and ranging. During data acquisition, a laser scanner emits pulses of monochromatic light. These laser pulses are dispersed and reflected from the ground surface. Reflected pulses that return to the scanner are detected, stopping the clock that measured the pulse's travel time while also recording the intensity. Given that the location of the Lidar scanner at the time of data acquisition is well known (due to on-board inertial navigation system and global positioning system) and given that the orientation of the outgoing laser pulse is known as well, the recorded pulse's travel time can be converted to distance and hence converted to absolute geographic coordinates of the reflecting object or surface. It is important to note that the laser pulse is not a point but has a certain width or foot print. As a result, an outgoing laser pulse may be reflected partially by objects at different distances when the footprint is larger than the object. One outgoing pulse may therefore generate multiple returning pulses. Presumably, the last returning pulse—representing the reflector with the largest distance to the instrument—represents the ground surface. Thus, Lidar-derived DEMs may be processed in a way to virtually remove vegetation cover to create a "bare-earth" digital terrain model, DTM. This feature of Lidar makes it a very powerful tool for Earth sciences and investigations of active tectonics, which aim to study the solid Earth surface.

Extensive data collection was initiated in the early 2000 first in western US, and later worldwide, aiming at building an archive dataset predating earthquakes to come to be in a position to measure earthquake deformation by differencing pre- and post-event Lidar acquisition. These data were also used to investigate recent surface rupture, such as the 1999 Hector Mine, Mw7.1, event that occurred in Southern California (Hudnut et al., 2002) and to study cumulative deformation associated to series of paste earthquakes (Arrowsmith & Zielke, 2009; Zielke et al., 2012). Eventually, occurrence of recent earthquakes in locations where pre-earthquake Lidar had been acquired, like in area of the El-Mayor event in Mexico in 2010, or in more vegetated area like the Kaikoura rupture in New Zealand in 2018, has shown the significant potential for such approach by allowing measurement of ground deformation at decimeter scale (Nissen et al., 2014; Oskin et al., 2012)

5.3.2.3 Structure-from-Motion

Although airborne and space imagery were first used for direct mapping of geological features, technologies that were initially developed by mapping agencies for the purpose of generating topography at a country scale have been revivified in the Earth science community during the last decade. As multi-stereoscopic acquisitions become standard, including for satellite imagery, several groups have developed codes based on photogrammetry and image correlation (adopting scale invariant feature transforms, SIFT; Lowe, 2004). These codes, sometimes referred to as Structure from Motion (SfM) technique, compute Digital Elevation Models (DEM) and orthorectify photos using large sets of overlapping images. Computational improvements enable object geometry to be reconstructed without a priori knowledge of camera position, relying instead on the sheer number of correlative pixels and their relative positions. Several software suites, including open-source and commercial software, provide tools to compute DEM with a resolution close or equal to the resolution of the input imagery. In turn, these DEM are used to compute orthophotos that can be used to quantify deformation. In the recent years, with the quick increase of high-resolution space imagery and also the emergence of low-cost easy-to-navigate drones, very high-resolution DEM for small areas are becoming the norm for any active tectonic or geomorphological studies, with DEM often at the centimetric scale (Gao et al., 2017).

In addition to offering exceptional support for mapping and a way to compute high-resolution topography, multiple acquisitions at different time periods also offer the opportunity to detect and quantify changes between successive acquisitions (e.g. Rosu et al., 2014). In that case, instead of having different image acquisitions in different geometries, to take advantage of the stereoscopic vision, images are brought into the most similar geometric configuration through orthorectification. Hence, in that case any change between images would be assigned to some deformation of the imaged landscape, for example due to an earthquake. Most of the studies using this kind of techniques are still only quantifying changes in 2D by measuring horizontal displacements. However, with more systematic multi-stereo acquisition this technique will soon be able to do routine measurements of deformation in 3D (Zhou et al., 2018).

One decisive input of these works using optical imagery to study earthquake deformation is that in addition to being able to quantify deformation directly in the fault zone, with no decorrelation issues such as with InSAR measurements, these techniques allow the measurement of distributed deformation in a zone that can span several hundreds of meters to few kilometers on each side of the fault zone. Thus, this kind of measurement can provide a better assessment of the full deformation budget at the surface than, for example, field investigations, especially in zones of fault complexity. Several studies have emphasized the importance of off-fault deformation (e.g., Antoine et al., 2021; Milliner et al., 2015; Vallage et al., 2015), clearly departing from the elastic rheology classically used to model crustal behavior, to the point where damage can, in part, control the way a rupture proceeds along its rupture path (Klinger et al., 2018). Similarly, such studies have highlighted the role of geometrical asperities in the modulation of the slip along strike for large strike-slip ruptures. What had been thought of as missing slip in fault jogs, because the amount of slip parallel to the strike would locally decrease, appears to be actually compensated by local increase on the component perpendicular to the fault strike, reflecting the volumetric deformation associated with the jog (Klinger et al., 2006). More generally, this method that allows measuring deformation directly in the fault zone, when combined into joint inversion with other geodetic measures, such as InSAR or GPS, has changed our understanding of the ground deformation. While it had been thought that slip would taper towards the free surface, generating some slip deficit at the surface compared to modeled slip at depth (Fialko et al., 2005), the addition of near field data into source models has shown that the slip deficit was mostly an artifact due to the lack of data close to the surface. Although the existence of slip deficit could not completely be ruled out at the free surface, likely it does not exceed 10–15% of the total slip, and the amount of distributed slip is likely often underestimated (Scott et al., 2019; Vallage et al., 2015).

5.3.3 Data Sets Gathered on the Ground

Getting access to high-resolution topography has long been an endeavor for scientists interested in active tectonics and geomorphological processes. Before generalization of high-resolution satellite imagery and use of drones, other techniques had been developed that are shortly reminded here for sake of completeness, although they tend to be quickly replaced by image-based techniques. Theodolite-distance meters, also named total stations, have been used widely in the 1990s. This tool, which casts an infrared beam to a specific target that is held by an operator, was used to measure topographic profiles across large geomorphological features such as alluvial fans or river terraces. The most efficient total stations are able to shoot a target located a few kilometers. Collecting data in this manner, especially if the goal is to build a DEM, was a slow process as it needs an operator to walk with the target across the landscape and to stop at any point of interest, while a second operator is triggering the shot and to register the measurement. Typical resolution is in millimeters.

This system has been progressively replaced in the 2000s by kinematic GPS, which consist of a mobile GPS unit mounted on a pole, a backpack, or a car, combined with a fixed station used as reference. This type of setting eases the collection of large batches of points to build local DEM, although it still requires to have an operator to move around the area of interest. Because the standard acquisition strategy with kinematic GPS is to follow a regular grid rather than specific geomorphological features, and then to interpolate a regular surface between points, it often results in smoother DEM leading to larger uncertainties in determination of geometry for landform features. Typical resolution is in one to few centimeters.

In parallel to airborne Lidar, some terrestrial Lidar system (TLS) have also been developed and used in the field of active tectonics and geomorphology (Haddad et al., 2012; Telling et al., 2017). The main advantage of the TLS compared to total station or kinematic GPS acquisition is the time and easiness of data acquisition. Like other Lidar-based systems, TLS does not require a specific target reflector (as is the case for the aforementioned total stations). Instead it uses the direct reflection of the laser beam from the ground surface. The most efficient TLS are able to get returns from natural reflectors located several kilometers away. Acquisition is fast and the density of points can be high, allowing post-processing of the point clouds to remove partial vegetation cover. Typical DEM acquired with this kind of instrument are centimetric in resolution for areas that could be rather extensive. This technique has also been used to document trenches as the TLS is often combined with a camera, allowing to build textured DEM of the trench walls (Haddad et al., 2012). The cost of such an instrument, however, is high and the TLS needs to be handled with special care. Further, for geomorphic investigations the shot angle of terrestrial Lidar systems is inferior when compared to airborne systems. Hence, use of TLS in the active tectonics community remains marginal.

Building point clouds of trench walls of even for small DEM of the surrounding area is also performed with photo acquisition from ground. In the case of trenching, the most classic strategy is to acquire a dense photo mosaic of the trench walls with the guaranty that each piece of the trench wall is covered several times with different view angles. In the case of small DEM, e.g., for trench site characterization, a camera is held at the top of 2–4 m-long pole with an automated image acquisition scheme. In some cases, to gain view, the camera can be fixed on a kite or a balloon, controlled from the ground. In all these configurations, because absolute location by GPS is not directly involved, a special attention should be paid to locate ground control points, which are later used to fix potential distortions in the combination of the images and to set absolute distances. Images acquired through these various systems are generally processed through SfM pipelines to generate textured point clouds.

5.3.4 Paleoseismic Excavation (Trenching in 2 or 3-D)

Previous sections focused mainly on different ways the Earth sciences community has been working on to measure earthquake-related deformation at ground surface, including those due to past earthquakes. Deciphering past-earthquake history, however, in addition to being able to measure past deformation, also involves to be able to propose timing for past earthquakes, and, in the best case, to associate timing with deformation per event.

To this purpose, paleoseismologists have developed various techniques, according to the style of deformation of the fault they are interested in. The most popular technique, which has been used for all kinds of faults, is 2D trenching. In that case, one or several trenches are excavated across a fault to retrieve evidence about past ground ruptures that were later buried by subsequent sediment deposits. Although the logic is rather simple, which has been applied many times in various contexts, not all paleoseismological trenches are equally successful and the number of sites where a large number of events could be documented remains limited (e.g., Daeron et al., 2007; Sieh et al., 1989; Weldon et al., 2005). In addition, it should be remembered that 2D trenching, especially in strike-slip environment, is mostly documenting secondary deformation such as vertical displacement along a fault whose motion is primarily horizontal. Thus, in most cases, although trenching provides enough evidence to constrain timing of past events, it is not usable to derive magnitude of earthquakes using classical scaling laws relating displacement to magnitude.

In specific cases, however, paleoseismological investigations can provide more information than only timing. In the case of strike-slip environment for example, where temporary landforms such as ephemeral channels or sediments spill-over get buried and offset through time, careful 3D trenching, with trenches perpendicular and parallel to the fault, might allow restoring landform features to their original shape and, thus, to derive slip-per-event for several successive earthquakes at one site (e.g., Wechsler et al., 2018). In case of normal faulting, because deformation is cumulative without being self-destructive, it might be easier to derive a slip-per-event history in trenches, although it would rely on strong assumption about geomorphological evolution of normal-fault scarps in alluvium (Klinger et al., 2003). In limestone environment a different strategy has been developed, based on dating successive exposures of a same fault scarp to identify successive earthquakes. In that case vertical displacement per event can be assessed based on vertical distribution of dates along the scarp (Benedetti et al., 2013).

5.4 Single-Event Rupture Characteristics

As mentioned in the previous section, a sound understanding of the rupture characteristics of recent earthquakes is important as it provides a framework in which to interpret evidence of faulting from past earthquakes. Adopting the aforementioned data sets and methods has provided insights about fault zone structure, fault segmentation, single-event fault slip distribution, and off-fault deformation.

5.4.1 Fault Structure and Segmentation

A key parameter for seismic hazard assessment is the lateral extent of past earthquake ruptures as it (the rupture length) provides an estimate for pre-instrumental earthquake magnitude (e.g., Field et al., 2015; Wesnousky, 2008). This information is crucial when aiming to assess a fault's or fault system's likely future behavior. One approach to determine the rupture length of past earthquakes is to correlate observations at successive paleoseismological sites to check if specific ruptures can be recognized in several places and therefore assign a minimum length for these earthquakes (DuRoss et al., 2016; Lefevre et al., 2018; Philibosian et al., 2017). This effort, however, is partly hampered by the lack of resolution of current dating techniques such as ^{14}C dating. Indeed, when a fault slip-rate is fast enough and leads to recurrence intervals of large earthquakes on the order of a few centuries, a typical uncertainty of ± 40 years in dating, combined with the reality that samples are not always available in units that closely bracket the event horizon, can quickly lead to strong ambiguities in correlating ruptures laterally from site to site. This effect may even be reinforced in cases where some doubt exists about having a unique very large rupture or a cluster of smaller events in a short interval of time (Lefevre et al., 2018). Hence, because paleoseismological data alone usually cannot unambiguously inform us about the lateral extent of past ruptures, they should be combined with observations of recent surface ruptures. Further, not only is it important to quantify the rupture length of past earthquakes but also identify the physical processes and parameters that caused a rupture to stop. The question becomes: Once an earthquake is initiated, what parameters control rupture propagation and rupture termination? Addressing this question is crucial as it directly affects the maximum magnitude earthquake size that a fault of fault system may generate.

Detailed mapping campaigns along the rupture traces of recent surface rupturing earthquakes, like for example the 2016, Mw7.8, Kaikoura event in New Zealand, indicate a high degree of fault geometric complexity. Faults are by no means planar. In fact, power-spectral density analysis revealed that the geometries of faults and surface ruptures exhibit a scale-invariant, fractal characteristics over a large range of spatial length scales (Candela et al., 2012). Recent progress in imaging (a) the earthquake rupture at the surface (via the various imaging techniques described earlier) and (b) the corresponding seismic source at depth, indicates that these geometric complexities at the ground surface might mirror similar complexities at depth (Wei et al., 2011). Thus, revisiting past surface-rupture maps, mostly for continental strike-slip earthquakes, has led to identify a couple of intrinsic fault-structure characteristics that might be key in controlling how a seismic rupture propagates through several fault segments and therefore control the final size of an earthquake.

Before continuing, it is important to address the term "fault segment". While there is currently no standard definition of what a fault segment is, the term typically refers to either (a) a section of a fault that is characterized by observed seismic behavior that distinctly differs from neighboring sections, or (b) a part of the fault that is geometrically consistent and different from neighboring sections or separated from them by distinct changes in fault geometry (e.g., fault bend, jogs, etc.; Fig. 5.7). The typical length of fault segments may differ distinctly, depending on which definition is considered and reported fault segment lengths range from more than 100 km (e.g., Schwartz, 2018) to only few kilometers (e.g., Manighetti et al., 2015a, 2015b). An attempt to clarify the concept of fault segmentation is to relate fault segmentation to actual, physical (i.e., geometrical) scaling. Compiling strike-slip surface rupture traces for large continental earthquakes during the last century, Klinger (2010) used geometrical features of the surface rupture, such as step-over and bends, to show that all strike-slip ruptures could be described as a succession of fault sections (i.e., segments), which have an average length around 15 ± 5 km. Thus, it has been proposed that there is a direct correlation between the average length of a singular fault segment and the thickness of the brittle continental crust. In fact, this relationship had already been inferred for southern California, based on the correlation between fault geometry and maximum depth of background seismicity (Bilham & Williams, 1985). Testing further this assumption, Cambonie et al. (2019) and Lefevre et al. (2018) have developed theoretical frameworks that explain this phenomenon and show that similar scaling relationships exist respectively for fractures in polymers-rupture experiments and for faults in sandbox experiments. Thus, this relationship might be universal for any fracture propagation into a brittle medium, including the crust of the Earth. The large ruptures may be seen as a succession of several failing segments that are separated by step-overs or other geometrical barriers. Eventually, such physics-based description of a fault segment might be useful by providing a building block to simplify discretization of complex fault systems into numerical models. While we discussed the effect of fault structure based on strike-slip faults here, it is plausible to assume corresponding relationships for reverse and normal faulting environments, where fault segment size might deviate from the aforementioned 15 ± 5 km average value, depending on a fault's overall dip angle and the thickness of the brittle crust.

The above-mentioned question however remains: What physical parameters or processes controls rupture propagation and termination and therefore the final size of an earthquake? What processes control the number of fault segments that may rupture in a given earthquake? One of these parameters is the scale of the geometric discontinuities between fault segments. In a compilation work, Wesnousky

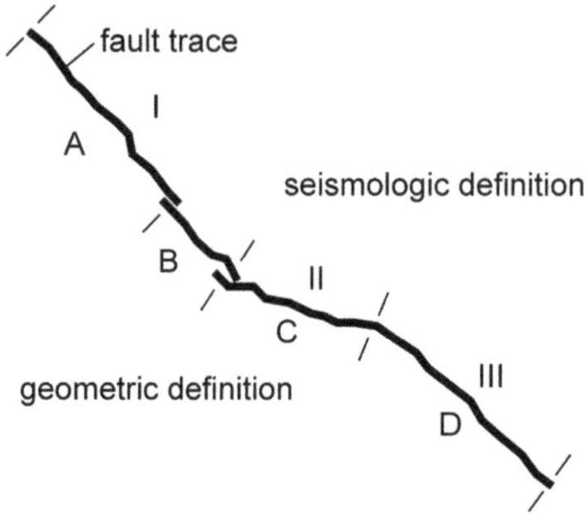

Fig. 5.7 Schematic representation of two definitions for fault segments. One is based on distinct and sufficiently large geometric complexities along a fault trace such as fault bends, or steps (segments with letters A to D). Another is based on a segment's distinct seismologic character, separating it from neighboring segments (segments with Roman numbers I to III). For example, segment I may commonly rupture entirely in large earthquakes. Segment II might be characterized by creep or by abundance of microseismicity. Segment III may rupture frequently in moderate-size earthquakes. Geometrically defined segments tend to be smaller (Klinger, 2010)

(2006) has shown that in a majority of cases a rupture would not propagate across a step-over wider than 3–4 km, and even stall in 40% of cases at narrower jogs. Although this empirical observation is not yet been formally justified by any theoretical work, to date it remains a solid empirical rule when it comes to envision rupture propagation through multiple fault sections and the estimation of probable maximum earthquake size.

5.4.2 Along-Fault Slip Distribution and Variability

Another indication that faults might be spatially structured (as indicated by fault geometric complexities) comes from observations of along-strike variability of co-seismic slip during large continental ruptures. In recent years, thanks to progress in space-based imagery, the capacity to measure dense slip distributions for large earthquakes has improved significantly. Thus, the lateral variability of co-seismic slip, which before was most often overlooked as related to noise and insufficient spatial sampling, can now being fully considered as a robust observation of a real physical phenomenon (Rockwell et al., 2002). Recent observations based on image correlation techniques have revealed that along-fault variability of surface slip is not purely random but exhibits fractal, self-affine characteristics (Milliner et al., 2016), where the wavelength of observed offset variability at the surface is proportional (as first order approximation) to the depth and magnitude of a corresponding slip variability at depth. In other words, high-wavenumber variations in surface slip are caused by shallow, spatially narrow variations whereas low-wavenumber variations are caused by deep and/or wide-spread variations (Zielke et al., 2015). This understanding goes in line with the comments made in the previous section—that rupture complexities observed at the surface may be mirrored by similar complexities at depth.

A persistent pattern in along-fault slip variability is associated with jogs. In most cases of strike-slip ruptures, the largest dataset available, one can show that the component of slip along strike decreases when the rupture reaches a jog, to pick up again as the rupture would pass the jog. Space imagery allowed to demonstrate that this decrease in displacement, in the component parallel to the strike of the fault, is usually due to distribution of the deformation over several fault branches inside the jog (Choi et al., 2018; Klinger et al., 2006). These faults do not have to be necessarily parallel to the main strike of the fault (King, 1986). Hence, careful examination of slip distributions allows deciphering fault segmentation, based on the small variation along strike of the slip distribution. This statement exemplifies the importance of constraining characteristics of the most recent rupture, preferably using differencing approaches, as it serves to identify segmentation and therefore distinctly influences further data interpretation.

As indicated in the previous paragraphs, it is relatively simple and straightforward the determine the along fault distribution of surface slip for recent earthquakes, given the availability of corresponding data sets and given that geomorphic overprinting has not (yet) been able to distinctly modify the shape of fault zone and displaced geomorphic markers so that their original pre-earthquake morphologies are well constrained. Hence, one might assume that corresponding offset measurement uncertainties are low, adding little to the aleatory variability of surface slip along the fault. However, a number of recent studies along recent surface ruptures indicated that measurement uncertainties of ± 10–15% (representing 2σ uncertainties) are not uncommon (e.g., Gold et al., 2013). That is, even for pristine faults and displaced geomorphic markers that are resolved at very high resolution, it is often difficult to precisely identify pre-earthquake morphologies. We want to raise this point here as it becomes relevant when performing offset measurements on displaced markers that have been displaced not recently, but in the past or by multiple earthquakes. In these cases, it becomes more and more difficult to assess the marker's original, pre-earthquake(s) morphology. Hence, we suggest to be cautious of studies that report offset measurement uncertainties that are (significantly) below this assumed minimum threshold.

5.4.3 Off-fault Deformation

Beyond the aleatory variability of slip along strike that can (in part) be directly tied to rupture geometry and step-over zones, detailed slip distributions have also shown variation of slip along strike unrelated to fault geometry. Several hypotheses, not necessarily exclusive of each other, have been proposed to explain such variation of slip along strike at the scale of several fault segments: Because faults do not always rupture in an identical fashion, with some fault segments rupturing either alone during moderate size earthquakes or in a cascade, during large magnitude events, the amount of accumulated elastic strain available along a specific segment may differ for consecutive segments, leading to slip variability over large distances (Rockwell & Klinger, 2013). An alternative hypothesis, often called the slip-deficit model, has been proposed by Fialko et al. (2005), who speculate that some part of the slip accommodated at depth during an earthquake does not reach the surface due to a deficit of accumulated strain in the upper crust. Instead, they proposed that strain in the upper crust would be accommodated through distributed failure during the interseismic period. Recent analyses have suggested, however, that a significant part of this alleged slip deficit might be due to biases related to limitations when using only InSAR data to derive slip distribution at surface (Marchandon et al., 2018; Scott et al., 2019; Vallage et al., 2015) and, even more fundamentally, to the modeling approach which in most cases is only considering elastic rheology for the entire brittle crust, which might not be suitable for the shallowest crust (Kaneko & Fialko, 2011). Eventually, with the emergence of higher resolution data sets during the last two decades, from the mid-2000s, the possibility of having a significant amount of deformation accommodated off of the primary fault is reconsidered. Several efforts in ground mapping ruptures after large earthquakes had already suggested that some part of the deformation, in specific geological configurations, is accommodated off-fault (Rockwell et al., 2002; Sieh et al., 1993). These data, however, were either only qualitative or very limited spatially. More comprehensive deformation surveys at high resolution, both for actual and for past earthquakes, have revealed that in many cases it is possible to find evidence of off-fault deformation (Antoine et al., 2021; Klinger et al., 2018; Scott et al., 2018; Vallage et al., 2015), which can reach up to 25% of the total slip accommodated during one event (Choi et al., 2018). Until now, this possibility has been mostly ignored by the paleoseismology community although it bears significant consequences when it comes to evaluate deformation budget and slip-rates in the framework of fault hazard assessment.

5.5 Earthquake Recurrence and Along-Fault Slip Accumulation

As is indicated in the introductory section to this chapter, the most important factor that motivates investigations of earthquake rupture and recurrence characteristics of large, potentially dangerous earthquakes is to gain some predictive capabilities in order to estimate what future earthquakes along a given fault or within a given region might be expected. Society and stakeholders are looking for information from the earthquake geology and geophysical communities that enables them to prepare for future occurrence of these disasters. Before being able to make predictive statements, however, it is important to first conduct careful descriptions of past earthquake ruptures (Marzocchi, 2018). Because the driving motivator is to understand how earthquake recurrence is characterized, one approach is to investigate the sequence of earthquakes along individual faults or fault segments. Here, the goal is to characterize the size and timing of earthquakes that occur along individual faults. Assuming that (all) active faults are known, this approach may give a good approximation/picture of the hazard that is posed by the fault system as a whole.

5.5.1 Along Fault Slip Accumulation Patterns

In order to resolve the accumulation of slip along a given fault through the occurrence of surface rupturing earthquakes, we need to be able to distinguish between the increments of slip produced by individual earthquakes. If it is the case that geomorphic markers form between successive surface rupturing earthquakes, then we can measure the corresponding displacement increments. Based on stratigraphic and geomorphic evidence, a number of end-member scenarios for earthquake recurrence and along-fault slip accumulation have been formulated, namely the uniform-slip, the variable-slip, and the characteristic slip model. These models have served as a framework to interpret paleoseismic and tectono-geomorphic data since the 1980s. Since their formulation, many more studies have been carried out to investigate this topic, indicating that the current models are not sufficient to explain some new observations. Following, we want to mention the existing apparent discrepancies and present a comprehensive model for earthquake recurrence and slip accumulation along individual faults or fault segments that can resolve these discrepancies.

While the data sets and analysis techniques have improved, the underlying concepts and assumptions have not changed: Researchers aim to identify and measure displaced geomorphic markers and stratigraphic units in order to constrain a fault's earthquake recurrence and slip accumulation pattern. Following this approach, numerous studies have identified that the accumulation of surface slip is dominated by the repetition of same-size offset increments as is prominently expressed via pronounced peaks in COPD plots (e.g., Haddon et al., 2016; Klinger et al., 2011; Kurtz et al., 2018; Manighetti et al., 2015a, 2015b; Ren et al., 2016; Zielke et al., 2010). For continental strike-slip faults, the reported offset increments are typically in the order of 3–6 m. While currently not sufficiently investigated, it is plausible to assume a correlation between the size of these offset increments and for example a fault's geometric characteristics such as the aforementioned structural or frictional segmentation or along-fault variations in other rupture-controlling parameters. That said, it is important to note that the existence of same-size offset increments, does not explicitly convey how many earthquakes have contributed to each of these increments or COPD peaks. The corresponding offset observations generally lack an age constraint and thus cannot be strictly attributed to a single discrete event. The implied repetition of similar-sized displacement increments is based on the assumption that marker formation is more frequent than marker displacement, so that clusters of offset observations represent the same number of surface-rupturing earthquakes. If, on the other hand, this assumption is not valid—i.e., earthquakes occur more frequently than the formation of geomorphic markers—then individual groups of offsets may represent more than one earthquake. In fact, for the Carrizo section of the San Andreas Fault in south-central California, it was found based on independent paleoseismic evidence that the number of earthquakes contributing to individual COPD peaks does not match the peaks' number: the 1st peak at ~5.5 m was formed by one earthquake, 2nd peak at ~10 m was formed by 2–3 earthquakes while the 3rd peak at ~15 m was formed by 5 earthquakes (Akçiz et al., 2010; Grant-Ludwig et al., 2010; Zielke et al., 2010). Furthermore, we note that recent studies have revealed a discrepancy between (a) the variability of inter-event time as derived from paleoseismic excavations, and (b) the variability of offset increments as derived from geomorphic investigations. The coefficient of variation (CV), which is the ratio of a dataset's standard deviation σ and mean μ $CV = \sigma/\mu$, provides a simple tool to quantitatively express this discrepancy. The earthquake inter-event time along a given fault segment exhibits a quasi-random behavior (with CV values typically between 0.5 and 1.0) whereas the observed accumulation pattern of slip along a fault segment is quasi-periodic (with CV values typically between 0.2 and 0.4; Zielke, 2018). How is it possible to explain these apparent discrepancies?

Numerical simulations and statistical considerations by Zielke (2018) and Zielke et al. (2015) have suggested that this discrepancy is an effect of bimodality in a fault's earthquake magnitude frequency distribution in combination with the different resolution potential of stratigraphic records (often used to constrain earthquake timing) and geomorphic records (often used to constrain offset increments). Other tested explanations were not able to consistently reproduce the observed behavior and to resolve this apparent discrepancy (Zielke, 2018). Here, bimodality in magnitude frequency distribution refers to a distribution that follows a Gutenberg-Richter relationship for earthquakes with magnitudes below M_1 while larger earthquakes are clustered around magnitude M_2 (Fig. 5.8; Ben-Zion, 2003; Wesnousky, 1994; Youngs & Coppersmith, 1985; Zielke & Arrowsmith, 2008).

Magnitudes M_1 and M_2 are separated by a gap. We want to note that a comprehensive data compilation by Wesnousky (1994) provided evidence for this bimodal behavior for faults in southern California while Ben-Zion (2003) and Zielke and Arrowsmith (2008) discussed the physical basis for its existence. For the following argument, we assume that some earthquakes with magnitude between $M_t <= M <= M_1$ are also capable of generating surface rupture. Alternatively, we could consider ruptures along another (neighboring) fault segment that penetrated into the studied segment but were not capable of causing it to fail completely.

Aside from a bimodal magnitude frequency distribution, we also mentioned the different resolution potential of stratigraphic and geomorphic records in order to explain the aforementioned discrepancies. The stratigraphic record is capable of storing evidence of surface rupturing earthquakes with as little as a few cm of surface displacement. In fact, these small-scale features are the ones that are commonly analyzed in paleoseismic investigations. The landscape however is—in response to the acting tectonic, climatic, and gravitational forces—constantly

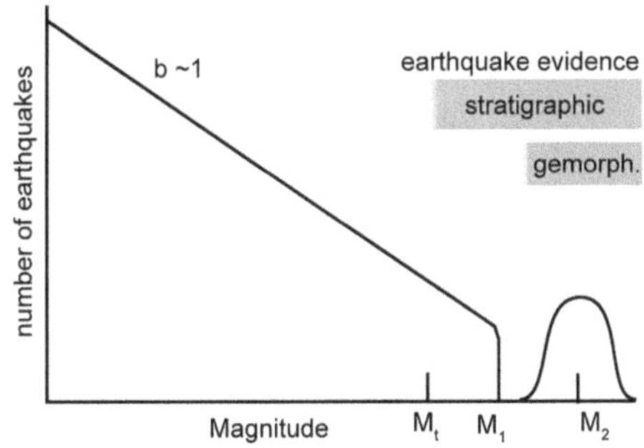

Fig. 5.8 Schematic representation of a discrete bimodal magnitude frequency distribution. Earthquakes up to magnitude M_1 follow the Gutenberg-Richter relationship. A peaked Gaussian-like distribution is observed around M_2. Following the discussion presented in the main text, we assume that earthquakes exceeding M_t are capable of generating surface ruptures. Due to the difference in resolution potential of stratigraphic and geomorphic record, we suggest that the latter is able to only resolve slip contributions from M_2 while stratigraphic records may contain signals form M_2 as well as smaller surface rupturing earthquakes. Modified after Zielke et al. (2015)

changing, making it difficult to preserve erosional geomorphic evidence of faulting at the same resolution as the depositional stratigraphic record is capable of. Hence, the stratigraphic record (i.e., the data set that typically constraints earthquake timing) is able to resolve details of earthquake recurrence that the geomorphic record is not able to sense or preserve. The most recent surface-rupturing earthquake of the Parkfield section of the San Andreas Fault in 2004 exemplifies this notion. This earthquake generated a surface rupture with centimeter-scale surface displacements (Rymer et al., 2006; Toké et al., 2006). While the corresponding disruptions of the stratigraphic record remain preserved in the sub-surface, the earthquake's geomorphic footprint has been erased in the last 20 years. Additionally, the distinct variability of slip along a fault during an earthquake along with the difficulties in assessing a marker's pre-earthquake morphology typically result in offset ranges (assigned to a displaced marker) of ±20–30% (2σ; e.g., 5.0 ± 1.2 m). These relatively large uncertainties make it practically impossible to resolve small-earthquake, centi- or decimeter-scale surface slip contributions (their offset contributions would be hidden in the measurement uncertainty and offset variability of dominating offset observations).

Figure 5.9 summarizes the conceptual model of earthquake recurrence and along-fault slip accumulation for individual fault segments that is capable of resolving the aforementioned discrepancies (Zielke, 2018; Zielke et al., 2015), further arguing that a fault segment can simultaneously exhibit "characteristic" as well as "variable-slip" behavior (Schwartz & Coppersmith, 1984). Slip accumulation along individual fault segments is dominated but not exclusively linked to the occurrence of similar-size, large, but relatively infrequent earthquakes—earthquakes commonly referred to as "characteristic earthquakes" in the sense that they produce a characteristic amount of slip over a characteristic length of the fault (M_2 in Fig. 5.8). In addition to those large and infrequent earthquakes, another group of distinctly smaller (but still surface-rupturing) and more randomly occurring earthquakes is required to satisfy the apparently disparate stratigraphic and geomorphic observations. Those latter earthquakes may be following the Gutenberg-Richter relation ($M_t <= M <= M_l$ in Fig. 5.8) but could also be related to slip leakage from nearby fault sections.

Hence, we may say that each COPD peak consists of the slip contribution of one large, inherently "characteristic" slip event and an additional number of smaller surface rupturing events (Fig. 5.9). We want to stress that the term "characteristic" as used here refers to the slip increment of a full-rupture earthquake along a fault segment (Zielke et al., 2015) and not to the magnitude or lateral extent of the corresponding entire earthquake rupture. While the former (characteristic slip events) dominate the geomorphic signal, the latter (smaller, randomly occurring surface ruptures) dominate the stratigraphic signal, resolving the discrepancy in CV estimates for slip accumulation and earthquake timing. If faults indeed behave in this manner, we would excavate a trench at a location along such a fault (Fig. 5.9) and observe inter-event times that are quasi-random (high CV value) due to the occurrence of smaller surface-rupturing earthquakes or slip-leakage from nearby segments. At the same time, detailed analysis of geomorphic evidence of faulting would provide quasi-periodic, similar-size offset increments (low CV value).

5.5.2 Fault Slip-Rates

Long-term (Holocene) geologic fault slip-rates are a manifestation of the strain accumulation rate and therefore constrain the timing and potential size of large and destructive earthquakes. Slip rates refer to the accumulated motion along a fault over a certain time frame as it is accommodated by numerous earthquakes and potentially also by aseismic phenomena, and hence are a time-averaged pseudo quantity. It does not refer to the "instantaneous" rate of motion at any given point in time as it would be resolved by geodetically-derived rates. Such long-term fault slip-rates are generally constrained using displaced geomorphic features whose time of formation or abandonment has been determined via geochronological methods. Some slip-rate studies have found that geologic fault slip-rates may vary along fault (spatial variation; e.g., Behr et al., 2010; Sieh & Jahns, 1984) as well as through time (temporal variation; e.g., Friedrich et al., 2003; Weldon et al., 2004). While spatial variation might be explained for example by along-fault changes of fault orientation relative to a regional tectonic stress-field or by existence of additional faults that take up some portion of the strain budget, it is not equally/readily clear what might be causing the reported temporal variations in geologic slip-rate. Possible explanations include temporal variations in frictional properties along the fault surface or in the applied far-field loading stresses. Another possibility is that these changes in long-term slip-rates are a representation of temporal earthquake clustering (suggesting the existence of earthquake super cycles), where the occurrence of earthquakes and strain-release are concentrated in time (high slip-rate),

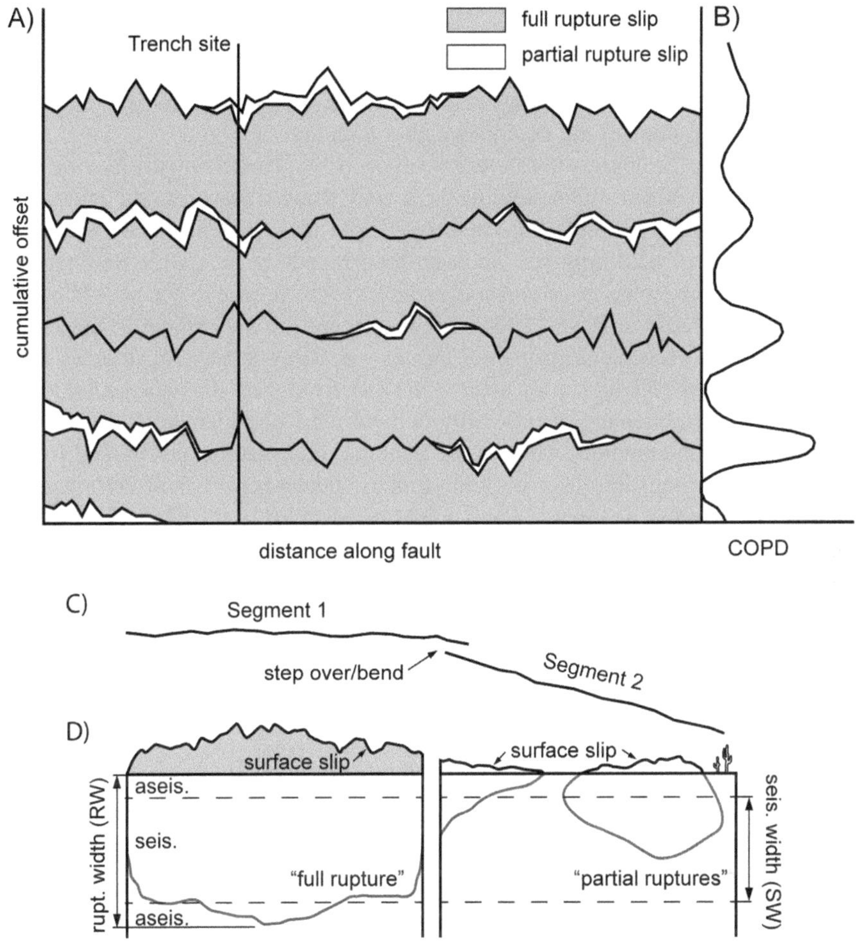

Fig. 5.9 Conceptual model for earthquake recurrence i.e., slip accumulation along individual fault segments. **A** Slip accumulation is dominated by full-rupture earthquakes and their signal is what is generally expressed and stored in the geomorphic records. As a result, corresponding CV values for slip accumulation (e.g., derived from COPD plots, **B**) are quasi-characteristic with values of 0.2–0.4. However, partial fault rupture, e.g., due to distinctly smaller surface-rupturing earthquakes or due to ruptures coming from neighboring segments may occur. While their disturbance of the morphology may be small and soon removed i.e., not recognizable, they also affect the stratigraphic record where they can persist for substantially longer periods of time. The randomness in occurrence time of these smaller earthquakes introduces the large variability in inter-event time typically with values of 0.5–1.0 as is often reported from paleoseismic trench investigations. Modified after Zielke et al. (2015)

followed by phases that are characterized by subdued seismic activity (low slip-rate) (e.g., Dolan et al., 2007; Friedrich et al., 2003). That said, at present the body of literature concerning slip-rates along various faults (e.g., Gold & Cowgill, 2011; Meade et al., 2013; Reilinger & McClusky, 2011) suggests that temporal variations in slip-rate are the exception rather than the rule.

In addition to geologic slip-rate estimates, geodetic rate estimates provide further information about strain accumulation and strain release during inter- and post-seismic phases. Comparisons of geologically and geodetically derived estimates often agree well with one another (e.g., Meade et al., 2013; Reilinger & McClusky, 2011). However certain studies have revealed large differences between the two (e.g., Meade & Hager, 2005). Some of these differences can be attributed to transient, post-seismic deformation signals—possibly lasting for decades after a large earthquake—which may affect GNSS-derived rates but not geologically derived rates. Additionally, geodetically derived rates consider the (interseismic) deformation of the medium whereas the geologic observations are limited to the displacements along the observed fault strands. Persistent off-fault deformation, or deformation along secondary faults might be resolved in the geodetic rates but might be missing in the geologic rates. Furthermore, if geologic rates are integrated over only a few earthquake cycles then it is likely that these rates are still sensitive to the time difference between a geomorphic markers' age and the time interval of marker displacement (because there is no reason for them to coincide in time). If these conditions exist or phenomena are persistent in space and time along a given fault, then we would expect that geodetic rates are consistently above the geologic rates. Thus such geologic—geodetic discrepancies do not require changes in slip-rate, although they also cannot be ruled out per se.

5.6 Variability in Earthquake Rupture and Uncertainty in Offset Measurements

Quantifying the amount of surface displacement that is associated with an individual earthquake or a sequence of earthquakes requires measuring the offset of displaced geomorphic markers. Like any other measurement of a physical quantity, these measurements are associated with epistemic uncertainties due to the measurement process and aleatory variability due to natural changes of the measured quantity in space or time. It is important to know the sources of uncertainty well in order to perform meaningful offset measurements. In this section we will address measurement uncertainty, offset variability, as well as the potential for biased measurements and the resolution potential of geomorphic records.

5.6.1 Slip Variation and Measurement Uncertainty

Let's assume that two closely-spaced geomorphic markers along a fault trace were displaced by the same number of earthquakes. Naively, we might expect that measuring their respective displacements provides identical offset amounts. However, actual offset measurements along real fault traces have shown that this expectation is commonly not met. There are two factors that control this discrepancy in offset measurement. One is the uncertainty in the offset measurement that relates to our ability to correctly infer the pre-earthquake morphology of the displaced marker and therefore our ability to uniquely constrain a reference datum against which to measure. The other is actual physically-driven variability along the fault.

Being able to make well-informed assumptions about the morphology is crucial for every offset measurement. In fact, the whole measurement hinges on it. While this statement may seem trivial, it is in fact not. The problem arises because geomorphic markers are not simple geometrical shapes such as straight or sub-linear lines, but instead have a spatial extent with a dimension, orientation, and shape that may vary substantially on either side of a fault. Geomorphic alteration subsequent to the displacement event further modifies the shape of these features, making it more and more difficult to make informed assessments of their pre-earthquake and pre-alteration shape. In analogy, it is a little like looking at a melting ice cube to then assess what its original shape has been before melting began. If not all of the cube is gone, we may be able to make a meaningful assessment about its initial shape, but at some point, we only know that melting took place. Going back to geomorphic features, at some point we only know that offset took place but cannot put meaningful numbers on the amount of offset that was taking place. Given that the aforementioned hypothetical two displaced geomorphic markers may exhibit different shapes, orientations to the fault trace, different exposure to erosional forces, or other differences, it becomes clear why identical offset measurements may not be expected, even if the actual displacement had been the same.

As to the factor of genuine along-fault variability in surface rupture and fault displacement, we can observe that most often the amount of surface slip is not constant along an earthquake's rupture trace. As discussed in a previous section, this along-fault variability of surface slip is not purely random but exhibits fractal, self-affine characteristics (Milliner et al., 2016). The observation of fractal characteristics in surface slip distribution on its own does not provide a physical explanation for its existence. Potential causes for the observed variability include along-fault spatial (and possibly temporal) variation in off-fault deformation, frictional properties, and material contrasts along the fault surface as well as the existence of (persistent) fault geometric complexities. Hence, it might be expected that the along-fault distribution of surface slip and therefore the offsets of geomorphic markers along a rupture trace, varies (a) along the fault, and (b) from one earthquake to the next.

One factor that reduces slip observed along a fault is off-fault deformation. As indicated in a previous section, off-fault deformation can take up a significant amount of strain release. The amount of displacement along the main fault that is generated by an earthquake is therefore lowered (compared to scenario where no off-fault deformation takes place). This can lead to an underestimation of past-earthquake size. Some of the observed variability in along-fault slip distribution is attributed to spatial variation in off-fault deformation. At the moment, measurements of off-fault deformation have been few, as they can only be done for earthquakes that happened recently enough to have good coverage with high-resolution 2-D areal data sets, i.e., remotely sensed displacement fields. The observed amount of off-fault deformation varies distinctly between earthquakes and within individual earthquakes for the few cases in which it has been measured. This variability and the limited number of observations makes it at the moment difficult to incorporate these important observations into the analysis of offset observations that were made for past earthquakes. Unfortunately, it seems not possible to simply apply a scaling factor between observed on-fault slip and reported off-fault deformation to determined overall strain release because it is currently unknown how off-fault deformation varies (a) along fault, (b) from earthquake to earthquake, (c) from one fault to another fault. Generally speaking, it is current consensus that off-fault deformation is stronger where fault geometry is more complex. Hence, it is important to record that geometry along with the measured offsets, to provide a framework in which to analyze/interpret these observations.

Both the variability and the uncertainty are important aspects to take into account when performing offset measurements and when interpreting them. As a result, we want to suggest taking offset measurements where the influence of fault geometry and lithology variations are minimized so that the actual (variation of) slip as it relates to the seismic source at depth may be resolved.

5.6.2 Potential for Measurement Bias

As noted in the previous section, an earthquake's displacement along the rupture surfaces at depth as well as the corresponding along-fault surface displacement exhibit an inherent variability that reflects a number of processes and phenomena. The goal of offset measurements is to quantify the slip distribution of a single earthquake and the accumulation of slip from a succession of earthquakes with the goal to identify potentially existing patterns within these observations. It is important to acknowledge that the underlying goal here may already affect offset measurements. If we are aiming to identify patterns then we are more likely to do so than the other way around. Hence, performing offset measurements to constrain slip-accumulation patterns has the potential of being biased. This bias may become fortified when the smallest offsets along a fault section all fall into the same offset range. In that case, we are generally more likely to identify multiples of that offset value (e.g., Weldon et al., 1996). The problem is directly related to the nature of geomorphic markers, being spatially extensive features (as opposed to simple lines) that change their shape over time. Every offset measurement requires from the researcher to make an assumption of the marker's pre-earthquake morphology. Depending on the specific shape and geomorphic age of the displaced marker, this assumption might or might not be well constrained. Hence, it is possible that the interpretation of a marker's pre-earthquake morphology is "modified" i.e., biased to the extent that the resulting offset measurement values fit into the expected/anticipated pattern. It is very hard for the human mind to switch off this inherent pattern recognition/anticipation. Making this statement does not mean that recurring increments of slip along a fault, like those mentioned in previous sections of the chapter, cannot or do not exist. It is, however, important to acknowledge the difficulties that arise from offset measurements that contain some subjective aspects (e.g., in form of the assumed pre-earthquake morphology and best displacement amount to reconstruct it).

While it is hard to avoid that bias, it is relatively easy to "trick ourselves" in a way that distinctly minimizes the potential for biased offset measurements: For example, if we measure a displacement of a geomorphic marker without knowing the scale of the ruler and/or changing its scale over and over again, then we are not affected by this bias because the numbers (measurement values in unknown units) themselves are meaningless unless we know the scaling term that converts these arbitrary units into the same one (e.g., into meters). Hence, a good (albeit not established) approach in the field might be to have a set of measuring tapes that all look the same but have different, arbitrary units on them (e.g., where 1m represents for example 0.6 m for tape 1, 0.8 m for tape 2, 1.5 m for tape 3, etc.). Measuring with these devices, recording the measured quantity as well as the tape's number, allows one to perform unbiased offset measurements. The same approach can be easily integrated into computer-aided measurement approaches, for example it is implemented in LaDiCaoz_v2 (Haddon et al., 2016; Zielke & Arrowsmith, 2012).

In addition to these manual offset measurements, semi-automatic offset measurement may be performed as well. Given the nature of this "hands-off" approach, these measurements can be assumed to be unbiased, although they are subject to the same assumptions and uncertainties about a landform's original geometry. A recent paper by Stewart et al. (2018) introduced such a semi-automatic measurement tool and compared offset measurements taken with this tool with those (among others) by Zielke et al. (2010) for the Carrizo segment of the southern San Andreas Fault who used LaDiCaoz. This comparison reported a good agreement between both approaches indicating (a) that manual offset measurements are not per-se biased and (b) that the semi-automatic approach is able to perform well. While semi-automatic measurement approaches provide a great alternative to manual approaches, it must be noted that they also require assumptions about the marker's pre-earthquake morphology just like manual measurements. Automated approaches also must assume a marker's shape as it existed before surface faulting occurred to disrupt it. For example, the aforementioned semi-automatic approach assumes that the marker was linear as it crossed the fault zone. This assumption can certainly be plausible, depending on local geological, hydrological, and geomorphic conditions. That said, they also limit the number of usable markers—although this may be for the better because they concentrate on those that are easier to interpret and where measurements are more defensible. The future may hold offset measurements that are guided by machine learning algorithms and heuristic landscape evolution models (e.g., Reitman et al., 2019) in order to provide the best possible constraints on earthquake slip accumulation.

With all of these approaches, it is important to acknowledge measurement bias and aim to use the tools that are available to minimize it in order to make meaningful measurements of offset for the reconstruction of slip-accumulation patterns.

5.6.3 Resolution Potential of Geomorphic Records and Residence Time of Markers in Land Scape

An important aspect of measuring earthquake surface slip by reconstructing geomorphic markers to their pre-earthquake morphology is to acknowledge the resolution potential of those markers. For example, a marker's physical dimension, the size of its drainage area (if the marker is a fluvial channel), and other parameters affect what amount of offset it can resolve (Wallace, 1968). Similarly, the presence of a geomorphic feature in a landscape (how long it remains) is dependent on its spatial scale and amplitude. Small features and/or steep slopes will be removed from the landscape or strongly modified more quickly than features that are larger and/or more gradual. Hence extending the slip-per-event history into the past is limited to earthquakes that are large enough to substantially and distinctly disrupt geomorphic features of sufficient size to remain in the landscape over multiple earthquake cycles. Additionally, given the aforementioned uncertainties that are associated with offset measurements and their variability both along-strike and between events, it is not realistic to resolve the offset-increments of more than ~5 successive large earthquakes (Zielke, 2018). Stratigraphic records may however be more sensitive. While this sensitivity is beneficial for resolving more details of rupture history, it also makes it challenging to combine the two types of data (stratigraphic and geomorphic records). This was discussed to some extend in the previous section. Here, we only want to raise awareness of this issue again, making the claim that, generally, it is not permissible to directly combine results from stratigraphic and geomorphic investigations because they do not represent the same history.

5.6.4 Rupture End Points and Overlapping Ruptures

While this topic is addressed in another chapter of this book, we still want to briefly comment on it here as it relates to offset measurements along continental faults and their behavior in terms of earthquake recurrence and slip accumulation. In the description mentioned above, we address scenarios where offset measurements were taken along well-defined fault segments, not considering their spatial extent or interaction with other, neighboring sections. Earthquake rupture into neighboring sections is likely to occur—at least occasionally—and might find its expression in the geomorphic as well as stratigraphic record. In fact, observations by Salisbury et al. (2012) show this quite clearly for the Clark strand of the San Jacinto fault in southern California. However, as suggested by the presence of similar-size slip increments and the statistical considerations by Zielke (2018) it appears that those partial/bleeding ruptures play a smaller role in terms of actual slip accumulation. That said, their signal may be well identified in the stratigraphic record (trenching) and therefore lead to distinct deviations in inter-event time intervals.

5.7 Guidelines for Interpretation of Tectono-Geomorphic Data

In the last sections we mentioned a number of times the lurking difficulties that are associated with investigations aiming to constrain earthquake recurrence and slip-accumulation history along continental faults. In order to avoid or minimize some of the difficulties and common pitfalls, we want to provide a few guidelines that we hope will help the reader when performing such investigations. These guidelines will concentrate on the interpretation and measurement of geomorphic markers. Please note that these guidelines can be considered equally valid and relevant for manual (in the field or with computer-based tools such as LaDiCaoz) or semi-automatic measurement approaches.

As a first step it is important to gain a comprehensive understanding of a site's morphology and the character of the fault zone (e.g., identifying potential existence of multiple fault strands or other complex geometries, including discrepant slip vectors or variable kinematics). This understanding can only partially be gained with optical imagery. High-resolution topographic data are important for this step. The additional information that they provide permit discerning for example whether a marker (such as a channel) is displaced tectonically or merely deflected by a topographic high. It further allows one to make more precise estimates of the pre-offset marker morphology and its likely geomorphic alteration since displacement. We further recommend using different representations of the landscape, in the form of contour plots as well as hillshade, slopeshade, aspect, and elevation maps in order to gain a detailed understanding of the site that surrounds a marker. Contour plots are particularly important, as they are able, when interpreted correctly, to convey information that may hard to identify in other visualization approaches. This includes assessing the plausibility that the assumed up-fault and down-fault marker sections are really part of the same (now-displaced) feature. Hence, we advocate for a detailed analysis of the landscape in which the identified geomorphic marker is located before performing offset measurements.

We recommend concentrating investigations on geomorphic markers and fault zones that are geometrically simple, i.e., for which the pre-earthquake morphology can be confidently assumed and the fault zone permits the simplifying approximation of a single fault plane. The goal is to limit the potential influence of other phenomena that might affect on-fault offset or our ability to measure it.

We further recommend adopting "blind measurement" routines as was discussed in Sect. 5.6.2 in an effort to avoid measurement bias. Doing so will greatly increase the trust and confidence in the presented results. Being able to state that offset measurements were free of potential measurement bias is a great step.

We further recommend performing offset measurements not on hillshade plots but rather on topo-slope plots or slope+aspect plots. Even better is the use of contour lines. Using hillshade representations (as well as optical imagery) can be problematic because the shadow that is cast may smear or hide important details of marker and fault trace. Using slope+aspect and contour line representations provides a clearer picture in that regard and therefore result in more reliable offset measurements.

Lastly, we want to advise that the researchers performing these types of analyses are aware of the limitations of the data set that they are using and the limited resolution potential of the geomorphic record. We made corresponding remarks in a previous section: The offset ranges that are assigned to a displaced marker are very important and it is important that these representations of measurement uncertainty are representative of the data and marker that were analyzed. For example, assigned offset ranges that are at or below $\pm 10\%$ (2σ) are dubious and might be related to biased offset measurement. Similarly, providing cm-scale precision for meter-scale offsets is nonsensical (e.g., 4.31 ± 0.23 m), i.e., fails to acknowledge the limitations of data resolution.

5.8 Outlook

Constraining the recurrence of earthquakes is important. Doing so requires high-resolution spatial data sets as well as detailed geomorphic investigations and precise as well as abundant geochronometers. The increasing availability of high-resolution (spatial) data sets is a great resource to active tectonics investigations and they promise to be able to resolve more and more details of recent earthquakes (for which differencing approaches and source inversion can be adopted). These data sets also help us to investigate earthquakes from the pre-instrumental era—which is crucial in the attempt to gain an understanding of earthquake recurrence and along-fault slip accumulation (i.e., strain release). That said, while the data sets become more abundant and of finer resolution, the resolution potential of stratigraphic and geomorphic records is limited and may at some point act as a hard boundary that we cannot overcome with the currently adopted approaches. Very importantly, in order to extract as much information as possible out of geomorphic and stratigraphic records, we need to gain an even deeper understanding of their formation and modification over geologic (Holocene) time scales. Analog as well as numerical modeling approaches may provide the means to further learn about the landscape's response to different climatic, tectonic, and gravitational forcing.

These approaches may help us to improve estimates of a marker's pre-earthquake morphology and even open avenues to investigate markers that cannot be analyzed with the current approaches.

Acknowledgements The preparation of this chapter is supported with funding from King Abdullah University of Science and Technology (KAUST), grant BAS/1/1339-01-01

References

Akçiz, S. O., Grant Ludwig, L., Arrowsmith, J. R., & Zielke, O. (2010). Century-long average time intervals between earthquake ruptures of the San Andreas Fault in the Carrizo Plain, California. *Geology, 38*(9), 787–790. https://doi.org/10.1130/G30995.1

Antoine, S. L., Klinger, Y., Delorme, A., Wang, K., Bürgmann, R., & Gold, R. D. (2021). Diffuse deformation and surface faulting distribution from submetric image correlation along the 2019 Ridgecrest, California, ruptures. *Bulletin of the Seismological Society of America, 111*(5), 2275–2302. https://doi.org/10.1785/0120210036

Arrowsmith, J. R., Rhodes, D. D., & Pollard, D. D. (1998). Morphologic dating of scarps formed by repeated slip events along the San Andreas Fault, Carrizo Plain, California. *Journal of Geophysical Research, 103*, 10141–10160.

Arrowsmith, J. R., & Zielke, O. (2009). Tectonic geomorphology of the San Andreas Fault Zone from high-resolution topography: An example from the Cholame segment. *Geomorphology, 113*(1), 70–81. https://doi.org/10.1016/j.geomorph.2009.01.002

Behr, W. M., Rood, D. H., Fletcher, K. E., Guzman, N., Finkel, R., Hanks, T. C., Hudnut, K. W., Kendrick, K. J., Platt, J. P., Sharp, W. D., Weldon, R. J., & Yule, J. D. (2010). Uncertainties in slip-rate estimates for the mission creek strand of the Southern San Andreas Fault at Biskra Palms Oasis, Southern California. *Geological Society of America Bulletin, 122*(9), 1360–1377. https://doi.org/10.1130/B30020.1

Benedetti, L., Manighetti, I., Gaudemer, Y., Finkel, R., Malavieille, J., Pou, K., Arnold, M., Aumaitre, G., Bourles, D., & Keddadouche, K. (2013). Earthquake synchrony and clustering on Fucino faults (Central Italy) as revealed from in situ 36Cl exposure dating. *Journal of Geophysical Research, 118*, 4948–4974. https://doi.org/10.1002/jgrb.50299

Ben-Zion, Y. (2003). Appendix 2-Key formulas in earthquake seismology. *International Handbook of Earthquake and Engineering Seismology, 81B*, 1857–1875.

Bevis, M., Hudnut, K., Sanchez, R., Toth, C., Grejner-Brzezinska, D., Kendrick, E., Caccamise, D., Raleigh, D., Zhou, H., Shan, S., Shindle, W., Yong, A., Harvey, J., Borsa, A., Ayoub, F., Shrestha, R., Carter, B., Sartori, M., Phillips, D., & Coloma, F. (2005). *The B4 Project: Scanning the San Andreas and San Jacinto Fault Zones, Abstract H34B–01 presented at 2005 Fall Meeting*. AGU.

Bilham, R., & Williams, P. (1985). Sawtooth segmentation and deformation processes on the southern San Andreas fault, California. *Geophysical Research Letters, 12*, 557–560.

Brothers, D. S., Miller, N. C., Vaughn, B. J., Haeussler, P. J., Greene, H. G., Andrews, B. D., Zielke, O., Watt, J., & Dartnell, P. (2020). Plate boundary localization, slip-rates and rupture segmentation of the Queen Charlotte Fault based on submarine tectonic geomorphology. *Earth and Planetary Science Letters, 530*, 115882. https://doi.org/10.1016/j.epsl.2019.115882

Burbank, D. W., & Anderson, R. S. (2012). *Tectonic geomorphology* (2nd ed.). Blackwell Publishing Ltd.

Burgmann, R., Rosen, P. A., & Fielding, E. J. (2000). Synthetic aperture radar interferometry to measure Earth's surface topography and its deformation. *Annual Review of Earth and Planetary Sciences, 28*, 169–209. https://doi.org/10.1146/annurev.earth.28.1.169

Cambonie, T., Klinger, Y., & Lazarus, V. (2019). Similarities between mode III crack growth patterns and strike-slip faults. *Philosophical Transactions of the Royal Society A,* 377–392. https://doi.org/10.1098/rsta.2017.0392

Candela, T., Renard, F., Klinger, Y., Mair, K., Schmittbuhl, J., & Brodsky, E. E. (2012). Roughness of fault surfaces over nine decades of length scales. *Journal of Geophysical Research, 117,* B08409. https://doi.org/10.1029/2011JB009041

Choi, J. H., Klinger, Y., Ferry, M., Ritz, J. F., Kurtz, R., Rizza, M., Bollinger, L., Davaasambuu, B., Tsend-Ayush, N., & Demberel, S. (2018). Geologic inheritance and earthquake rupture processes: The 1905 M $\geq$ 8 Tsetserleg-Bulnay strike-slip earthquake sequence, Mongolia. *Journal of Geophysical Research, 123*(2), 1925–1953. https://doi.org/10.1002/2017JB013962

Daeron, M., Klinger, Y., Tapponnier, P., Elias, A., Jacques, E., & Sursock, A. (2007). A 12,000-year-long record of up to 14 paleo-earthquakes on the Yammouneh fault (Levant fault system). *Bulletin of the Seismological Society of America, 97*(3), 749–771.

Dolan, J. F., Bowman, D. D., & Sammis, C. G. (2007). Long-range and long-term fault interactions in Southern California. *Geology, 35*(9), 855–858. https://doi.org/10.1130/G23789A.1

DuRoss, C., Personius, S., Crone, A., Olig, S., Hylland, M., Lund, W., & Schwartz, D. (2016). Fault segmentation: New concepts from the Wasatch fault zone, Utah, USA. *Journal of Geophysical Research, 121*(2), 1131–1157.

Field, E. H., Biasi, G. P., Bird, P., Dawson, T. E., Felzer, K. R., Jackson, D. D., Johnson, K. M., Jordan, T. H., Madden, C., Michael, A. J., Milner, K. R., Page, M. T., Parsons, T., Powers, P. M., Shaw, B. E., Thatcher, W. R., Weldon, R. J., II, & Zeng, Y. (2015). Long-term time-dependent probabilities of the third Uniform California Earthquake Rupture Forecast (UCERF3). *Bulletin of the Seismological Society of America, 105*(2A), 511–543. https://doi.org/10.1785/0120140093

Fialko, Y., Sandwell, D., Simons, M., & Rosen, P. (2005). Three-dimensional deformation caused by the Bam, Iran, earthquake and the origin of shallow slip deficit. *Nature, 435,* 295–299.

Friedrich, A., Wernicke, B. P., Niemi, N. A., Bennett, R. A., & Davis, J. L. (2003). Comparison of geodetic and geologic data from the Wasatch region, Utah, and implications for the spectral character of Earth deformation at periods of 10 to 10 million years. *Journal of Geophysical Research, 108*(4), 1–23. https://doi.org/10.1029/2001JB000682

Gao, M., Xu, X., Klinger, Y., Van der Woerd, J., & Tapponier, P. (2017). High-resolution mapping based on an unmanned aerial vehicle (UAV) to capture paleoseismic offsets along the Altyn-Tagh fault, China. *Scientific Reports, 7,* 8281. https://doi.org/10.1038/s41598-017-08119-2

Gold, P. O., Oskin, M. E., Elliott, A. J., Hinojosa-Corona, A., Taylor, M. H., Kreylos, O., & Cowgill, E. (2013). Coseismic slip variation assessed from terrestrial lidar scans of the El Mayor-Cucupah Surface Rupture. *Earth and Planetary Science Letters, 366,* 151–162. https://doi.org/10.1016/j.epsl.2013.01.040

Gold, R. D., & Cowgill, E. (2011). Deriving fault-slip histories to test for secular variation in slip, with examples from the Kunlun and Awatere faults. *Earth and Planetary Science Letters, 301,* 52–64.

Grant-Ludwig, L., Akçiz, S. O., Noriega, G. N., Zielke, O., & Arrowsmith, J. R. (2010). Climate modulated channel incision and rupture history of the San Andreas fault in the Carrizo Plain. *Science, 327,* 1117–1119. https://doi.org/10.1126/science.1182837

Haddad, D. E., Akciz, S. O., Arrowsmith, J. R., Rhodes, D. D., Oldow, J. S., Zielke, O., Toke, N. A., Haddad, A. G., Mauer J., & Shilpakar, P. (2012). Application of airborne and terrestrial laser scanning to paleoseismology. *Geosphere, 8*(4), 771–786. https://doi.org/10.1130/GES00701.1

Haddon, E. K., Amos, C. B., Zielke, O., Jayko, A. S., & Burgmann, R. (2016). Surface slip during large Owens Valley earthquakes. *Geochemistry, Geophysics, Geosystems, 17*(6), 2239–2269.

Hanks, T. C., & Andrews, D. J. (1989). Effect of far-field slope on morphologic dating of scarplike landforms. *Journal of Geophysical Research, 94,* 565–573.

Hudnut, K. W., Borsa, A., Glennie, C., & Minster, J.-B. (2002). High-resolution topography along surface rupture of the 16 October 1999 Hector Mine, California Earthquake (Mw 7.1) from airborne laser swath mapping. *Bulletin of the Seismological Society of America, 92*(4), 1570–1576. https://doi.org/10.1785/0120000934

Ide, S. (2007). Slip inversion. In G. Schubert (Ed.), *Treatise on geophysics earthquake seismology* (Vol. 4). Elsevier.

Kaneko, Y., & Fialko, Y. (2011). Shallow slip deficit due to large strike-slip earthquakes in dynamic rupture simulations with elasto-plastic off-fault response. *Geophysical Journal International, 186*(3), 1389–1403.

King, G. (1986). Speculations on the geometry of the initiation and termination processes of earthquake rupture and its relation to morphology and geological structures. *Pure and Applied Geophysics, 124*(3), 567–585. https://doi.org/10.1007/BF00877216

King, G., Klinger Y., Bowman, D., & Tapponnier, P. (2005). Slip partitioned surface breaks for the 2001 Kokoxili earthquake, China (Mw 7.8). *Bulletin of the Seismological Society of America, 95*(2), 731–738.

Klinger, Y. (2010). Relation between continental strike-slip earthquake segmentation and thickness of the crust. *Journal of Geophysical Research, 115*, B07306. https://doi.org/10.1029/2009JB006550

Klinger, Y., Avouac, J. P., Bourles, D., & Tisnerat, N. (2003). Alluvial deposition and lake-level fluctuations forced by late Quaternary climate change: The Dead-Sea case example. *Sedimentary Geology, 162*, 119–139.

Klinger, Y., Etchebes, M., Tapponier, P., & Narteau, C. (2011). Characteristic slip for five great earthquakes along the Fuyun Fault in China. *Nature Geoscience, 4*, 389–392. https://doi.org/10.1038/NGEO1158

Klinger, Y., Michel, R., & King, G. C. P. (2006). Evidence for an earthquake barrier model from MW ~7.8 Kokoxili (Tibet) earthquake slip-distribution. *Earth and Planetary Science Letters, 242*, 354–364.

Klinger, Y., Okubo, K., Vallage, A., Champenois, J., Delorme, A., Rougier, E., Lei, Z., Knight, E., Munjiza, A., Satriano, C., Baize, S., Langridge, R., & Bhat, H. (2018). Earthquake damage patterns resolve complex rupture processes. *Geophysical Research Letters, 45*. https://doi.org/10.1029/2018GL078842

Klinger, Y., Xu, X., Tapponnier, P., Van der Woerd J., Lasserre C., & King G. (2005). High-resolution satellite imagery mapping of the surface rupture and slip distribution of the Mw7.8, 14 November 2001 Kokoxili earthquake, Kunlun fault, northern Tibet, China. *Bulletin of the Seismological Society of America, 95*(5), 1970–1987.

Kurtz, R., Klinger, Y., Ferry, M., & Ritz, J.-F. (2018). Horizontal surface-slip distribution trough several seismic cycles: The eastern Bogd fault, Goby Altai, Mongolia. *Tectonophysics, 734*(735), 167–182. https://doi.org/10.1016/j.tecto.2018.03.011

Lefevre, M., Klinger, Y., Al-Qaryouti, M., Le Beon, M., & Moumani, K. (2018). Slip deficit and temporal clustering along the Dead Sea fault from paleoseismological investigations. *Scientific Reports, 8*(4511). https://doi.org/10.1038/s41598-018-22627-9

Lowe, D. G. (2004). Distinctive image features from scale-invariant keypoints. *International Journal of Computer Vision, 60*(2), 91–110.

Mai, P. M., Schorlemmer, D., Page, M., Ampuero, J.-P., Asano, K., Causse, M., Custodio, S., Fan, W., Festa, G., Galis, M., Gallovic, F., Imperatori, W., Kaeser, M., Malytskyy, D., Okuwaki, R., Pollitz, F., Passone, L., Razafindrakoto, H., Sekiguchi, H., & Zielke, O. (2016). The earthquake-source inversion validation (SIV) project. *Seismological Research Letters, 87*(3), 690–708. https://doi.org/10.1785/0220150231

Manighetti, I., Caulet, C., De Barros, L., Perrin, C., Cappa, F., & Gaudemer, Y. (2015b). Generic along-strike segmentation of Afar normal faults, East Africa: Implications on fault growth and stress heterogeneity on seismogenic fault planes. *Geochemistry, Geophysics, Geosystems, 16*(2), 443–467.

Manighetti, I., Perrin, C., Dominguez, S., Garamois, S., Gaudemer, Y., Malavieille, J., Matteo, L., Delor, E., Vitard, C., & Beaupretre, S. (2015a). Recovering paleoearthquake slip record in a highly dynamic alluvial and tectonic region (Hope Fault, New Zealand) from airborne lidar. *Journal of Geophysical Research, 120*, 1–26. https://doi.org/10.1002/2014JB011787

Marchandon, M., Vergnolle, M., Sudhaus, H., & Cavalié, O. (2018). Fault geometry and slip distribution at depth of the 1997 Mw7.2 Zirkuh earthquake: Contribution of near-field displacement data. *Journal of Geophysical Research, 123*(2), 1904–1924.

Marzocchi, W. (2018). Predictive seismology. *Seismological Research Letters, 89*(6), 1998–2000. https://doi.org/10.1785/0220180238

Massonnet, D., & Feigl, K. L. (1998). Radar interferometry and its application to changes in the earth's surface. *Reviews of Geophysics, 36*(4), 441–500. https://doi.org/10.1029/97RG03139

McGill, S. F., & Sieh, K. E. (1991). Surficial offsets on the central and eastern Garlock Fault associated with prehistoric earthquakes. *Journal of Geophysical Research, 96*(B13), 21597–21621.

Meade, B. J., & Hager, B. H. (2005). Block models of crustal motion in southern California constrained by GPS measurements. *Journal of Geophysical Research, 110*, B03403. https://doi.org/10.1029/2004JB003209

Meade, B. J., Klinger, Y., & Hetland, E. A. (2013). Inference of multiple earthquake-cycle relaxation timescales from irregular geodetic sampling of interseismic deformation. *Bulletin of the Seismological Society of America, 103*(5), 2824–2835. https://doi.org/10.1785/0120130006

Milliner, C. W. D., Dolan, J. F., Hollingsworth, J., Leprince, S., Ayoub, F., & Sammis, C. G. (2015). Quantifying near-field and off-fault deformation patterns of the 1992 Mw 7.3 Landers earthquake. *Geochemistry, Geophysics, Geosystems, 16*(5), 1,577–1,598.

Milliner, C. W. D., Sammis, C., Allam, A. A., Dolan, J. F., Hollingsworth, J., Leprince, S., & Ayoub, F. (2016). Resolving fine-scale heterogeneity of coseismic slip and the relation to fault structure. *Science and Reports, 6*, 27201. https://doi.org/10.1038/srep27201

Nissen, E., Maruyama, T., Arrowsmith, J. R., Elliott, J. R., Krishnan, A. K., Oskin, M. E., & Saripalli, S. (2014). Coseismic fault zone deformation revealed with differential lidar: Examples from Japanese Mw ~7 intraplate earthquakes. *Earth and Planetary Science Letters, 405*, 244–256. https://doi.org/10.1016/j.epsl.2014.08.031

Oskin, M. E., Arrowsmith, J. R., Hinojosa-Corona, A., Elliott, A. J., Fletcher, J. M., Fielding, E. J., Gold, P. O., Javier Gonzalez Garcia, J., Hudnut, K. W., Liu-Zeng, J., & Teran, O. J. (2012). Near-field deformation from the El Mayor-Cucupah earthquake revealed by differential lidar. *Science, 335*, 702–705. https://doi.org/10.1126/science.1213778

Philibosian, B., Sieh, K., Avouac, J. P., Natawidjaja, D., Chiang, H., Wu, C., Shen, C., Daryono, M., Perfettini, H., Suwargadi, B., Lu, Y., & Wang, X. (2017). Earthquake supercycles on the Mentawai segment of the Sunda megathrust in the seventeenth century and earlier. *Journal of Geophysical Research, 122*, 642–676. https://doi.org/10.1002/2016JB013560

Power, W. L., & Tullis, T. E. (1991). Euclidean and fractal models for the description of rock surface roughness. *Journal of Geophysical Research, 96*(B1), 415–424.

Reilinger, R., & McClusky, S. (2011). Nubia-Arabia-Eurasia plate motions and the dynamics of Mediterranean and Middle East tectonics. *Geophysical Journal International, 186*, 971–979. https://doi.org/10.1111/j.1365-246X.2011.05133.x

Ren, Z., Zhang, Z., Chen, T., Yan, S., Yin, J., Zhang, P., Zheng, W., Zhang, H., & Li, C. (2016). Clustering of offsets on the Haiyuan fault and their relationship to paleoearthquakes. *Geological Society of America Bulletin, 128*(1–2), 3–18.

Rockwell, T. K., Lindvall, S., Dawson, T., Langridge, R., Lettis, W., & Kinger, Y. (2002). Lateral offsets on surveyed cultural features resulting from the 1999 İzmit and Düzce Earthquakes, Turkey. *Bulletin of the Seismological Society of America, 92*(1), 79–94.

Rockwell, T. K., & Klinger, Y. (2013). Surface rupture and slip distribution of the 1940 Imperial Valley Earthquake, Imperial Fault, Southern California: Implications for rupture segmentation and dynamics. *Bulletin of the Seismological Society of America, 103*(2A), 629–640. https://doi.org/10.1785/0120120192

Rosu, A. M., Pierrot-Deseilligny, M., Delorme, A., Binet, R., & Klinger, Y. (2014). Measurement of ground displacement from optical satellite image correlation using the free open-source software MicMac. *Journal of Photogrammetry and Remote Sensing*. https://doi.org/10.1016/j.isprsjprs.2014.03.002

Rymer, M. J., Tinsley III, J. C., Treiman, J. A., Arrowsmith, J. R., Clahan, K. B., Rosinski, A. M., ... & Bawden, G. W. (2006). Surface fault slip associated with the 2004 Parkfield, California, earthquake. *Bulletin of the Seismological Society of America, 96*(4B), S11–S27. https://doi.org/10.1785/0120050830

Sagy, A., Brodsky, E. E., & Axen, G. J. (2007). Evolution of fault-surface roughness with slip. *Geology, 35*(3), 283–286. https://doi.org/10.1130/G23235A.1

Salisbury, J. B., Haddad, D. E., Rockwell, T., Arrowsmith, J. R., Madugo, C., Zielke, O., & Scharer, K. (2014). Validation of meter-scale surface faulting offset measurements from high-resolution topographic data. *Geosphere, 11*(6), 1884–1901. https://doi.org/10.1130/GES01197.1

Salisbury, J. B., Haddad, D. E., Rockwell, T., Arrowsmith, J. R., Madugo, C., Zielke, O., & Scharer, K. M. (2015). Validation of meter-scale surface faulting offset measurements from high-resolution topographic data. *Geosphere, 11*(6), 1–19. https://doi.org/10.1130/GES01197.1

Salisbury, J. B., Rockwell, T. K., Middelton, T. J., & Hudnut, K. W. (2012). Lidar and field observations of slip distribution for the most recent surface ruptures along the Central San Jacinto Fault. *Bulletin of the Seismological Society of America, 102*(2), 598–619. https://doi.org/10.1785/0120110068

Scharer, K. M., Salisbury, J. B., Arrowsmith, J. R., & Rockwell, T. K. (2014). Southern San Andreas Fault evaluation field activity: Approaches to measuring small geomorphic offsets—Challenges and recommendations for active fault studies. *Seismological Research Letters, 85*(1), 68–76. https://doi.org/10.1785/0220130108

Schwartz, D. P. (2018). Past and future fault rupture lengths in seismic source characterization—The long and short of it. *Bulletin of the Seismological Society of America, 108*(5A), 2493–2520.

Schwartz, D. P., & Coppersmith, K. J. (1984). Fault behavior and characteristic earthquakes: Examples from the Wasatch and San Andreas Fault Zones. *Journal of Geophysical Research, 89*(B7), 5681–5698.

Scott, C., Arrowsmith, R., Nissen, E., Lajoie, L., Maruyama, T., & Chiba, T. (2018). The M7 2016 Kumamoto, Japan, earthquake: 3-D deformation along the fault and within the damage zone constrained from differential Lidar topography. *Journal of Geophysical Research, 123*(7), 6138–6155.

Scott, C., Champenois, J., Klinger, Y., Nissen, E., Maruyama, T., Chiba, T., & Arrowsmith, R. (2019). 2016 M7 Kumamoto, Japan, earthquake slip field derived from a join inversion of differential Lidar topography, optical correlation, and InSAR surface displacements. *Geophysical Research Letters*. https://doi.org/10.1029/2019GL082202

Shimazaki, K., & Nakata, T. (1980). Time-predictable recurrence model for large earthquakes. *Geophysical Research Letters, 7*(4), 279–282.

Sieh, K. E., & Jahns, R. H. (1984). Holocene activity of the San Andreas Fault at Wallace Creek, California. *Geological Society of America Bulletin, 95*, 883–896.

Sieh, K. E., Jones, L., Hauksson, E., Hudnut, K., Eberhart-Phillips, D., Heaton, T., Hough, S., Hutton, K., Kanamori, H., Lilje, A., Lindvall, S., McGill, S., Mori, J., Rubin, C., Spotila, J., Stock, J., Kie Thio, H., Treiman, J., Wernicke, B., & Zachariasen, J. (1993). Near-field investigations of the Landers earthquake sequence. *Science, 260*, 171–176.

Sieh, K. E., Stuiver, M., & Brillinger, D. (1989). A more precise chronology of earthquakes produces by the San Andreas Fault in Southern California. *Journal of Geophysical Research, 94*, 603–623.

Stewart, N., Gaudemer, Y., Manighetti, I., Serreau, L., Vincendeau, A., Dominguez, S., Matteo, L., & Malavieille, J. (2018). "3D_Fault_Offsets", a Matlab code to automatically measure lateral and vertical fault offsets in topographic data: Application to San Andreas, Owens Valley, and Hope Faults. *Journal of Geophysical Research, 123*, 815–835. https://doi.org/10.1002/2017JB014863

Tapponnier, P., & Molnar, P. (1977). Active faulting and tectonics in China. *Journal of Geophysical Research, 82*, 2905–2930.

Telling, J., Lyda, A., Hartzell, P., & Glennie, C. (2017). Review of Earth science research using terrestrial laser scanning. *Earth-Science Reviews, 169*, 35–68. https://doi.org/10.1016/j.earscirev.2017.04.007

Toké, N. A., Arrowsmith, J. R., Young, J. J., & Crosby, C. J. (2006). Paleoseismic and postseismic observations of surface slip along the Parkfield segment of the San Andreas fault. *Bulletin of the Seismological Society of America, 96*(4B), S221–S238. https://doi.org/10.1785/0120050809

Vallage, A., Klinger, Y., Grandin, R., Bhat, H. S., & Pierrot-Deseilligny, M. (2015). Inelastic surface deformation during the 2013 Mw7.7 Balochistan, Pakistan, earthquake. *Geology, 43*(12), 1,079–1,082. https://doi.org/10.1130/G37290.1

Wallace, R. E. (1968). Notes on stream channels offset by the San Andreas Fault, Southern Coast Ranges, California. In W. R. Dickson & A. Grantz (Eds.), *Proceedings of Conference on Geologic Problems of the San Andreas fault system* (Vol. 11, pp. 6–21). Stanford Univ. Publ., Geol. Sci., Univ. Ser.

Wang, T., Wei, S., & Jónsson, S. (2015). Coseismic displacements from SAR image offsets between different satellite sensors: Application to the 2001 Bhuj (India) earthquake. *Geophysical Research Letters, 42*, 7022–7030. https://doi.org/10.1002/2015GL064585

Wechsler, N., Rockwell, T., & Klinger, Y. (2018). Variable slip-rate and slip-per-event on a plate boundary fault: The Dead Sea fault in northern Israel. *Tectonophysics, 722*, 210–226. https://doi.org/10.1016/j.tecto.2017.10.017

Wei, S., Fielding, E., Leprince, S., Sladen, A., Avouac, J.-P., Helmberger, D., Hauksson, E., Chu, R., Simons, M., Hudnut, K., Herring, T., & Briggs, R. (2011). Superficial simplicity of the 2010 El Mayor-Cucapah Earthquake of Baja California in Mexico. *Nature Geoscience, 4*(9), 615–618. https://doi.org/10.1038/ngeo1213

Weldon, R. J., Fumal, T., & Biasi, G. (2004). Wrightwood and the earthquake cycle. *Geological Society of America Today, 14*(9). https://doi.org/10.1130/1052-5173

Weldon, R. J., Fumal, T. E., Biasi, G. P., & Scharer, K. M. (2005). Past and future earthquakes on the San Andreas Fault. *Science, 308*, 966–967.

Weldon, R. J., II., McCalpin, J. P., & Rockwell, T. K. (1996). Paleoseismology of strike-slip tectonic environments. In J. McCalpin (Ed.), *Paleoseismology* (pp. 271–329). Academic Press.

Wesnousky, S. G. (1994). The Gutenberg-Richter or characteristic earthquake distribution, which is it? *Bulletin of the Seismological Society of America, 84*(6), 1940–1959.

Wesnousky, S. G. (2006). Predicting the endpoints of earthquake ruptures. *Nature, 444*, 358–360. https://doi.org/10.1038/nature05275

Wesnousky, S. G. (2008). Displacement and geometrical characteristics of earthquake surface ruptures: Issues and implications of seismic-hazard analysis and the process of earthquake rupture. *Bulletin of the Seismological Society of America, 98*(4), 1609–1632. https://doi.org/10.1785/0120070111

Youngs, R. R., & Coppersmith, K. J. (1985). Implications of fault slip rates and earthquake recurrence models to probabilistic seismic hazard estimates. *Bulletin of the Seismological Society of America, 75*(4), 939–964.

Zhou, Y., Parson, B., & Walker, R. (2018). Characterizing complex surface rupture in the 2013 Mw 7.7 Balochistan earthquake using three-dimensional displacements. *Journal of Geophysical Research, 123*(11).

Zielke, O. (2018). Earthquake recurrence and the resolution potential of tectono-geomorphic records. *Bulletin of the Seismological Society of America, 108*(3A), 1399–1413. https://doi.org/10.1785/0120170241

Zielke, O., & Arrowsmith, J. R. (2008). Depth variation of coseismic stress drop explains bimodal earthquake magnitude–frequency distribution. *Geophysical Research Letters, 35*, L24301. https://doi.org/10.1029/2008GL036249

Zielke, O., & Arrowsmith, J. R. (2012). LaDiCaoz and LiDARimager–MATLAB GUIs for LiDAR data handling and lateral displacement measurement. *Geosphere, 8*(1), 206–221. https://doi.org/10.1130/GES00686.1

Zielke, O., & Strecker, M. R. (2009). Recurrence of large earthquakes in magmatic continental rifts: insights from a paleoseismic study along the Laikipia–Marmanet Fault, Subukia Valley, Kenya Rift. *Bulletin of the Seismological Society of America, 99*(1), 61–70 https://doi.org/10.1785/0120080015

Zielke, O., Arrowsmith, J. R., Grant Ludwig, L., & Akçiz, S. O. (2010). Slip in the 1857 and earlier large earthquakes along the Carrizo Plain, San Andreas Fault. *Science, 327*, 1119–1122. https://doi.org/10.1126/science.1182781

Zielke, O., Arrowsmith, J. R., Grant Ludwig, L., & Akçiz, S. O. (2012). High-resolution topography-derived offsets along the 1857 Fort Tejon Earthquake Rupture Trace, San Andreas Fault. *Bulletin of the Seismological Society of America, 102*(3), 1135–1154. https://doi.org/10.1785/0120110230

Zielke, O., Klinger, Y., & Arrowsmith, J. R. (2015). Fault slip and earthquake recurrence along strike-slip faults—Contributions of high-resolution geomorphic data. *Tectonophysics, 638*, 43–62. https://doi.org/10.1016/j.tecto.2014.11.004

Paleoseismology and Paleogeodesy Using Coral Microatolls

6

Belle Philibosian

Abbreviations

ELW	Extreme low water
ENSO	El Niño-Southern Oscillation
GIA	Glacial isostatic adjustment
GNSS	Global navigation satellite system
HLG	Highest level of growth
HLS	Highest level of survival
InSAR	Interferometric synthetic aperture radar
IOD	Indian Ocean Dipole
RSL	Relative sea level
SLA	Sea-level anomaly
UAV	Uncrewed aerial vehicle

6.1 The Need for Precision and Data Distribution in Paleoseismic Techniques

The growing realization that patterns of earthquake recurrence are complex has made it ever more important to establish the detailed parameters that characterize the extent (rupture length, width, slip distribution) in addition to timing of past earthquakes. Due to the brevity of historical records in comparison to tectonic processes, most fault sections have ruptured in no more than one well-documented

B. Philibosian (✉)
U.S. Geological Survey Earthquake Science Center, Moffett Field, California, USA
e-mail: bphilibosian@usgs.gov

A. Elliott and C. Gruetzner (eds.), *Understanding Past Earthquakes*,
https://doi.org/10.1007/978-3-031-73580-6_6

recent historical earthquake. It was formerly common practice to assume that future ruptures were likely to resemble these historically documented events, and furthermore that ruptures would necessarily terminate at structural discontinuities. While these assumptions are not without merit, various lines of paleoseismic evidence and a few examples of multiple historical ruptures illustrate that not all earthquakes repeat the lengths, endpoints, and slip magnitudes of prior events (Philibosian & Meltzner, 2020, and references therein). Therefore, observations over multiple repeat ruptures are necessary to accurately define the range of behavior for a particular fault (and thus anticipate future events). Rather than expecting all ruptures on a given fault to be similar and recur quasi-periodically, researchers now seek to discover the maximum, minimum, and typical rupture areas of large earthquakes; the distribution, variability, and spatial applicability of recurrence intervals; and patterns of earthquake clustering in space and time. These detailed studies require tools with high temporal and geodetic precision as well as wide distribution of study sites.

In order to assess earthquake recurrence behavior and rupture segmentation along a given fault, the known earthquake chronology must be long enough to span several full-system seismic cycles (in which the entire fault is re-ruptured), and must be established at many locations along the fault. Very long earthquake chronologies (~10 events or more) that are limited to a single point along the fault (e.g., Clark et al., 2013; Klinger et al., 2015) are useful for calculating site-specific recurrence intervals but less informative about fault behavior overall because the rupture extent of each event is not known. Further, the data must be sufficiently temporally precise and spatially dense to delineate individual ruptures. Technically speaking, determinations of rupture extent based on correlating geologic observations from site to site are able to establish only the maximum rupture extent of an individual earthquake, with the caveat that any "single event" could in theory be composed of multiple non-overlapping smaller events within the time resolution of the geologic data. Occam's Razor is generally applied to interpret the smallest number of individual ruptures that are permitted by the data, forming a "maximum rupture model" (e.g., Scharer & Yule, 2020). However, with the emerging knowledge that rupture cascades—temporally clustered series of earthquakes on neighboring fault sections—are common (Philibosian & Meltzner, 2020, and references therein), this assumption may not be strongly justified. As the time resolution of an earthquake chronology becomes more precise, the likelihood increases that an apparent single instance of ground deformation along a fault truly represents a single event, so high-resolution techniques are necessary (with historical records being ideal). Finally, for any detailed chronology, it is also important to evaluate the strength of the evidence of absence of earthquakes; i.e., how confidently the occurrence of earthquakes can be excluded during periods when there is no positive evidence for earthquakes. Geologic or historical records that are likely to have significant temporal gaps do not permit reliable assessment of event recurrence as is possible with continuous records.

Coral microatoll growth is one of the best types of geologic record for paleoseismology and paleogeodesy given these needs. Studying the patterns and rates of microatoll growth offers long, continuous, widely distributed records of centimeter-scale vertical tectonic motion with potentially annual-level temporal precision. The primary limitation is, of course, that a fault must produce significant vertical deformation in a tropical coastal area in order to be studied using this technique. Nevertheless, there are numerous such fault zones worldwide, many of them woefully under-studied since there are few other paleoseismic techniques that may be applied when a fault lies underwater. Furthermore, many of these fault zones are subduction megathrusts, which produce the world's largest earthquakes as well as regionally destructive tsunamis, thereby having an outsize significance in global earthquake hazard. This chapter describes how coral microatolls may be used to understand past earthquakes, with the note that a thorough guide to the practical techniques and potential pitfalls of coral microatoll sampling and interpretation was published by Meltzner and Woodroffe (2015) in a chapter of the *Handbook of Sea-Level Research.* Many considerations mentioned only briefly herein, particularly those in Sect. 6.2, are covered in greater detail in that chapter.

6.2 Microatoll Growth Height as a Proxy for Relative Sea Level

Coral upward growth is ultimately limited by subaerial exposure. The upper surfaces of corals growing in the intertidal zone will eventually die when they are exposed for a sufficient period of time during low tide. Such intertidal reef-building coral colonies, particularly those with massive morphologies that grow radially outward starting from a single polyp, tend to attain a shape with a subhorizontal dead top surrounded by a ring of living tissue (Fig. 6.1). The term "microatoll" was coined for these specimens based on their resemblance to ring-shaped coral reef atoll islands. While coral growth is in actuality limited by duration and intensity of exposure (a coral lifted arbitrarily high above the water for a short time will likely suffer no harm), in practice this limit translates to an elevation (termed the highest level of survival, or HLS) relative to sea level. The significance of coral microatolls as sea-level recorders was first recognized by Scoffin and Stoddart (1978), and since then microatolls have become widely used in climate and oceanography studies and occasionally used in tectonic studies.

Microatolls that are formed in radially growing species that produce annual density bands analogous to tree rings are particularly useful, as these produce records with annual precision. These density bands are often not clearly visible to the naked eye but usually form stark contrasts in x-ray transmission images of thin coral slices. Studies in the Indo-Pacific have predominantly relied upon the common, widely distributed, fast-growing species *Porites lutea* and *lobata* which can grow for hundreds of years, producing colonies meters in diameter. Slower-growing and typically shorter-lived species such as *Goniastrea retiformis* and others formerly classified in the family Faviidae (now reassigned to the families Merulinidae or Lobophylliidae) are also used occasionally. In the Caribbean, *Pseudodiploria strigosa* (brain coral), *Siderastrea siderea*, and *Orbicella sp.* are

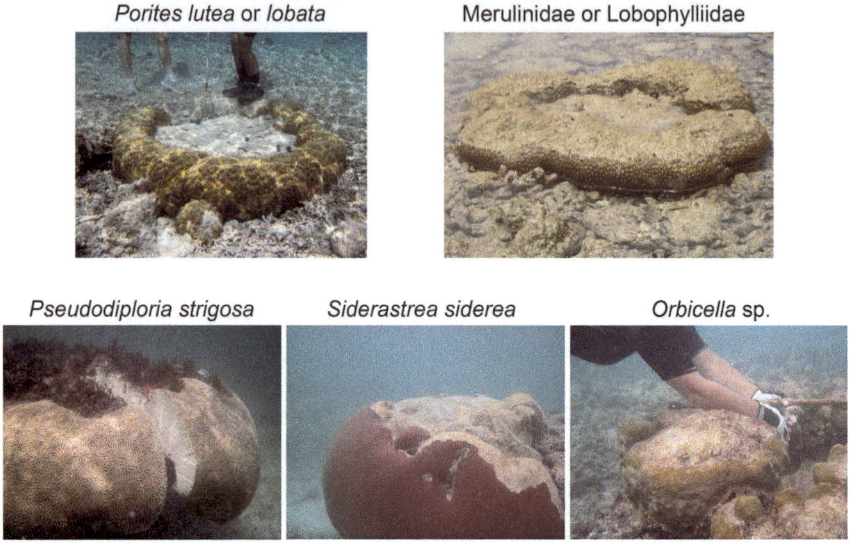

Fig. 6.1 Examples of living microatolls formed from radially growing coral species commonly used in relative sea level (RSL) studies. In addition to *Porites lutea/lobata*, studies in the Indo-Pacific have also used larger-polyp species within the family Merulinidae and the former family Faviidae, the latter of which have been reassigned to either Merulinidae or Lobophylliidae. *Pseudodiploria strigosa* (brain coral) was formerly classified as *Diploria strigosa* and genus *Orbicella* as part of genus *Montastrea* (Budd et al., 2012). Genus *Orbicella* contains the three species *O. annularis, franksi*, and *faveolata*, but these are difficult to distinguish in the field. In general, taxonomic identification is not necessary (and has often not been done rigorously) for tectonic studies—the morphology of any radially-growing microatoll can be interpreted in terms of RSL without knowing the species or genus. *Porites* photo by A. J. Meltzner, Merulinidae / Lobophylliidae photo by K. Sieh, others from Philibosian et al. (2022), licensed under CC-BY 4.0

the most frequently used species. Specimens of each are shown in Fig. 6.1. Each of these species is common and widely distributed throughout its region (Indo-Pacific or Caribbean), so appropriate specimens are likely to be found throughout the global range of reef-building corals (Fig. 6.2).

The basic procedure of interpreting RSL from a microatoll begins with identifying a specimen that tracked sea level over the course of its growth and preserves this record. Such microatolls have subhorizontal dead upper surfaces that have not experienced significant erosion, typically indicated by the presence of multiple concentric rings produced by minor sea-level fluctuations (Fig. 6.3). A microatoll which has had the upper surface planed off by wave erosion or extensively altered by burrowing animals will not accurately preserve an RSL record. A further important consideration is to ensure that the specimen grew in an environment that experienced the lowest tides of the open ocean, rather than in a pool where water remained trapped above low tide level. The HLS of corals that grew in such a pool will not be sensitive to RSL fall below the sill of the pool. Such "ponding" or "moating" can be identified and avoided by observing the site at low tide and/or

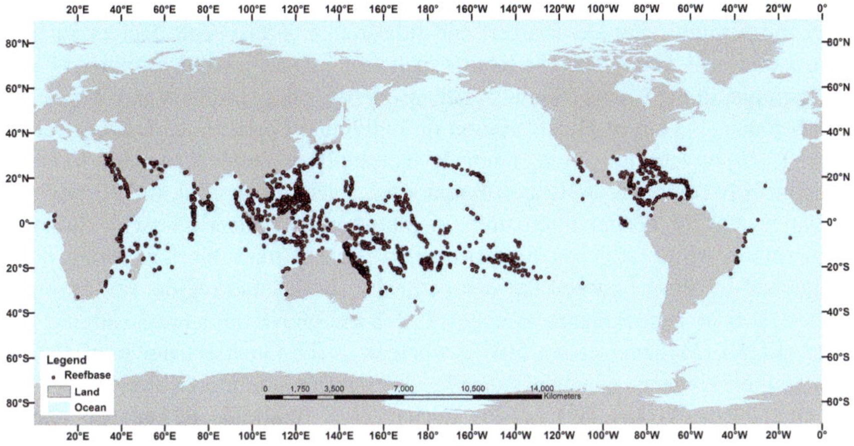

Fig. 6.2 Global distribution of reef-building corals from the National Oceanic and Atmospheric Administration (NOAA), https://oceanservice.noaa.gov/facts/coralwaters.html

Fig. 6.3 Radial slab (~1.5 m long) cut through a microatoll with several preserved concentric rings. Whereas the microatoll has sustained some erosion, the rings indicate the RSL history is preserved. This specimen turned out to be 2500 years old, illustrating that microatolls can preserve RSL records over long periods

surveying the elevation of the reef substrate to confirm that an outlet to the open ocean exists at a level well below the observed HLS.

Provided the above conditions are met, the elevation of the upper surface of the microatoll, from the center to the outer rim, can be interpreted as a time series of HLS, a proxy for RSL. This can be done either approximately by surveying a transect of elevation along a radius and estimating the time component from coral growth rate, or more precisely by collecting a radial slice of the microatoll so that the annual bands may be imaged allowing the height in each year to be measured (Fig. 6.4). However, like most proxy signals, coral HLS is an imperfect recorder of sea level and its particular limitations must be considered for accurate interpretation. First, because the actual controlling factor is exposure, the relation between

HLS and sea level is complex and varies depending on regional factors such as water temperature, climate/weather, and tidal range, site-specific factors such as wave action, and colony-specific factors such as coral species and specimen health. For example, in the Cocos (Keeling) Islands of Australia, Smithers and Woodroffe (2000) found 2–3 cm of HLS variation on individual *Porites lutea/lobata* microatolls, ~10 cm of variation among multiple microatolls at a typical site, and 40 cm of variation between microatolls at different sites. Zachariasen et al. (2000) reported similar results for *Porites* microatolls in the Mentawai Islands west of Sumatra. Furthermore, while HLS can be expected to roughly track the low spring tides, the amount of offset between the two varies by species and region. For example, *Porites* HLS in west Sumatra averages 19 ± 8 cm above the annual extreme low water (ELW) (Meltzner et al., 2010), whereas in the Great Barrier Reef *Porites* HLS is approximately at annual mean low spring tide (Scoffin & Stoddart, 1978), and in the Cocos (Keeling) Islands *Porites* HLS lies midway between annual mean low spring tide and mean low neap tide (Smithers & Woodroffe, 2000). In west Sumatra, the HLS of *Goniastrea retiformis* is typically ~ 10 cm above that of *Porites* (Natawidjaja et al., 2006). In the Lesser Antilles, *Pseudodiploria strigosa* HLS averages 5 ± 5 cm below annual ELW, while *Siderastrea siderea* HLS typically lies ~5 cm below that of *Pseudodiploria* (Philibosian et al., 2022).

The end result of these complexities is that HLS is best interpreted in a relative sense. In order of decreasing precision and accuracy, differences in HLS may be measured over time on an individual coral colony, between colonies of the same/similar species at an individual site, and between colonies of the same/similar species in a given region with similar tidal range and climate. Direct comparisons of HLS between different genera and/or different regions are unlikely to be meaningful. Further, coral HLS can be considered a more reliable and robust proxy for relative sea level when it is measured on a microatoll that is part of a field-documented population of specimens with a similar elevation and growth morphology. The growth of any single coral colony may be affected by individual health or microenvironmental factors, and it is therefore impossible to determine whether a single isolated specimen is representative of sea-level history at that site.

A second major consideration is that coral responds much more quickly to relative sea-level (RSL) fall than RSL rise, since the former kills existing coral (leaving the dead coral skeleton as evidence, termed a "die-down") and the latter merely provides more vertical space for coral to grow. The growth rates of reef-building corals range from a few millimeters to a few centimeters per year. Thus, RSL increases typically must persist for several years or more before they will have a noticeable impact, whereas RSL decreases may have an impact within days or weeks depending on magnitude. Even in a stable RSL environment, this asymmetric response has significant effects due to sea-level anomalies (SLAs). SLAs are additional short-term fluctuations superimposed on the harmonic tidal cycle, which occur due to oceanographic phenomena such as the interannual El Niño-Southern Oscillation (ENSO) and the Indian Ocean Dipole (IOD), extreme storms, or other non-tidal factors. Typical SLA amplitude varies regionally and is generally < 10 cm, though extreme ENSO and IOD events can reach tens of cm (e.g.,

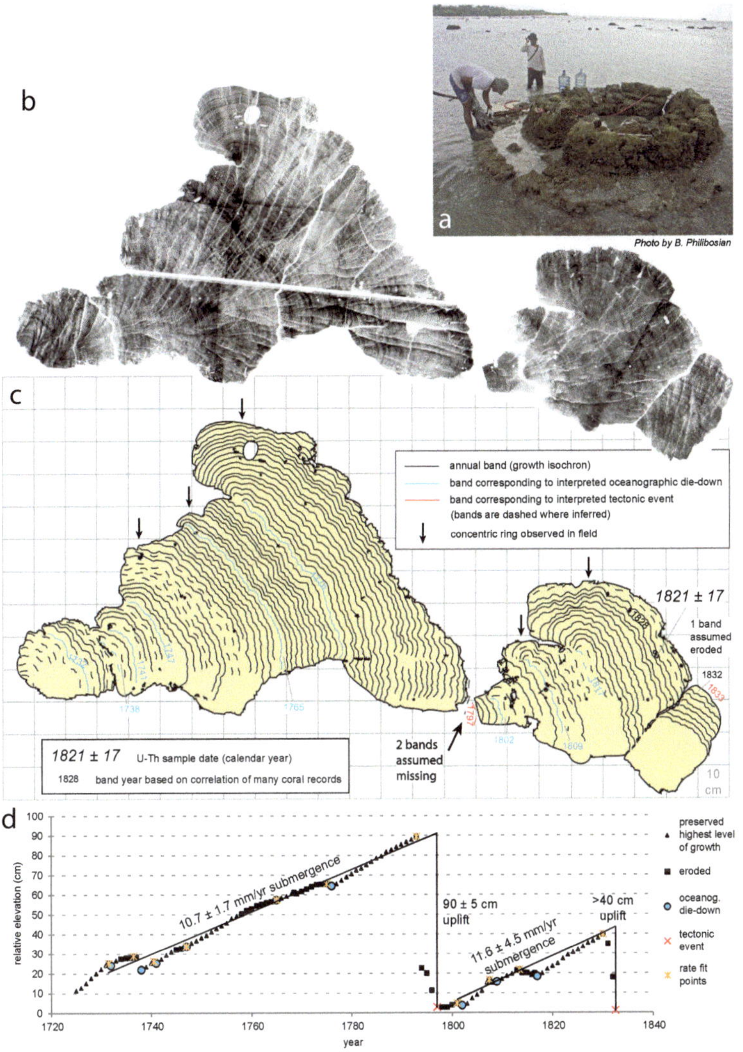

Fig. 6.4 Example microatoll interpretation from Betumonga, North Pagai Island, west Sumatra (adapted from Philibosian et al., 2014). **a** Field photograph of "hat"-shaped fossil microatoll indicating multiple uplift events. Note the lower, outer "brim" of the "hat," just at the water line, on which a person is standing. Slab is being cut with a hydraulic chainsaw. Field of similar microatolls can be seen in the background, indicating that this colony is representative of a population and not an isolated individual specimen. **b** X-ray transmission image mosaic of the collected slice showing annual density banding. Darker areas indicate denser growth. Small holes along horizontal lines were drilled to mark orientation; larger holes were drilled to collect samples for U-Th dating. **c** Interpreted microatoll cross-section. **d** Plot of annual coral growth height with measured uplifts and submergence rates estimated by linear fits to points preceding die-downs. The colony survived the first 90-cm uplift so this amount can be measured directly. The second uplift killed the colony, so only a 40-cm minimum can be measured (total 1833 uplift was estimated using the method shown in Fig. 6.6)

Becker et al., 2012; Webster et al., 1999). These temporary fluctuations typically persist for less than a year, ranging from days (storms) to months (ENSO/IOD). Due to the response asymmetry, it is generally the troughs (but not the peaks) of this quasi-random sea-level fluctuation that affect the annual ELW experienced by corals and thus the annual HLS. A sufficiently strong negative SLA will cause a die-down to a lower-than-normal level if the SLA persists long enough and/or occurs during spring tide.

The coral growth response to RSL fluctuation translates to an asymmetric record of submergence vs. emergence, and the impacts of this asymmetry must be considered when interpreting changes in RSL from changes in microatoll growth height. Because of this asymmetric response, the time series of coral growth height measured from the upper surface of a microatoll is not a straightforward annual record of coral HLS. In reality, growth in each year is limited either by reaching/exceeding HLS (and experiencing a die-down) or by the coral growth rate if HLS is not reached. These latter points (designated HLG for highest level of growth) are identified by convex-upward annual bands (see Fig. 4b and c) and represent lower bounds for HLS. Interannual SLA fluctuation typically produces a coral growth pattern of alternating upward growth periods and die-downs. If the coral surface has experienced some erosion, it may not be possible to distinguish HLG years from die-down years, but preserved concentric rings indicate the approximate timing and magnitude of die-downs (Fig. 4c). Each ring is produced by a die-down, so the elevation of the ring approximates the pre-die-down height while the trough immediately outward from the ring represents the approximate time and elevation of that die-down.

Figure 6.5 schematically illustrates coral growth height in response to sudden and gradual RSL rise and fall in the presence of random fluctuation from SLAs. Each annual height is classified as an HLG point or a die-down. Large die-downs immediately record the magnitude of sudden emergence events (years 17 and 66), but sudden submergence events are initially marked only by unconstrained upward growth (years 34 and 83). The magnitude of a submergence event is not constrained until the coral growth catches up to HLS, which may take a number of years in an environment of long-term emergence (years 83–90) and even longer in a submerging environment (years 34–47). When measuring rates of RSL change over many years, arbitrarily high rates of emergence will be recorded precisely, as the coral will experience a die-down nearly every year (unless and until the entire coral colony dies). In contrast, the precision of recorded submergence rates decreases as the submergence rate increases, because the coral will experience die-downs progressively less frequently (i.e., only during sufficiently strong negative SLA events). Note that in Fig. 6.5, during years 18–34 (gradual submergence) there are 6 die-downs, whereas during years 67–83 (an equal period of emergence) there are 13 die-downs. At the extreme, if the sustained submergence rate outpaces the coral growth rate, the coral will never experience a die-down and will simply grow radially without forming a microatoll. In such a situation, coral records can only place a lower bound (the coral growth rate) on the rate of sea-level rise.

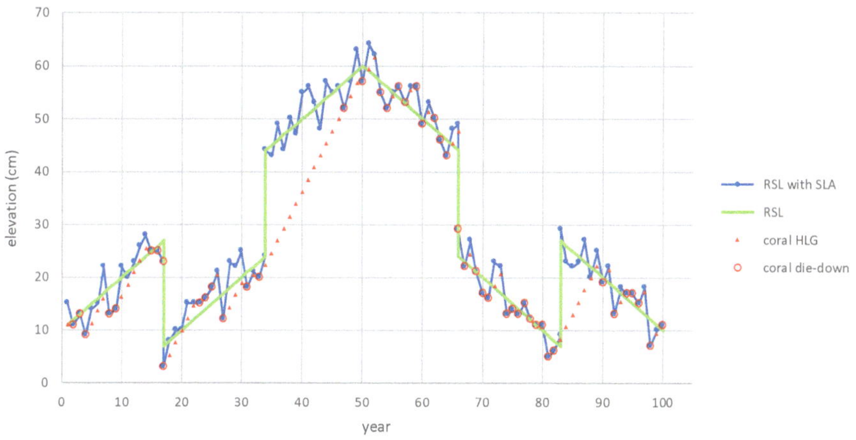

Fig. 6.5 Synthetic illustrative relative sea level (RSL) history and resultant microatoll growth record. Green line shows RSL history with steady 1 cm/yr submergence over the first 50 years and equivalent emergence over the second 50 years, with a sudden 20-cm emergence and submergence event during each period. Blue line with markers shows the same RSL curve with random ±5 cm interannual sea-level anomaly (SLA) fluctuation added. Red markers show response of a coral with 2.3 cm/yr upward growth rate; triangles indicate years in which height is limited by growth rate (highest level of growth, HLG) and circles show years when it is limited by sea level (die-downs). Note that in reality, the coral growth curve would be vertically offset from mean sea level by some amount dependent on the tidal range, coral species, etc.

The ultimate impact of the "noise" from SLA fluctuation is that a rate of gradual RSL change can be measured accurately only as an average over a period with many HLS hits (several years in emerging environments and typically a decade or more in submerging environments). Various strategies have been employed for deriving these rates using linear fits to coral growth height records, including fitting all points on the growth curve (HLGs, die-downs, and eroded points), fitting all non-eroded points, or fitting only die-downs since these are the actual HLS hits. However, as the die-down years typically do not represent mean HLS but actually negative SLA events, the emerging preferred method (first devised by Meltzner et al., 2010) is to fit the highest points immediately preceding die-downs, i.e. the tops of the concentric rings found on the upper surface of the microatoll (see Fig. 4d for an illustration). While these points are often slightly eroded, they represent the best estimates for mean HLS. The point preceding the first die-down in a coral growth history may be excluded if it is below the general trend, as the coral may not have reached mean HLS prior to being affected by a large negative SLA.

A final complicating factor to consider when estimating RSL trends are harmonic tidal cycles, which occur at various intervals due to resonances in the respective orbits of the Earth and Moon. Cycles with periods ≤ 1 year generally have little or no impact on coral-based RSL histories since the temporal precision is roughly annual. However, longer-period harmonics such as the 18.61-year lunar nodal cycle may have noticeable effects. Although tidal cycles do not alter mean sea level, they affect the amplitude of the tidal fluctuation which controls annual ELW, making coral die-downs more likely during the phase of the cycle where tidal range is increasing or at its peak (Meltzner et al., 2017). Along with SLAs, this is a factor that adds "noise" to the signal from monotonic trends in RSL change, with the result that such rates are most accurately measured by averaging over decades.

6.3 Extracting Tectonic Vertical Deformation from Microatoll RSL Records

Coral-based RSL records are not only excellent recorders of sudden, presumably coseismic vertical deformation events, but due to their continuous nature they also demonstrate that no significant coseismic deformation occurred at any other time, as well as tracking gradual (interseismic) tectonic deformation. However, there are several interpretive steps that must be taken to extract tectonic signals from an RSL history, in addition to the considerations associated with translating coral growth history into RSL described in Sect. 6.2. An RSL history represents the summation of tectonic and non-tectonic land-level changes with oceanographic and climatic sea-level changes. To isolate the tectonic signal, all other factors must be removed, including land-level change due to volcanic processes or loading, coastal subsidence due to sediment compaction or groundwater withdrawal, glacial isostatic adjustment (GIA), trends in eustatic/steric sea level, and SLAs. I will discuss each of these in turn, indicating how they may be disambiguated from tectonic signals.

If the coral study site is located on a volcanic island or near a volcanic edifice, an effort must be made to estimate and remove the land-level change due to active volcanic inflation/deflation and long-term island subsidence due to volcanic loading. Coastal subsidence related to sediment compaction or groundwater extraction must similarly be addressed; however, this factor is usually not a major concern for coral studies since compaction is most significant in the thick sedimentary sequences found in large river delta complexes, and corals are generally intolerant of freshwater input. Such non-tectonic land-level changes, at least in the modern period, may be independently estimated using terrestrial geodetic instruments such as global navigation satellite system (GNSS) receiver stations and subsequently removed from the measured RSL trend. However, this technique requires the signal to be sufficiently regionally homogeneous to affect geodetic instruments outside the influence of the expected tectonic signal. Volcanic deformation and delta subsidence tend to be more local and may be challenging to isolate in that manner. If non-tectonic land-level change is expected to be significant but local, another method must be used, such as estimating volcanic deflation based on eruptive volumes or the likely compaction rate based on sediment thickness. Rates for periods long before present may be very difficult to determine; an order-of-magnitude estimate that increases the uncertainty of the tectonic rate may be the best that can be done.

GIA-driven RSL change is unevenly distributed around the globe with the highest modern rates (up to ~1 cm/yr) concentrated in the polar regions (e.g., Peltier, 1999). Coral microatolls of course do not occur in polar regions, but even in the tropics GIA of a few mm/yr may be a concern. GIA includes a complex suite of processes that affect both absolute sea level and land level, with transitions along continental margins (Mitrovica & Milne, 2002). Thus, the GIA-driven RSL change at a given site is the net summation of these factors, which is less straightforward to estimate than pure land-level change and may be especially difficult to determine for a site in the vicinity of a continental margin. The modern land-level change component of GIA may be estimated in a given region using geodetic instruments outside the influence of the expected tectonic signal, but this method will not provide estimates of the sea-level change component of GIA (nor can this method be applied to periods long before present). Order-of-magnitude estimates of total GIA may be obtained from global models (e.g., Mitrovica & Milne, 2002) and removed from RSL records if their magnitude is not insignificant. More accurate estimates of GIA-driven RSL change, particularly for time periods > 1 ka, require data compilation and modeling efforts far beyond the scope of a typical tectonic study and are generally done as part of regional total RSL-change studies that also include eustatic/steric contributions (e.g., Khan et al., 2017). In summary, GIA is non-trivial to determine independently, but in the modern period is limited to a few mm/yr in the regions where coral reefs grow and is a broad-scale signal that will typically affect all sites in a study area essentially equally. Thus, GIA can be confidently excluded as a cause of disparate rates of RSL change between nearby sites.

Eustatic/steric sea-level change is generally a more significant concern than GIA or non-tectonic land-level change. This issue is most significant after the year 1900, as anthropogenic global sea-level rise averaged 1.7 mm/yr over the twentieth century (Church & White, 2006) and reached ~3 mm/yr in the late 20th and early twenty-first century, with ongoing acceleration (e.g., Nerem et al., 2018). These rates far exceed all other estimated rates of global sea-level change since post-glacial sea levels stabilized in the mid-Holocene. Rates of sea-level change also vary regionally (e.g., Church et al., 2004) and regional trends can be estimated from tide gauge data (again, outside the tectonically deforming area, unless land-level change is independently estimated with a co-located GNSS instrument) or satellite ocean altimetry data. The temporal extent of tide gauge data varies and some regions still have few stations, but many records cover the twentieth century and a few extend back into the nineteenth century, whereas satellite ocean altimetry data are available only since the early 1990s. Coral microatoll records from outside the deforming area may also be used to estimate sea-level change (e.g., Majewski et al., 2022). For a tectonic study, the regional rate of sea-level change should be estimated as precisely as possible given the available data and then removed from the RSL trend to obtain the tectonic deformation rate. For time periods between the mid-Holocene (~6 ka) sea-level stabilization and the year 1900, regional constraints are generally poor and it is reasonable to assume zero sea-level change in the absence of other information, as global rates averaged < 1 mm/yr since 6 ka and even lower since 2 ka (Fleming et al., 1998).

Due to corrections for perhaps imprecisely known rates of non-tectonic land-level change or absolute sea-level change, absolute rates of tectonic land-level change may have relatively large uncertainties. However, if the non-tectonic contributions to RSL change are expected to be homogeneous over large regions (e.g., hundreds of km), as is often the case, spatial variations between contemporaneous rates over shorter length scales (e.g., ~100 km) can be attributed to tectonic deformation. Furthermore, the differences between rates can be computed more precisely, without any contribution to the uncertainty from corrections for non-tectonic effects.

As noted in Sect. 6.2, interannual "noise" from SLAs means that rates of RSL change can be measured only as decade-long averages. An additional common challenge in interpreting coral microatoll slabs is distinguishing small to moderate die-downs (less than ~15 cm) caused by extreme negative SLAs from those due to tectonic uplift. Larger die-downs almost always require uplift, as SLAs are rarely large enough to cause die-downs of > 15 cm (Meltzner & Woodroffe, 2015), but the cause of smaller die-downs can be ambiguous. Specifically, a sudden (<1 yr) tectonic uplift, such as from an earthquake or slow earthquake, that is followed by a years-long rapid subsidence (e.g., postseismic relaxation), may be indistinguishable from a negative SLA of equivalent amplitude, since in either case the coral will take some time to recover at its growth rate. For example, a major die-down in 1961–2 and subsequent recovery observed in many west Sumatran microatolls was initially interpreted as a paired aseismic tectonic uplift and subsidence event, though the coincidence with a known major IOD event was noted (Natawidjaja et al., 2004). However, with a more spatially distributed microatoll dataset, it became clear that the 1961–2 event affected nearly every specimen in the entire region and in general the corals each recovered to the same HLS as before the event, indicating that the observed die-down was likely due to the SLA alone with no tectonic contribution necessary (Philibosian et al., 2014). Uninterrupted, complete recovery after a die-down is the best distinguishing factor of an SLA vs. a tectonic uplift. While it is not uncommon for a period of rapid postseismic subsidence to follow a coseismic uplift, the magnitude of the subsidence is usually significantly less than the initial uplift, meaning that the coral will hit a lower HLS in the aftermath of a permanent sudden uplift rather than having unimpeded upward growth until it attains its pre-event HLS.

In comparison to gradual (e.g., interseismic) tectonic deformation rates, sudden (e.g., coseismic) tectonic deformation is relatively straightforward to measure from coral microatolls. Sudden uplifts are the simplest to interpret, as the coral will die down immediately and the uplift can be measured directly by comparing the pre- and post-die-down elevations. If the uplift is large enough that the entire coral colony dies, the height of the previously living outer perimeter surface provides a minimum uplift constraint. If post-uplift microatoll records exist, the total

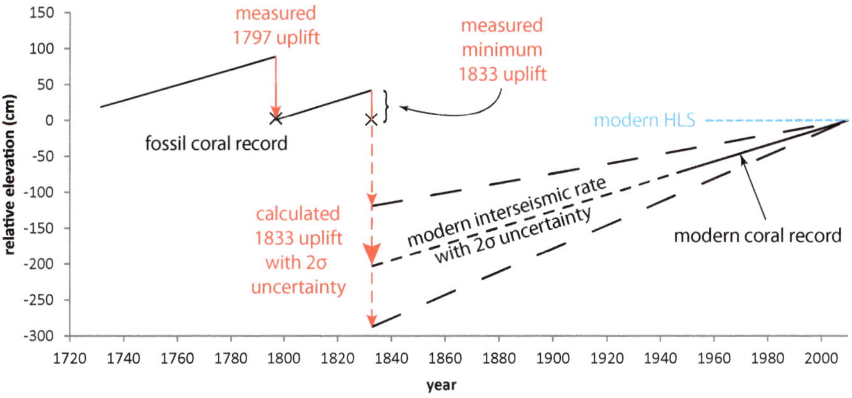

Fig. 6.6 Example schematic calculation of total coseismic uplift for a case where the coral colony was killed (in 1833). Adapted from Philibosian et al. (2014)

uplift may be estimated by projecting the later submergence/emergence rate from a known HLS back to the time of the uplift, although this technique assumes the rate remained unchanged during the record hiatus. This is not an unreasonable first-order assumption in the absence of other information, but interseismic deformation rate changes in response to earthquake ruptures (e.g., Philibosian et al., 2014) or even in the absence of any obvious trigger (e.g., Meltzner et al., 2012, 2015) have been documented and must always be considered possible. Figures 6.4 and 6.6 show an illustrative example of how tectonic uplifts that did and did not kill the colony, as well as interseismic subsidence rates, were measured from a *Porites* microatoll in the west Sumatra region.

One potential pitfall of interpreting large uplift events from mass coral death bears noting: if all microatolls at a particular site die completely at a particular time, the cause may be an environmental mass mortality event unrelated to RSL change, such as the extreme red tide in west Sumatra in 1997–8 (Abram et al., 2003). Such events may be distinguished from sudden tectonic uplift by a few characteristics: dead coral surfaces may be immediately recolonized at their prior living elevation after an environmental mortality event, whereas a tectonic uplift will reset HLS to a lower elevation for years following. Also, complete reef death at a particular site is very likely due to tectonic uplift if it is contemporaneous with clearly tectonic uplift (large die-downs that reset HLS but do not completely kill the microatoll) at other sites in the region; complete death may occur in the peak uplift area but lesser amounts of uplift will occur around the periphery. Environmental mass mortality events were likely quite rare in the past, though they are becoming more common due to climate change (e.g., Hughes et al., 2018).

Sudden subsidence events can be identified based on the initiation of uninhibited upward growth from a former HLS, but the amount of subsidence cannot be measured until the coral reaches HLS again, which may take a very long time in an environment of gradual submergence. (For a recent subsidence event, as in an immediate post-earthquake survey, the amount can be estimated by comparing the current sea level to the pre-event HLS as in Briggs et al., 2006, but this cannot be done for paleo-events.) Furthermore, the measured subsidence will include whatever deformation (interseismic, postseismic, or other coseismic events) occurred during the intervening period and will require correction for an assumed rate of non-tectonic RSL change over that period. The latter rate (and interseismic contribution) can reasonably be estimated from adjoining periods during which the coral was tracking RSL, but there is necessarily greater resultant uncertainty for subsidence events than for uplifts. See Philibosian et al. (2012) for an example of a microatoll study focusing on a tectonic subsidence event; in that case the coral growth did not re-attain HLS before a subsequent uplift event killed the colony 35 years later. Therefore, the measured subsidence was merely a minimum and had additional uncertainty due to necessary assumptions about the interseismic RSL change during that period.

One final consideration to note is that a large sudden uplift or subsidence event that respectively kills or submerges entire reef platforms may produce a hiatus in recording. In the case of uplift, if all formerly intertidal microatolls are killed, it may be decades or longer before new colonies grow and form microatolls that track RSL. In the case of subsidence, if all formerly intertidal microatolls become subtidal, a sufficiently large subsequent uplift might still be recorded (e.g., Philibosian et al., 2012) but otherwise the hiatus will persist until drowned microatolls grow back up to HLS, or new colonies are established at shallower depths. These effects may mask the occurrence of later earthquakes in a local sequence. For example, the 1797 Sumatran Sunda Megathrust rupture killed reefs in the northern Mentawai Islands which were then unavailable to record the second major event in the sequence which occurred in 1833 (Philibosian et al., 2014). In that case the precise earthquake dates are known from historical records and it was possible to estimate the 1833 uplift in the northern Mentawais based on the uplift gradient observed in the southern Mentawais. In contrast, a nearly 500-year hiatus with no preserved microatolls followed a large c. 1450 uplift on northern Simeulue Island, and the lack of historical or other constraints leaves many possible scenarios for tectonic events and deformation during that period (Meltzner et al., 2010).

6.4 Temporal Precision

Coral microatolls have several major advantages in terms of temporal precision. First, for coral colonies that are sampled many years after death (termed "fossil" corals though they are generally not recrystallized as true fossils would be), absolute age uncertainties of a decade or less can commonly be obtained through uranium–thorium disequilibrium dating (Chiang et al., 2023; Edwards et al., 1988; Shen et al., 2002, 2008, 2012), which is generally much more precise than radiocarbon dating since there is no need to account for atmospheric calibration or the marine reservoir effect. Second, provided a cross-sectional slice of the microatoll is collected and x-rayed, the annual bands can be identified and counted, such that time intervals between events recorded in a single continuous coral record typically have uncertainties of less than a year. For a coral sampled with a living outer surface (or known date of death), the ages of events earlier in its growth history can therefore be determined with annual precision, subject only to uncertainties in the band counting and not to any uncertainties associated with radiometric analysis. If no slice is collected, it is still possible to estimate the coral growth chronology by measuring an elevation transect across the top of the microatoll and assuming an average radial growth rate for the species (or interpolating between the ages of two radiometrically dated samples), but this method delivers far less precision in both the coral growth height and the age.

The precisely known *relative* ages of RSL-changing events provide an additional tool for *absolute* age determination. Large-amplitude SLAs are typically regional in extent, causing notable contemporaneous die-downs on many corals throughout a study area. Tectonic die-downs occur less frequently but similarly typically affect corals at numerous sites contemporaneously. Therefore, fossil coral records which overlap in absolute age may be more precisely correlated with each other by matching the "bar code" sequence of regional oceanographic and tectonic die-downs (Meltzner et al., 2012, 2015; Philibosian et al., 2014). Provided the microatoll upper surfaces preserve enough die-downs, this correlation method is in general more reliable than relying on the assumption that the outer perimeters of fossil corals with similar radiometric ages represent contemporaneous living surfaces at the time of death (as was done by Natawidjaja et al., 2006). While the coral deaths are indeed likely contemporaneous, most fossil microatolls are at least slightly eroded and some unknown number of annual bands are missing from the outer surface of the sampled slice. Therefore, the youngest preserved band on a microatoll may have formed several years before the colony died—a critical distinction when counting bands with annual precision.

It is important to note that even with the best practices and circumstances, the coral technique, though extremely precise by geologic standards, would likely conflate earthquakes on neighboring fault segments that occurred within a few years of each other. Thus, a rupture cascade within a short time interval (such as the 2004 M9.1 Aceh–Andaman and 2005 M8.6 Nias–Simeulue earthquakes on the Sunda Megathrust) might be interpreted as a single event, and those very closely spaced in time (such as the M8.4 and M7.9 Mentawai Sunda Megathrust earthquakes in 2007, which were separated by only 12 h) almost certainly would be. However, cascading events separated by at least five years, such as the 1797/1833 doublet and the five-event 17th-century cascade in the Mentawais, can generally be distinguished using the microatoll technique (Philibosian et al., 2014, 2017). Five years of growth is typically sufficient to exceed uncertainties in coral record correlation and interpretation due to erosion or unclear annual banding; events separated by as little as 4 years have been distinguished in some records (e.g., Philibosian et al., 2017).

6.5 Inferring Fault Slip and Coupling Parameters from a Vertical Deformation Field

Coral microatoll-based observations provide not only remarkably precise ages of tectonic events, but are well-suited to record the vertical deformation field for these events as well as during other stages of the earthquake cycle. Synthesis of contemporaneous vertical surface deformation across many sites in a region can be used to constrain fault slip parameters at depth. These methods of inference are by no means unique to coral data and entail the same assumptions and limitations that apply to interpretation of deformation fields from GNSS station networks, interferometric synthetic aperture radar (InSAR), etc. As for any type of geodetic data, forward or inverse modeling of the surface deformation field can infer the extent, distribution, and magnitude of slip or coupling on an underlying fault interface. The broader the spatial distribution of sites, the better constrained the earthquake rupture or fault coupling parameters will be. Although coral-based observations are limited to the vertical component of deformation and are less temporally precise and have greater measurement uncertainty than modern geodetic instruments, their great value lies in their potentially broad natural distribution across an area of tectonic deformation. The rupture extent and slip distribution of recent Sumatran subduction earthquakes were far better constrained with the addition of distributed microatoll data than they could have been with the few instrumental geodetic records alone (Briggs et al., 2006; Konca et al., 2008; Morgan et al., 2017, 2020). For paleoevents, of course, there are no instrumental geodetic records at all, yet coral microatoll data can provide a deformation field and resultant fault rupture parameters with almost the same level of detail (Philibosian et al., 2014, 2017).

It also bears reiterating that unlike most other types of paleoseismic data, absence of evidence for vertical deformation in a microatoll record is effective evidence of absence. For example, in a paleoseismic trench study, there is generally a possibility of missed earthquakes due to a hiatus in sedimentation or strata being removed by erosion, so it is often difficult to demonstrate definitively that a particular earthquake rupture did *not* extend to a given site. This inhibits the determination of rupture extent for individual events. In contrast, the continuous growth of a microatoll will definitively preserve evidence of every vertical deformation event that occurs during its lifetime. Therefore, microatoll records exhibiting zero deformation at the time of an earthquake recognized at other sites provide definitive limits on rupture extent for that event.

Coral microatoll techniques may be applied in any tectonic setting which involves significant vertical deformation, including compression/thrust faulting, extension/normal faulting, and oblique strike-slip faulting. Subduction zones are the most abundant such environment in the proximity of coastlines and therefore the vast majority of tectonic microatoll studies have been in subduction settings, although some have been done in an oblique strike-slip setting in Haiti (Hayes et al., 2010; Weil-Accardo et al., 2016b). The expected vertical deformation patterns for the interseismic and coseismic phases of the subduction earthquake cycle

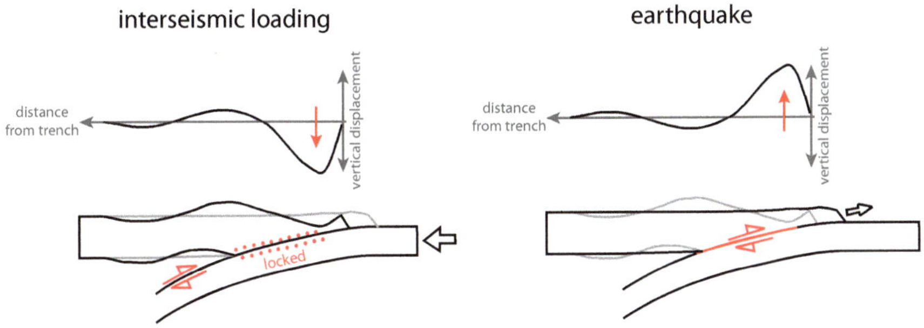

Fig. 6.7 Schematic diagram of elastic deformation during the interseismic and coseismic phases of a subduction earthquake cycle (after Savage, 1983). Postseismic deformation is more variable, but often includes short-term subsidence in the coseismically uplifted area due to afterslip at the periphery of the rupture area and longer-term low-amplitude regional deformation due to viscoelastic relaxation. From Philibosian et al. (2022), licensed under CC-BY 4.0

are shown in Fig. 6.7. During interseismic loading, the deformation field is dominated by subsidence above the updip edge of the locked seismogenic zone, with more minor uplift above the downdip edge farther from the trench. More detailed modeling can be used to infer the distribution and degree of coupling along both strike and dip. If deformation is fully elastic, the coseismic deformation, dominated by uplift nearer the trench with more minor subsidence farther away, will be the inverse of the interseismic field. These peaks of uplift and subsidence roughly outline the along-dip rupture extent, with the along-strike rupture extent indicated by the limits of deformation along that axis. Thus, the long-term gradual subsidence and uplift measured from corals yields the interseismic fault behavior while sudden deaths or drownings of coral define earthquake ruptures. Transient post-seismic deformation, typically logarithmically decaying with time after a rupture, is often too brief to be captured well by annual-resolution coral records with SLA noise, but longer-term effects such as viscoelastic relaxation or alteration of the fault coupling distribution may be recorded by coral microatolls.

6.6 Logistical Advantages and Disadvantages

One major advantage of coral microatolls as a paleoseismic tool is the relative logistical simplicity of establishing chronologies at many locations along a fault, highly desirable for establishing the extent and slip distribution of individual earthquake ruptures. Studies of the Sumatran Sunda Megathrust by a single research group, over a period of ~20 years including four Ph.D. theses, established 96 paleoseismic study sites distributed along ~1000 km of the fault (Fig. 6.8), with an additional 32 sites that documented 20th-century interseismic deformation only. In comparison, numerous groups working along the 1100-km-long San Andreas fault in California over the past ~40 years have established 37 paleoseismic trench study sites, with each site requiring at least one graduate thesis or equivalent analysis

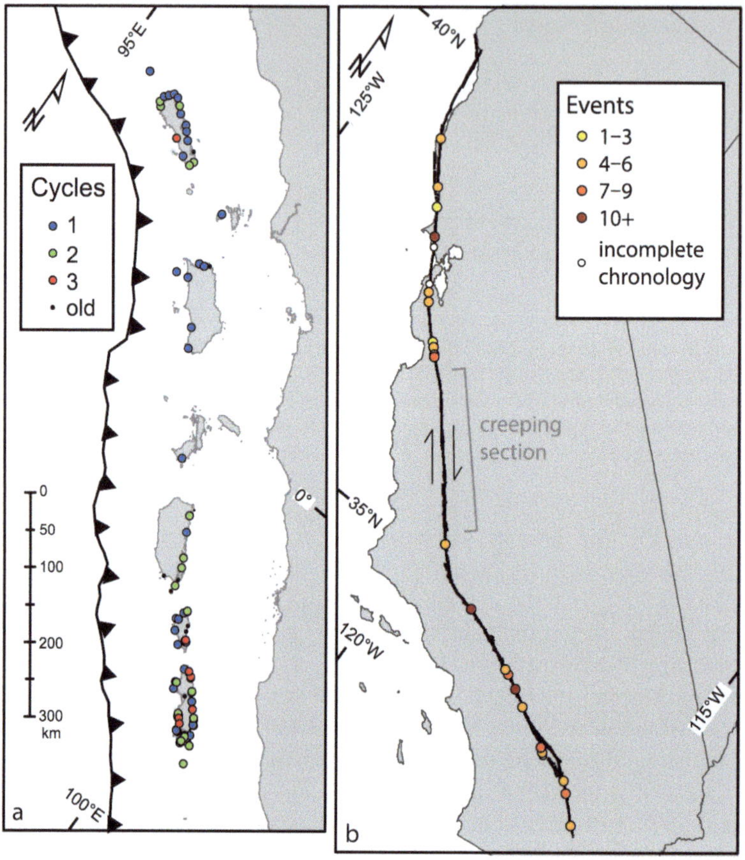

Fig. 6.8 Site density and earthquake chronology length comparison for **a** coral microatolls in west Sumatra and **b** trench studies on the San Andreas fault in California, each the best-developed example of the respective paleoseismic technique. Both maps are shown at the same spatial scale. Each "cycle" along the Sumatran margin includes multiple individual earthquakes that dominantly rupture different sections of the fault. Data at "old" Sumatra sites is limited to times prior to the most recent three cycles. San Andreas event tallies (compiled by Scharer & Streig, 2018) at some sites may overestimate the number of full system cycles if tail ends of neighboring ruptures overlap. Figure adapted from Philibosian and Meltzner (2020), licensed under CC-BY 4.0

effort. While the San Andreas trench sites typically have longer records covering more earthquake cycles than the Sumatran microatoll study sites (recurrence intervals on the two faults are similar), record length cannot substitute for the site density and precision required to confidently establish rupture extent for individual earthquakes.

Major disadvantages of the technique are the logistical complexity and expense associated with sampling sites that may be accessible only by boat, and with the series of specialized equipment and facilities required for cutting, thin-slicing, and x-raying coral slabs (see Meltzner & Woodroffe, 2015 for details). There are significant economies of scale with these procedures such that for a study involving many sites the cost per site is lower than for fault trench studies, but for one or two sites the coral microatoll technique will likely be more expensive and logistically difficult than an equivalent number of trench studies. Additionally, sampling coral (even dead coral rock) and transporting coral across international borders both typically require specific government permits. It is generally simplest if the coral slab processing can be done in the same country as the slabs are collected.

Finally, it must be noted that the primary factors controlling the success of the method are the abundance of coral microatolls and the degree of preservation of fossil colonies. Microatolls are most abundant on fringing reef platforms or lagoons that are no more than a few meters below mean sea level; other coastal environments are not conducive to microatoll formation. Additionally, even among areas with appropriate environments, microatoll abundance varies. Fields containing dozens of microatolls (either living or fossil) are relatively common in the west Sumatra region, whereas in the Lesser Antilles microatoll fields typically have fewer individual colonies and fossil microatolls seem to be essentially absent (Philibosian et al., 2022). The cause of this discrepancy is not known; possibilities include weaker long-term resilience of the skeletons of Caribbean coral species in comparison to Indo-Pacific species, more frequent destruction/erosion of Caribbean reefs by tropical cyclones (which do not occur in Sumatra), or greater human modification of Caribbean reefs (Antillean islands have greater population density than those in the Sumatran outer arc, and coral rock is often harvested as a building material). Differing sea-level trends may also be implicated: the continuously rising late Holocene sea level in most of the Caribbean (~0.5 mm/yr in the Lesser Antilles; Khan et al., 2017) may have inhibited the formation of microatolls if coral growth was unable to keep pace, and any older microatolls that did form would now be underwater where they are difficult to find as well as subject to disturbance by waves. In contrast, equatorial regions such as Sumatra generally experienced a mid-Holocene highstand with subsequently stable or falling sea level (e.g., Khan et al., 2015), although the pervasive tectonic deformation inhibits direct measurement of Holocene paleo sea level in Sumatra specifically (Briggs et al., 2008).

Even in regions where microatolls are abundant, extensive reconnaissance of coastlines is typically required to locate promising study sites. As they are inherently submerged most of the time, microatolls are most easily spotted and marked from above rather than from boat-based or land-based surveys, and at no more than a few meters' diameter they cannot be resolved on most satellite and aerial imagery (though shallow reef platforms and lagoons that potentially host microatolls can easily be identified that way). Very high-resolution aerial imagery or surveys from low-velocity, low-altitude aircraft such as helicopters are the most effective ways to find microatoll study sites. Human-crewed flights, though very expensive and logistically challenging, are the most efficient as researchers can observe potential study sites from multiple angles and make repeat passes to gather more information. However, photographic surveys via uncrewed aerial vehicles (UAVs) have emerged as a potentially much cheaper viable alternative as UAV availability has expanded, technology has improved, and costs have dropped. Small hand-transportable UAVs can easily survey kilometers of coastline in a single flight, though larger craft are still necessary for long-range regional canvassing.

6.7 The Impact of the Coral Microatoll Technique on Tectonic Studies

The coral microatoll technique was first applied to tectonic research by Taylor et al. (1987), who used microatolls to constrain coseismic deformation that had occurred during several 20th-century earthquakes along the Vanuatu subduction zone. The tectonic application of microatolls was subsequently further developed and used extensively to establish the history of coseismic and interseismic deformation along the Sumatran Sunda Megathrust (Meltzner et al., 2010, 2012, 2015; Morgan et al., 2017, 2020; Natawidjaja et al., 2004, 2006, 2007; Philibosian et al., 2012, 2014, 2017; Sieh et al., 1999, 2008; Tsang et al., 2015a, 2015b; Zachariasen et al., 1999, 2000). The majority of the best practices described in this chapter and by Meltzner and Woodroffe (2015) were established over the course of the Sumatran tectonic studies. The technique is gradually spreading to other tropical tectonic study areas, having been applied with varying degrees of success to subduction paleoseismology and paleogeodesy along the northernmost Sunda (Arakan) margin (Mondal et al., 2018; Shyu et al., 2018), the Solomon Islands (Thirumalai et al., 2015), the Lesser Antilles (Philibosian et al., 2022; Weil-Accardo et al., 2016a), and the Ryukyu arc (Debaecker et al., 2023; Weil-Accardo et al., 2019), with nascent application in the Philippines (e.g., Li et al., 2021; Sarkawi et al., 2021, 2022). Microatolls have also served as vital recorders for vertical deformation during recent earthquakes in the Solomon Islands (Taylor et al., 2008) and Haiti (Hayes et al., 2010; Weil-Accardo et al., 2016b), as well as those along the Sumatran Sunda Megathrust.

In conclusion, successful coral microatoll studies ultimately provide maps of tectonic vertical deformation with annual precision. This precision allows individual earthquakes that occurred only a few years apart to be distinguished in the geologic record, and demonstrates that no earthquakes with significant vertical deformation occurred during parts of the record that lack sudden RSL changes. For each individual earthquake rupture, the distribution of deformation can be modeled to obtain the rupture extent and fault slip distribution, using the same procedures as would be done for any other type of geodetic data such as from GNSS stations. Similarly, gradual interseismic deformation can be modeled to estimate the distribution of fault coupling. The potentially broad spatial distribution of sites is particularly useful for studying subduction megathrust faults, as fault plane properties and earthquake rupture patterns often vary significantly along dip (Kanamori, 2014; Lay et al., 2012). Obtaining such precise parameters for individual paleo-earthquakes is vital for assessing fault behavior including the persistence of rupture segmentation, recurrence patterns, and similarity of successive events, key issues in both fault mechanics and hazard assessment.

Acknowledgements I thank the editors for inviting me to contribute this chapter. Numerous improvements were made based on suggestions from the editors and a thorough review by Aron Meltzner.

References

Abram, N. J., Gagan, M. K., McCulloch, M. T., Chappell, J., & Hantoro, W. S. (2003). Coral reef death during the 1997 Indian Ocean Dipole linked to Indonesian wildfires. *Science, 301*, 952–955.

Becker, M., Meyssignac, B., Letetrel, C., Llovel, W., Cazenave, A., & Delcroix, T. (2012). Sea level variations at tropical Pacific islands since 1950. *Global and Planetary Change, 80–81*, 85–98. https://doi.org/10.1016/j.gloplacha.2011.09.004

Briggs, R.W., Sieh, K., Amidon, W.H., Galetzka, J., Prayudi, D., Suprihanto, I., Sastra, N., Suwargadi, B., Natawidjaja, D., & Farr, T. G. (2008). Persistent elastic behavior above a megathrust rupture patch: Nias island, West Sumatra. *J Geophys Res 113*, Article B12406. https://doi.org/10.1029/2008JB005684

Briggs, R. W., Sieh, K., Meltzner, A. J., Natawidjaja, D., Galetzka, J., Suwargadi, B., Hsu, Y., Simons, M., Hananto, N., Suprihanto, I., Prayudi, D., Avouac, J.-P., Prawirodirdjo, L., & Bock, Y. (2006). Deformation and slip along the Sunda megathrust in the great 2005 Nias-Simeulue earthquake. *Science, 311*, 1897–1901. https://doi.org/10.1126/science.1122602

Budd, A. F., Fukami, H., Smith, N. D., & Knowlton, N. (2012). Taxonomic classification of the reef coral family Mussidae (Cnidaria: Anthozoa: Scleractinia). *Zoological Journal of the Linnean Society 166*, 465–529. https://doi.org/10.1111/j.1096-3642.2012.00855.x

Chiang, H. -W., Philibosian, B., Meltzner, A. J., Wu, C. -C., Shen, C. -C., Edwards, R. L., Chuang, C. -K., Suwargadi, B. W., Natawidjaja, D. H. (2023). Investigating spatio-temporal variability of initial 230Th/232Th in intertidal corals. *Quaternary Science Reviews 307*, Article 108005. https://doi.org/10.1016/j.quascirev.2023.108005.

Church, J. A., & White, N. J. (2006). A 20th century acceleration in global sea-level rise. *Geophys Res Lett 33*. https://doi.org/10.1029/2005GL024826

Church, J. A., White, N. J., Coleman, R., Lambeck, K., & Mitrovica, J. X. (2004). Estimates of the regional distribution of sea level rise over the 1950–2000 period. *Journal of Climate, 17*, 2609–2625.

Clark, K. J., Cochran, U. A., Berryman, K. R., Biasi, G., Villamor, P., Bartholomew, T., Litchfield, N., Pantosti, D., Marco, S., Van Dissen, R., Turner, G., & Hemphill-Haley, M. (2013). Deriving a long paleoseismic record from a shallow-water Holocene basin next to the Alpine Fault, New Zealand. *Geological Society of America Bulletin, 125*, 811–832. https://doi.org/10.1130/B30693.1

Debaecker, S., Feuillet, N., Satake, K., Sowa, K., Yamada, M., Watanabe, A., Saiki, A., Saurel, J.-M., Nakamura, M., Occhipinti, G., Yu, T. -L., & Shen, C. -C. (2023). Recent relative sea-level changes recorded by coral microatolls in Southern Ryukyus islands, Japan: Implication for the seismic cycle of the megathrust. *Geochemistry, Geophysics, Geosystems 24*, Article e2022GC010587. https://doi.org/10.1029/2022GC010587

Edwards, R. L., Taylor, F. W., & Wasserburg, G. J. (1988). Dating earthquakes with high-precision thorium-230 ages of very young corals. *Earth and Planetary Science Letters, 90*, 371–381.

Fleming, K., Johnston, P., Zwartz, D., Yokoyama, Y., Lambeck, K., & Chappell, J. (1998). Refining the eustatic sea-level curve since the Last Glacial Maximum using far- and intermediate-field sites. *Earth and Planetary Science Letters, 163*, 327–342. https://doi.org/10.1016/S0012-821X(98)00198-8

Hayes, G. P., Briggs, R. W., Sladen, A., Fielding, E. J., Prentice, C., Hudnut, K., Mann, P., Taylor, F. W., Crone, A. J., Gold, R., Ito, T., & Simons, M. (2010). Complex rupture during the 12 January 2010 Haiti earthquake. *Nature Geoscience, 3*, 800–805. https://doi.org/10.1038/ngeo977

Hughes, T. P., Anderson, K. D., Connolly, S. R., Heron, S. F., Kerry, J. T., Lough, J. M., Baird, A. H., Baum, J. K., Berumen, M. L., Bridge, T. C., Claar, D. C., Eakin, C. M., Gilmour, J. P., Graham, N. A. J., Harrison, H., Hobbs, J.-P.A., Hoey, A. S., Hoogenboom, M., Lowe, R. J., … Wilson, S. K. (2018). Spatial and temporal patterns of mass bleaching of corals in the Anthropocene. *Science, 359*, 80–83. https://doi.org/10.1126/science.aan8048

Kanamori, H. (2014). The diversity of large earthquakes and its implications for hazard mitigation. *Annual Review of Earth and Planetary Sciences, 42*, 7–26. https://doi.org/10.1146/annurev-earth-060313-055034

Khan, N. S., Ashe, E., Horton, B. P., Dutton, A., Kopp, R. E., Brocard, G., Engelhart, S. E., Hill, D. F., Peltier, W. R., Vane, C. H., & Scatena, F. N. (2017). Drivers of Holocene sea-level change in the Caribbean. *Quaternary Science Reviews, 155*, 13–36. https://doi.org/10.1016/j.quascirev.2016.08.032

Khan, N. S., Ashe, E., Shaw, T. A., Vacchi, M., Walker, J., Peltier, W. R., Kopp, R. E., & Horton, B. P. (2015). Holocene relative sea-level changes from near-, intermediate-, and far-field locations. *Current Climate Change Reports, 1*, 247–262. https://doi.org/10.1007/s40641-015-0029-z

Klinger, Y., Le Beon, M., & Al-Qaryouti, M. (2015). 5000 yr of paleoseismicity along the southern Dead Sea Fault. *Geophysical Journal International, 202*, 313–327. https://doi.org/10.1093/gji/ggv134

Konca, A. O., Avouac, J.-P., Sladen, A., Meltzner, A. J., Sieh, K., Fang, P., Li, Z., Galetzka, J., Genrich, J., Chlieh, M., Natawidjaja, D., Bock, Y., Fielding, E. J., Ji, C., & Helmberger, D. V. (2008). Partial rupture of a locked patch of the Sumatra megathrust during the 2007 earthquake sequence. *Nature, 456*, 631–635. https://doi.org/10.1038/nature07572

Lay, T., Kanamori, H., Ammon, C. J., Koper, K. D., Hutko, A. R., Ye, L., Yue, H., & Rushing, T. M. (2012). Depth-varying rupture properties of subduction zone megathrust faults. *J Geophys Res 117*, Article B04311. https://doi.org/10.1029/2011JB009133

Li, X., Lim, J. T. Y., Gopal, A., Mitchell, A. P., Sarkawi, G. M., Meltzner, A. J., Pang, I., Wang, Y., Sarmiento, L. F., Komori, J., Maxwell, K. V., Weil-Accardo, J., Yu, T.-L., Hu, H.-M., Shen, C.-C., Gong, S.-Y., Lin, K., Lu, Y., Wang, X., & Ramos, N. T. (2021). *Coral evidence for Mid-Holocene sea-level rise and an Abrupt Uplift event at Badoc, Ilocos Norte, Philippines, Abstract T45F-06, American Geophysical Union Fall Meeting.*

Majewski, J. M., Meltzner, A. J., Switzer, A. D., Shaw, T. A., Li, T., Bradley, S., Walker, J. S., Kopp, R. E., Samanta, D., Natawidjaja, D. H., Suwargadi, B. W., & Horton, B. P. (2022). Extending instrumental sea-level records using coral microatolls, an example from Southeast Asia. *Geophysical Research Letters*. https://doi.org/10.1029/2021GL095710

Meltzner, A. J., Sieh, K., Chiang, H.-W., Shen, C.-C., Suwargadi, B.W., Natawidjaja, D.H., Philibosian, B., & Briggs, R. W. (2012). Persistent termini of 2004- and 2005-like ruptures of the Sunda megathrust. *J Geophys Res 117*, Article B04405. https://doi.org/10.1029/2011JB008888

Meltzner, A. J., Sieh, K., Chiang, H. -W., Shen, C. -C., Suwargadi, B. W., Natawidjaja, D. H., Philibosian, B. E., Briggs, R. W., & Galetzka, J. (2010). Coral evidence for earthquake recurrence and an A.D. 1390–1455 cluster at the south end of the 2004 Aceh-Andaman rupture. *J Geophys Res 115*, Article B10402. https://doi.org/10.1029/2010JB007499

Meltzner, A. J., Sieh, K., Chiang, H.-W., Wu, C.-C., Tsang, L. L. H., Shen, C.-C., Hill, E. M., Suwargadi, B. W., Natawidjaja, D. H., Philibosian, B., & Briggs, R. W. (2015). Time-varying interseismic strain rates and similar seismic ruptures on the Nias-Simeulue patch of the Sunda megathrust. *Quaternary Science Reviews, 122*, 258–281. https://doi.org/10.1016/j.quascirev.2015.06.003

Meltzner, A. J., Switzer, A. D., Horton, B. P., Ashe, E., Qiu, Q., Hill, D. F., Bradley, S. L., Kopp, R. E., Hill, E. M., Majewski, J. M., Natawidjaja, D. H., Suwargadi, B. W. (2017). Half-metre sea-level fluctuations on centennial timescales from mid-Holocene corals of Southeast Asia. *Nature Communs, 8*, Article 14387. https://doi.org/10.1038/ncomms14387

Meltzner, A. J., & Woodroffe, C. D. (2015). Coral microatolls. In I. Shennan, A. J. Long & B. P. Horton (Eds.), *Handbook of sea-level research*. Wiley.

Mitrovica, J. X., & Milne, G. A. (2002). On the origin of late Holocene sea-level highstands within equatorial ocean basins. *Quaternary Science Reviews, 21*, 2179–2190. https://doi.org/10.1016/S0277-3791(02)00080-X

Mondal, D. R., McHugh, C. M., Mortlock, R. A., Steckler, M. S., Mustaque, S., & Akhter, S. H. (2018). Microatolls document the 1762 and prior earthquakes along the southeast coast of Bangladesh. *Tectonophysics, 745*, 196–213. https://doi.org/10.1016/j.tecto.2018.07.020

Morgan, P. M., Feng, L., Meltzner, A. J., Lindsey, E. O., Tsang, L. L. H., & Hill, E. M. (2017). Sibling earthquakes generated within a persistent rupture barrier on the Sunda megathrust under Simeulue Island. *Geophysical Research Letters, 44*, 2159–2166. https://doi.org/10.1002/2016GL071901

Morgan, P.M., Feng, L., Meltzner, A.J., Mallick, R., Hill, E.M., 2020. Diverse slip behavior of the Banyak Islands sub-segment of the Sunda megathrust in Sumatra, Indonesia. J. Geophys. Res. Solid Earth. https://doi.org/10.1029/2020JB020011

Natawidjaja, D. H., Sieh, K., Chlieh, M., Galetzka, J., Suwargadi, B. W., Cheng, H., Edwards, R. L., Avouac, J.-P., Ward, S. N. (2006). Source parameters of the great Sumatran megathrust earthquakes of 1797 and 1833 inferred from coral microatolls. *J Geophys Res 111*, Article 6403. https://doi.org/10.1029/2005JB004025

Natawidjaja, D. H., Sieh, K., Galetzka, J., Suwargadi, B. W., Cheng, H., Edwards, R. L., Chlieh, M. (2007). Interseismic deformation above the Sunda Megathrust recorded in coral microatolls of the Mentawai islands, West Sumatra. *J Geophys Res 112*, Article 2404. https://doi.org/10.1029/2006JB004450

Natawidjaja, D.H., Sieh, K., Ward, S.N., Cheng, H., Edwards, R.L., Galetzka, J., & Suwargadi, B.W. (2004). Paleogeodetic records of seismic and aseismic subduction from central Sumatran microatolls, Indonesia. *J Geophys Res 109*, Article 4306. https://doi.org/10.1029/2003JB002398

Nerem, R. S., Beckley, B. D., Fasullo, J. T., Hamlington, B. D., Masters, D., & Mitchum, G. T. (2018). Climate-change–driven accelerated sea-level rise detected in the altimeter era. *Proceedings of the National Academy of Sciences of the United States of America, 115*, 2022–2025. https://doi.org/10.1073/pnas.1717312115

Peltier, W. R. (1999). Global sea level rise and glacial isostatic adjustment. *Global and Planetary Change, 20*, 93–123. https://doi.org/10.1016/S0921-8181(98)00066-6

Philibosian, B., Feuillet, N., Jacques, E., Weil-Accardo, J., Guihou, A., Mériaux, A.-S., Anglade, A., Saurel, J.-M., & Deroussi, S., 2022. 20th-century strain accumulation on the Lesser Antilles Megathrust based on coral microatolls. *Earth and Planetary Science Letters 579*, Article 117343. https://doi.org/10.1016/j.epsl.2021.117343

Philibosian, B., & Meltzner, A. J. (2020). Segmentation and supercycles: A catalog of earthquake rupture patterns from the Sumatran Sunda Megathrust and other well-studied faults worldwide. *Quaternary Science Reviews 241*. https://doi.org/10.1016/j.quascirev.2020.106390

Philibosian, B., Sieh, K., Avouac, J.-P., Natawidjaja, D. H., Chiang, H.-W., Wu, C.-C., Perfettini, H., Shen, C.-C., Daryono, M. R., & Suwargadi, B. W. (2014). Rupture and variable coupling behavior of the Mentawai segment of the Sunda megathrust during the supercycle culmination of 1797 to 1833. *J Geophys Res 119*. https://doi.org/10.1002/2014JB011200

Philibosian, B., Sieh, K., Avouac, J.-P., Natawidjaja, D. H., Chiang, H.-W., Wu, C.-C., Shen, C.-C., Daryono, M. R., Perfettini, H., Suwargadi, B. W., Lu, Y., & Wang, X. (2017). Earthquake supercycles on the Mentawai segment of the Sunda megathrust in the seventeenth century and earlier. *J. Geophys. Res. Solid Earth, 122*, 642–676. https://doi.org/10.1002/2016JB013560

Philibosian, B., Sieh, K., Natawidjaja, D.H., Chiang, H.-W., Shen, C.-C., Suwargadi, B.W., Hill, E.M., Edwards, R.L., 2012. An ancient shallow slip event on the Mentawai segment of the Sunda megathrust, Sumatra. J. Geophys. Res. 117, Article B05401. https://doi.org/10.1029/2011JB009075

Sarkawi, G.M., Meltzner, A.J., Lim, J., Gopal, A., Mitchell, A.P., Sarmiento, L.F., Weil-Accardo, J., Maxwell, K.V., Komori, J., Li, X., Peng, D., Ramos, N.T., 2022. A coral microatoll record

of sea-level rise, interseismic subsidence, and El Niño in La Union, Philippines since 1902 CE, Abstract GC22B-01, American Geophysical Union Fall Meeting.
Sarkawi, G.M., Meltzner, A.J., Lim, J., Gopal, A., Mitchell, A.P., Sarmiento, L.F., Weil-Accardo, J., Maxwell, K.V., Komori, J., Li, X., Wang, Y., Pang, I., Ang, S., Lin, K., Wang, X., Hu, H.-M., Shen, C.-C., Gong, S.-Y., Lu, Y., Ramos, N.T., 2021. Tectonic uplift and subsidence inferred from coral archives of relative sea level in Balaoan, La Union, Philippines, Abstract T45F-09, American Geophysical Union Fall Meeting.
Savage, J. C. (1983). A dislocation model of strain accumulation and release at a subduction zone. *Journal of Geophysical Research, 88*, 4983–4996.
Scharer, K., Streig, A.. (2018). The San Andreas Fault system: Complexities along a major transform fault system and relation to earthquake hazards. in Duarte, J. (Ed.) *Transform plate boundaries and fracture zones*. Elsevier.
Scharer, K.M., Yule, D., 2020. A maximum rupture model for the southern San Andreas and San Jacinto Faults California, derived from paleoseismic earthquake ages: observations and limitations. Geophysical Research Letters 47, e2020GL088532. https://doi.org/10.1029/2020GL088532
Scoffin, T. P., & Stoddart, D. R. (1978). The nature and significance of microatolls. *Philos. Trans. r. Soc. London Ser. B, 284*, 99–122.
Shen, C.-C., Edwards, R. L., Cheng, H., Dorale, J. A., Thomas, R. B., Moran, S. B., Weinstein, S. E., & Edmonds, H. N. (2002). Uranium and thorium isotopic and concentration measurements by magnetic sector inductively coupled plasma mass spectrometry. *Chemical Geology, 185*, 165–178.
Shen, C.-C., Li, K. S., Sieh, K., Natawidjaja, D., Cheng, H., Wang, X., Edwards, R. L., Lam, D. D., Hsieh, Y. T., Fan, T. Y., Meltzner, A. J., Taylor, F. W., Quinn, T. M., Chiang, H. W., & Kilbourne, K. H. (2008). Variation of initial $^{230}Th/^{232}Th$ and limits of high precision U-Th dating of shallow-water corals. *Geochimica Et Cosmochimica Acta, 72*, 4201–4223.
Shen, C.-C., Wu, C.-C., Cheng, H., Edwards, R. L., Hsieh, Y.-T., Gallet, S., Chang, C.-C., Li, T.-Y., Lam, D. D., Kano, A., Hori, M., & Spötl, C. (2012). High-precision and high-resolution carbonate ^{230}Th dating by MC-ICP-MS with SEM protocols. *Geochimica Et Cosmochimica Acta, 99*, 71–86.
Shyu, J. B. H., Wang, C.-C., Wang, Y., Shen, C.-C., Chiang, H.-W., Liu, S.-C., Min, S., Aung, L. T., Than, O., & Tun, S. T. (2018). Upper-plate splay fault earthquakes along the Arakan subduction belt recorded by uplifted coral microatolls on northern Ramree Island, western Myanmar (Burma). *Earth and Planetary Science Letters, 484*, 241–252. https://doi.org/10.1016/j.epsl.2017.12.033
Sieh, K., Natawidjaja, D. H., Meltzner, A. J., Shen, C.-C., Cheng, H., Li, K.-S., Suwargadi, B. W., Galetzka, J., Philibosian, B., & Edwards, R. L. (2008). Earthquake supercycles inferred from sea-level changes recorded in the corals of west Sumatra. *Science, 322*, 1674–1678. https://doi.org/10.1126/science.1163589
Sieh, K., Ward, S. N., Natawidjaja, D., & Suwargadi, B. W. (1999). Crustal deformation at the Sumatran subduction zone revealed by coral rings. *Geophysical Research Letters, 26*, 3141–3144. https://doi.org/10.1029/1999GL005409
Smithers, S. G., & Woodroffe, C. D. (2000). Microatolls as sea-level indicators on a mid-ocean atoll. *Marine Geology, 168*, 61–78.
Taylor, F. W., Briggs, R. W., Frohlich, C., Brown, A., Hornbach, M., Papabatu, A. K., Meltzner, A. J., & Billy, D. (2008). Rupture across arc segment and plate boundaries in the 1 April 2007 Solomons earthquake. *Nature Geoscience, 1*, 253–257. https://doi.org/10.1038/ngeo159
Taylor, F. W., Frohlich, C., Lecolle, J., & Strecker, M. R. (1987). Analysis of partially emerged corals and reef terraces in the central Vanuatu Arc; comparison of contemporary coseismic and nonseismic with Quaternary vertical movements. *Journal of Geophysical Research, 92*, 4905–4933. https://doi.org/10.1029/JB092iB06p04905
Thirumalai, K., Taylor, F.W., Shen, C.-C., Lavier, L.L., Frohlich, C., Wallace, L.M., Wu, C.-C., Sun, H., & Papabatu, A.K. (2015). Variable Holocene deformation above a shallow subduction

zone extremely close to the trench. *Nature Communications 6*. https://doi.org/10.1038/ncomms8607

Tsang, L. L. H., Meltzner, A. J., Hill, E. M., Freymueller, J. T., & Sieh, K. (2015a). A paleogeodetic record of variable interseismic rates and megathrust coupling at Simeulue Island. *Sumatra. Geophys. Res. Lett., 42*, 10585–10594. https://doi.org/10.1002/2015GL066366

Tsang, L. L. H., Meltzner, A. J., Philibosian, B., Hill, E. M., Freymueller, J. T., & Sieh, K. (2015b). A 15 year slow-slip event on the Sunda megathrust offshore Sumatra. *Geophysical Research Letters, 42*, 6630–6638. https://doi.org/10.1002/2015GL064928

Webster, P. J., Moore, A. M., Loschnigg, J. P., & Leben, R. R. (1999). Coupled ocean-atmosphere dynamics in the Indian Ocean during 1997–98. *Nature, 401*, 356–360.

Weil-Accardo, J., Feuillet, N., Jacques, E., Deschamps, P., Beauducel, F., Cabioch, G., Tapponnier, P., Saurel, J. M., & Galetzka, J. (2016a). Two hundred thirty years of relative sea level changes due to climate and megathrust tectonics recorded in coral microatolls of Martinique (French West Indies). *J. Geophys. Res. Solid Earth, 121*, 2873–2903. https://doi.org/10.1002/2015JB012406

Weil-Accardo, J., Feuillet, N., Jacques, E., Deschamps, P., Saurel, J.-M., Thirumalai, K., Demeza, S., & Anglade, D. (2016b). Relative sea-level changes during the last century recorded by coral microatolls in Belloc, Haiti. *Global and Planetary Change, 139*, 1–14. https://doi.org/10.1016/j.gloplacha.2015.12.019

Weil-Accardo, J., Feuillet, N., Satake, K., Goto, T., Goto, K., Harada, T., Kayanne, H., Nakamura, M., Ramos, N., Saurel, J.-M., Sowa, K., Liu, S.-C., Yu, T.-L., Shen, C.-C., 2019. Relative sea-level changes over the past centuries in the central Ryukyu Arc inferred from coral microatolls. Journal of Geophysical Research: Solid Earth 125, e2019JB018466. https://doi.org/10.1029/2019JB018466

Zachariasen, J., Sieh, K., Taylor, F. W., Edwards, R. L., & Hantoro, W. S. (1999). Submergence and uplift associated with the giant 1833 Sumatran subduction earthquake: Evidence from coral microatolls. *Journal of Geophysical Research, 104*, 895–919. https://doi.org/10.1029/1998JB900050

Zachariasen, J., Sieh, K., Taylor, F. W., & Hantoro, W. S. (2000). Modern vertical deformation above the Sumatran subduction zone: Paleogeodetic insights from coral microatolls. *Bulletin of the Seismological Society of America, 90*, 897–913. https://doi.org/10.1785/0119980016

Lacustrine Records of Past Seismic Shaking

7

Jasper Moernaut, Jamie Howarth, Katrina Kremer, and Katleen Wils

7.1 The Need for Long and Continuous Archives of Past Seismic Shaking

The recurrence of large earthquakes in a region is typically characterized by pronounced variability in time and space. From a temporal perspective, it is often believed that their occurrence at plate boundaries follows a "seismic cycle" pattern where seismic hazard is time-dependent, i.e. stress on the ruptured fault (and thus hazard) is generally low after a major earthquake and slowly rises over elapsed time. Simple recurrence models based on a renewal process, such as periodic ("characteristic"), slip-predictable or time-predictable however do not fit well to many long-term paleoseismological records as these commonly show a range of rupture modes and types of "supercycles". Such "supercycles" can involve sequences of short interevent times separated by long quiescence periods or superimposed cycles of high magnitude events on top of more frequent smaller ruptures.

Supplementary Information The online version contains supplementary material available at https://doi.org/10.1007/978-3-031-73580-6_7.

J. Moernaut (✉)
Department of Geology, University of Innsbruck, Innsbruck, Austria
e-mail: Jasper.Moernaut@uibk.ac.at

J. Howarth
School of Geography, Environment and Earth Sciences, Victoria University of Wellington, Wellington, New Zealand

K. Kremer
Swiss Seismological Service, ETH Zürich, Zürich, Switzerland

K. Wils
Department of Geology, Ghent University, Ghent, Belgium

A. Elliott and C. Gruetzner (eds.), *Understanding Past Earthquakes*,
https://doi.org/10.1007/978-3-031-73580-6_7

These patterns are better described with a Long-Term Fault Memory model, where the occurrence probability of large earthquakes reflects the accumulated elastic strain over multiple seismic cycles, instead of the time elapsed since the last event (Salditch et al., 2020).

From a spatial perspective, rupture variability relates to possible differences in location, length and coseismic slip. Variability in rupture mode has been inferred for most studied plate boundary faults with long paleoseismic records. Various terminology has been used to describe rupture and fault segmentation in different settings, forming a complex and strongly debated topic. In this chapter, we follow the terminology concerning subduction megathrust segmentation used in the recent review of Philibosian and Meltzner (2020). This includes "master segments" that occur between two persistent rupture barriers and represent the maximum-length rupture that can plausibly occur on that portion of the fault. These master segments consist of smaller "asperities" (or "unitary segments") which can rupture individually, in combination or as cascades in which adjacent asperities rupture within a short time (hours to a few years). In transform fault settings, fault "segments" or "sections" have been identified as the fault surface between outstanding structural and geometric features. As ruptures can start, halt or even continue at such section boundaries, their role for rupture segmentation is often unclear. An important research field deals with the evaluation of the physical nature of potential rupture barriers and how these may link to structural and geometric fault features (Bilek & Lay, 2018; Lefevre et al., 2020; Shrivastava et al., 2019).

In contrast to plate boundaries, intraplate settings are characterized by low deformation rates (< few mm/yr of relative motion) and often exhibit broad zones of seismicity distributed over many potentially active faults in complex configurations. Large earthquakes typically have long millennia-scale average recurrence times and are found to be strongly episodic, spatially migrating and clustered (Stein et al., 2017). Therefore, recurrence patterns are poorly constrained by historical and instrumental data. Moreover, for many intraplate settings, it remains unclear how large the maximum credible earthquake can be (Vanneste et al., 2016) and which faults are capable of hosting significant events.

The listed types of spatiotemporal rupture variability in different tectonic regions pose major challenges for constraining recurrence parameters and selecting adequate models to calculate the probability of future large earthquakes. This implies that the often-cited mean recurrence rate obtained from paleoseismology only becomes meaningful when the variability in interevent times can be adequately constrained (McCalpin, 2009; Styron, 2019). To adequately estimate both the mean interevent time and their coefficient of variation (CoV: mean-normalized standard deviation), paleoseismic records must hold at least ~6–16 interevent times (Kempf et al., 2019). Logically, the required number is much lower for a more regular underlying pattern (e.g. CoV < 0.5) than for a bursty pattern (CoV > 1.1) (Moernaut, 2020). To calculate occurrence probability, a common method is to fit a probability function to the distribution of interevent times. However, this requires an even larger number of events (>30, but ideally > 100; see e.g. Parsons, 2008;

Clare et al., 2015). Besides statistical goodness-of-fit to e.g. exponential, lognormal, inverse Gaussian or weibull distributions, also theoretical considerations can guide the choice of best-fitting probability function (Abaimov et al., 2008).

Obtaining a sufficient number of events for descriptive recurrence statistics is often not possible for on-fault trench-based studies as it becomes technically difficult to expose older surface ruptures and the fast rate of near-fault landscape evolution precludes long records (e.g. Berryman et al., 2012a, b). In coastal marshes, long-term evolution of relative sea level can affect the stratigraphic record of tsunamis and coseismic elevation (Dura et al., 2016). Moreover, dynamic erosional and depositional processes in such terrestrial settings can hamper record completeness. In contrast, marine and lacustrine sedimentation in sufficiently deep basins typically occurs continuously and changes in sedimentation rate and type may be modest over several millennia (Sect. 7.3). Marine turbidite paleoseismology was primarily developed along the Cascadia Margin (Adams, 1990; Goldfinger, 2011; Goldfinger et al., 2003) and has been applied along many other active margins, such as Sumatra (Patton et al., 2015), Japan (Ikehara et al., 2016; Kioka et al., 2019), Hikurangi (Pouderoux et al., 2014), Chile (Bernhardt et al., 2015) and in other subaqueous basins such as in the Mediterranean Sea (Polonia et al., 2013; Ratzov et al., 2015) and its semi-enclosed basins (Corinth Gulf: Sergiou et al., 2017; Marmara Sea: Drab et al., 2012; Yakupoğlu et al., 2019). At Cascadia, comprehensive along-margin mapping of seismo-turbidites led to a 10 kyr-long reconstruction of megathrust earthquakes in space and time, encompassing partial ~300–600 km-long ruptures and full segment ~860 km-long ruptures (Goldfinger et al., 2012) which has been used to inform seismic hazard models (Kulkarni et al., 2013; Park et al., 2017; Petersen et al., 2014b). The reliability of marine turbidite paleoseismology has been heavily debated (Atwater et al., 2014; Goldfinger et al., 2014; Sumner et al., 2013; Talling, 2021). This has directed recent investigations towards testing the essential assumptions of the approach based on comprehensive mapping of event deposits related to recent events and/or quantitative considerations on seismic ground motions (Gomberg, 2018; Howarth et al., 2021b; McHugh et al., 2020). A comprehensive review on the basic principles and application of marine turbidite paleoseismology can be found in Goldfinger et al. (2012), whereas the present book chapter is entirely focused on lacustrine paleoseismology.

Many glacigenic lakes originated after the last general deglaciation and can thus hold 11–18 kyr long paleoseismic records. This is also true for fjords, which have likewise been studied for paleoseismic purposes (e.g. St.-Onge et al., 2004; Blais-Stevens et al., 2011; Bellwald et al., 2019; Wils et al., 2020). These studies mainly follow the lacustrine paleoseismic approach, as especially the inner-fjord areas can be considered as confined basins with only small influence of marine currents and sediment dynamics that are dominated by clastic input via large tributaries. Sedimentary sequences in crater or tectonic lakes may produce even longer records (>100 kyr) when investigated through deep drilling projects (e.g. Dead Sea: 220 kyr; Lu et al., 2020). For moderate-to-high seismicity settings, most lake paleoseismic records generally exhibit sufficient events for an appropriate estimation of basic recurrence parameters (Moernaut, 2020). Some of these can even be used for robustly fitting probability distributions to interevent times and thus calculation of the conditional probability that a certain shaking strength will be exceeded within a given time span. Besides constraining temporal recurrence, a more challenging aspect in lacustrine paleoseismology is the reconstruction of rupture location, extent and earthquake magnitudes through site-to-site correlations (Sect. 7.5). Because lacustrine records can exhibit a high temporal precision (Sect. 7.3) and allow for quantitative considerations on shaking strength (Sect. 7.4), these can significantly contribute to the reconstruction of past earthquake scenarios (Sects. 7.6, 7.7 and 7.8).

In this chapter, we briefly introduce the different types of earthquake-triggered lacustrine imprints, how these imprints can be dated and calibrated to ground motion information from recent and historical earthquakes and how to use multiple lake records to infer rupture location, extent and magnitude of past events. Moreover, we present the application and development of quantitative lake paleoseismology for representative case studies in intraplate, transform and subduction areas.

7.2 Identifying Earthquake Imprints in Lacustrine Sediments

Several pioneer studies in the '70–90 s' (Doig, 1986; Hibsch et al., 1997; Marco et al., 1996; Sims, 1975) inferred that strong seismic shaking can produce distinct deformation structures and event deposits preserved in lacustrine sedimentary sequences. In the last 20 years, lacustrine paleoseismology has significantly advanced and a steeply rising number of lake sequences have been studied over the world for use as "natural seismographs" (Fig. 7.1, Supplementary Table 1). The main lacustrine earthquake fingerprints are classified as follows (Figs. 7.2 and 7.3).

In-situ soft sediment deformation structures (SSDS): These mostly develop directly at the sediment–water interface and consist of liquefaction features or brittle and ductile deformation structures (Agnon et al., 2006; Monecke et al., 2004). Liquefaction features, such as sand blows and injection structures, occur due to very high excess pore pressure and require a pronounced stratigraphic variability with a relatively coarse-grained source layer capped by a fine-grained layer (Kluger et al., 2023; Obermeier, 1996). Fluid escape structures on top of large, buried mass-transport deposits have been identified on seismic profiles and may be used as a complementary paleoseismic proxy (Moernaut et al., 2009). Brittle and ductile SSDS are considered a direct consequence of seismically-induced transient shear stress between near-surface layers of different density and rheology. Comprehensive identification and mapping of lake outcrops in and around the Dead Sea (Ken-Tor et al., 2001; Marco et al., 1996; Migowski et al., 2004) documented different deformation types such as linear waves, asymmetric billows,

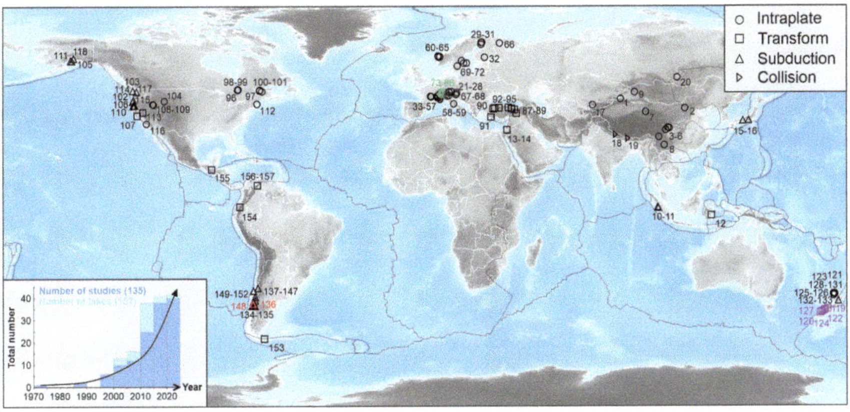

Fig. 7.1 World map with indication of lacustrine paleoseismic shaking records for different tectonic settings. More details of each of these lakes (numbered 1–157) are provided in the Supplementary Table 1 (compilation date: June 2024). Sites used to illustrate the use of lacustrine paleoseismology for earthquake source characterization in intraplate, transform and subduction zones (Sects. 7.6, 7.7, and 7.8) are indicated in green, purple and red, respectively. Lower left inset shows the increasing number of studied lakes and publications through time, binned into 5-year intervals

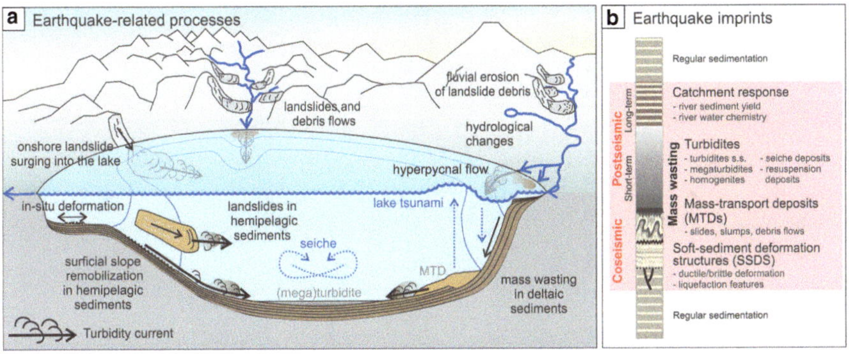

Fig. 7.2 Overview of earthquake-related processes and their lacustrine imprints. **a** Sedimentary processes in and around lakes that can be induced by strong earthquake shaking (Praet, 2020, modified after Van Daele et al., 2015). **b** Earthquake-induced imprints in sedimentary sequences which can be identified in sediment cores or outcrops. These range from coseismic in-situ deformation structures to early postseismic deposits related to subaqueous mass-wasting and signatures of long-term catchment response (Praet, 2020, modified after Avşar et al., 2014b)

coherent vortices and intra-clast breccias. Strong stratigraphic variability with contrasting mechanical properties can facilitate the development of slip horizons and aids in the identification of SSDS, but is not considered a prerequisite for their occurrence.

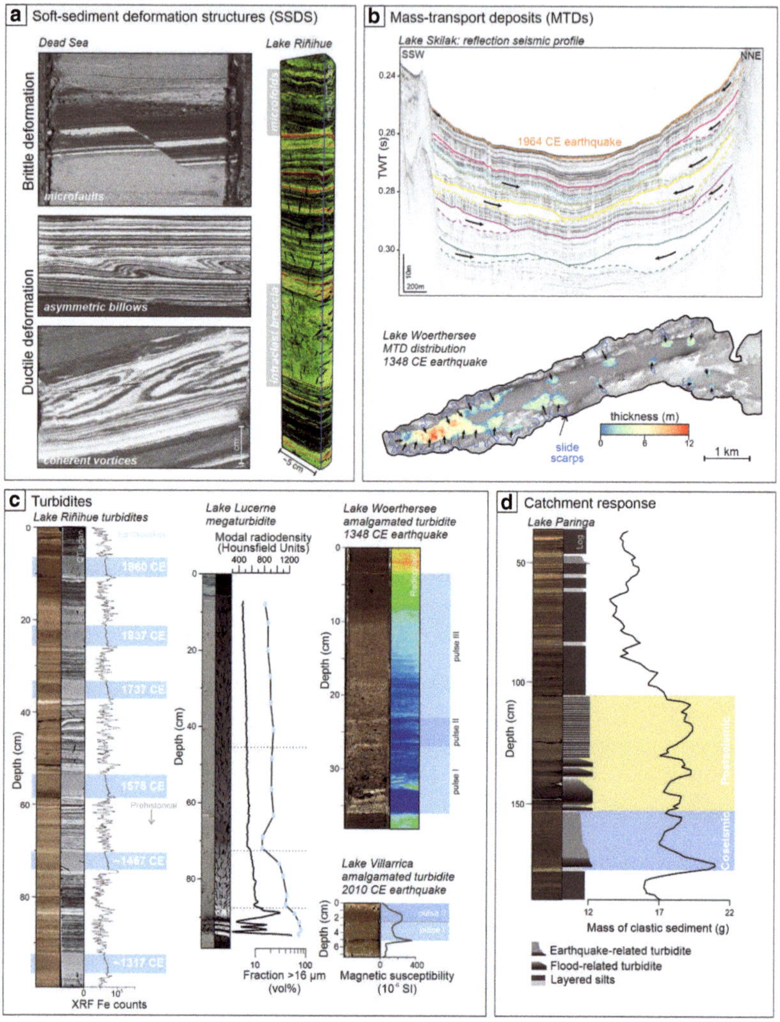

Fig. 7.3 Representative examples for the different types of earthquake imprints in lacustrine sequences (see Fig. 7.2). **a** A In-situ SSDS comprise e.g. brittle deformation structures such as micro-faults, ductile deformation structures such as different types of folded laminae (Dead Sea core: Lu et al., 2021a) and intra-clast breccia layers (micro-CT 3D reconstruction of slope sediments from Lake Riñihue, Chile: Molenaar et al., 2021). **b** Multiple coeval mass-transport deposits as mapped on subbottom profiles in Lake Skilak, Alaska (Praet et al., 2017) and Lake Wörthersee, Austria (Daxer et al., 2022b). **c** Sequence of seismo-turbidites in Lake Riñihue, Chile, linked to four strong historical megathrust earthquakes through a varve-based age-depth model (Moernaut et al., 2014). Seismic triggering often leads to (i) megaturbidites due to extensive mass-wasting and seiche action, which can be reconstructed by detailed analysis of sub-units in the megaturbidite (Lake Lucerne, Switzerland; Vermassen et al., 2023), and (ii) amalgamated turbidites indicating the synchronous remobilization of sediments on several slopes (Lake Wörthersee, Austria: Daxer et al., 2022b; Lake Villarrica, Chile: Van Daele et al., 2015). **d** Sedimentological signature of catchment response derived from Lake Paringa, New Zealand, involving a stack of flood-related turbidites above a seismo-turbidite induced by subaqueous mass-wasting (Howarth et al., 2014)

Deposits related to subaqueous mass-wasting: Strong seismic shaking can result in gravitational mass movements due to both transient shear stress and increasing pore pressure leading to failure of lacustrine sedimentary slope sequences. The resulting mass-transport deposits (MTDs) form heterogenous wedges with a positive topography and a generally erosive base in the sedimentary sequence (Sammartini et al., 2019; Schnellmann et al., 2002). The moving mass can incorporate lake water, dilute and eventually evolve into a turbidity current, which leads to turbidite deposits in the more distal, deeper part of the basin (e.g. Waldmann et al., 2011). Thick turbidites that are imaged on reflection seismic data are often called "megaturbidites" (Leithold et al., 2019). Recently, it was inferred that seismic shaking can also remobilize a thin veneer (few cm) of surficial slope sediments, which can be archived as stratigraphic gaps in the slope sequence and produces relatively thin turbidites in basin sequences (Moernaut et al., 2017a; Molenaar et al., 2021). The sedimentary facies of seismo-turbidites is strongly dependent on the failure process, source slope material, bottom morphology and transport distance (Daxer et al., 2022b; Gastineau et al., 2023; Moernaut et al., 2014; Vermassen et al., 2023; Wils et al., 2021b).

Deposits related to terrestrial processes: Seismically-triggered terrestrial mass movements can propagate into a lake (or fjord) and become incorporated in the sedimentary sequence. This mass-movement impact process can generate devastating impulse waves (e.g. Lituya Bay: Miller, 1960; Franco et al., 2020; Lake Lauerz: Bussmann & Anselmetti, 2010) that leave extensive sedimentary traces in and around the basin. Moreover, earthquake-triggered mass movements in the lake catchment can lead to a prolonged period of altered sedimentation ("catchment response") over several years to decades (Sect. 7.7; Howarth et al., 2012; Avşar et al., 2016; Van Daele et al., 2019; Oswald et al., 2021b). This involves enhanced fluviatile transport of terrestrial sediments to the lake caused by vegetation clearance and supply of loose landslide debris to the fluvial system, leading to a stack of hyperpycnal flow deposits in the lacustrine sequence above other types of earthquake imprints (Figs. 7.2 and 7.3d). Landslide-induced loose sediments can also be transported to lakes by eolian transport (dust storms) and be preserved in the sedimentary sequence (Jiang et al., 2017; Shi et al., 2022). In addition, earthquake-induced changes in the hydrology of the catchment, i.e. changes in the activity and composition of springs, have been reported and documented to affect the rate and type of lacustrine sedimentation (Archer et al., 2019; Daxer et al., 2024).

Deposits related to seiche action: Typical for closed (lake) and semi-closed (fjord) basins is the possible development of standing waves of the water body (seiches) initiated by (i) earthquake-triggered gravitational mass movements that displace lake water, as was observed on several local lakes caused by the 1960 M_W 9.5 Chile earthquake (Van Daele et al., 2015) or (ii) by seismic ground motion on its own, such as observed on Norwegian fjords caused by the 2011 M_W 9 Japan earthquake (Bondevik et al., 2013). Seiches can modulate the depositional signature of seismo-turbidites by sustaining fine-grained sediments in suspension, resulting in high segregation between finer and coarser sediments (Beck, 2009; Larsen et al., 2019). Moreover, seiching can generate significant bottom currents that remobilize near-coastal surface sediments and transport these to the deeper lake basins, leading to "seiche deposits". The oscillating bottom currents may lead to diagnostic alternation in grain size sorting at the base of seiche deposits (Chapron et al., 1999; Vermassen et al., 2023) and/or a strong foliation for the anisotropy of magnetic susceptibility for the muddy middle part of the deposit (Petersen et al., 2014a; Rapuc et al., 2018). Shaking-induced water column oscillations and turbulent flows at the lake bed may also facilitate vertical mixing of the water body (de la Fuente et al., 2010), which may be archived in the sedimentary sequence as peaks in redox-sensitive elements, such as Mn (Wils et al., 2021a).

Multibeam bathymetric sounding is a useful technique to accurately map relatively young mass-movement related deposits on the lake bottom and that allows for full coverage with a typical vertical resolution of a few decimeters and spatial resolution of 1–2 m (Hilbe et al., 2011; Sammartini et al., 2021; Strupler et al., 2019). Older MTDs in the stratigraphy are best identified by reflection seismic profiling. High-frequency seismic systems (>3 kHz: subbottom profilers) are capable of visualizing MTDs and megaturbidites in high vertical resolution (<20 cm), which allows these event deposits to be assigned to specific seismostratigraphic levels. Basin-wide MTD stratigraphies can then be established using a dense grid of 2D seismic profiles throughout the lake basin (Brooks, 2018; Karlin et al., 2004; Kremer et al., 2015; Moernaut et al., 2019; Praet et al., 2017; Strasser et al., 2006; Waldmann et al., 2011). In addition, seismic-stratigraphic analysis supports selecting the most suitable coring sites. For paleoseismic studies, the common research strategy involves the acquisition of numerous short gravity cores (0.5–1.5 m) to adequately trace the imprint of recent and historical earthquakes, followed by a few long piston cores (5–15 m) at key locations to construct long-term paleoseismic records. Core processing involves non-destructive core scanning for geophysical (e.g. magnetic susceptibility, gamma-density, P-wave velocity) and geochemical parameters (XRF-scanning). X-Ray Computed Tomography (CT) scanning is useful for a 3D visualization of the radiodensity patterns of sedimentary structures and thus facilitating the identification of subtle in-situ SSDS (Molenaar et al., 2021; Oswald et al., 2021b) and thin MTDs or turbidites (Vandekerkhove et al., 2020). Lab analysis of discrete samples can include any method used in paleolimnological and limnogeological research depending on the specific site characteristics and research objectives.

As most of the above-mentioned structures and deposits could also be induced by aseismic processes, every lacustrine paleoseismic study requires a detailed evaluation of possible trigger mechanisms (Sabatier et al., 2022). In many regions, temporal correlation to large earthquakes documented by instrumental records or historical reports is established (Archer et al., 2019; Migowski et al., 2004; Moernaut et al., 2014; Wilhelm et al., 2016), in this way providing a good idea on the potential sedimentary earthquake imprints within a specific lacustrine basin. A fully independent approach for seismic trigger evaluation builds upon the "synchronicity criterion", as initially developed for marine turbidites (Adams, 1990) and adapted for lacustrine MTDs and turbidites (Schnellmann et al., 2002). This concept assumes that simultaneous failure of different slopes in or around lakes hints at a regional triggering mechanism, i.e. earthquake shaking. Support for failure synchronicity can be gathered by the identification of (i) multiple MTDs on a single stratigraphic level, potentially overlain by a single (mega-)turbidite (Schnellmann et al., 2002), and/or (ii) an amalgamated turbidite consisting of different pulses—sometimes with indication for multiple flow directions—and overlain by a single fine-grained cap (Van Daele et al., 2014a, 2017; Vermassen et al., 2023; Wils et al., 2021b). The first argumentation requires a comprehensive seismic-stratigraphic mapping of the lake basin, whereas the second one requires detailed sediment core analyses. Correlation of synchronously-generated deposits across multiple independent lake basins further strengthens a seismic triggering mechanism (Brooks, 2018; Howarth et al., 2021a; Kremer et al., 2017; Moernaut et al., 2014; Oswald et al., 2022; Strasser et al., 2013). The importance of the synchronicity criterion implies that the spatial distribution of (multiple) event deposits is the most important paleoseismic "proxy" in lacustrine studies.

In situations where a spatial mapping approach is not possible, as is the case for small lakes or deep drilling projects in a single basin, the coexistence of different types of structures and deposits that are potentially earthquake-related can increase the confidence level for assigning a seismic trigger. For example, the long drill core in the Dead Sea shows many in-situ SSDS that are directly overlain by turbidites (Lu et al., 2021a). More commonly, the distinction between a seismic or other origin for turbidites in a single core is made by adopting a multi-proxy compositional analysis. Turbidites with an in-lake composition (e.g. mimicking the hemipelagic slope sediments) are commonly attributed to seismic shaking, because deep-water hemipelagic slopes with gradients below ~ 15–20° are stable under static conditions when no elevated excess pore pressure is present (e.g. Urlaub et al., 2015). In contrast, well-sorted turbidites with a clastic composition and mainly terrestrial organic matter indicate sediment input during river floods and debris flows (Arnaud et al., 2002; Fan et al., 2023; Glur et al., 2013; Sabatier et al., 2017; Schillereff et al., 2014; Simonneau et al., 2013; Wilhelm et al., 2015; Wirth et al., 2013) or gravitational delta collapses. The latter may occur due to earthquake shaking (Sect. 7.4.2; Praet et al., 2017), sudden sediment loading during flood events (Vandekerkhove et al., 2020), terrestrial mass movements impacting the delta plain (Kremer et al., 2012) or even spontaneously (Girardclos et al., 2007; Hilbe & Anselmetti, 2014).

In so-called "clastic lakes", where most sediment is delivered by fluviatile input (e.g. proglacial lakes), a systematic distinction between background sedimentation and small turbidites may require a statistical approach (Praet et al., 2020). Identifying an in-lake or allochthonous sediment source for turbidites mainly relies on flow-related parameters such as grain-size evolution and deposit geometry over the basin (Beck, 2009; Vandekerkhove et al., 2020). It is proposed that a sharp fining upwards trend in the basal part of turbidites indicates a rapidly decreasing flow velocity with time, which is diagnostic for subaqueous mass-wasting processes. A strongly basin-focused deposition can further support this interpretation. In contrast, flood-related turbidites show a more gradual fining upwards trend due to longer-lasting turbidity currents and a more even deposition throughout the lacustrine basins, the latter explained by fine-grained sediment transfer via interflows. Given these considerations on the source material and gravitational transport mechanisms of sediments, a detailed positioning of the core(s) with respect to inflowing rivers and basin morphology is crucial:

(i) Ideally, cores are located distal to large tributaries and their dynamic delta systems. In this way, large event deposits related to river floods and potential aseismic delta failures (Clare et al., 2016; Girardclos et al., 2007) can be avoided. Nevertheless, core locations near (smaller) inflows may form very sensitive recorders of seismic shaking as alluvial fan slopes can produce sandy turbidity currents for seismic intensities as low as ~ V½ (Sect. 7.4.2; Moernaut et al., 2014; Van Daele et al., 2020).

(ii) Locations with evidence for large slope failures and resulting MTDs generally lead to complex and incomplete paleoseismic records (Moernaut et al., 2019). This is because major slope failure events remove thick sequences of soft slope sediments and expose more consolidated and stable sediments. Sufficient time is required to recharge such slopes with soft sediments and produce potentially unstable sequences (Praet et al., 2017; Strasser et al., 2011). Large mass-wasting events also deform or erode the basin floor sediments they are deposited on, locally reducing the completeness of the record (Sammartini et al., 2021; Schnellmann et al., 2005). Moreover, large MTDs affect basin morphology, potentially changing the pathways of turbidity currents and leading to a migration of turbidite depocenters.

Deep-water SSDS in fine-grained soft sediments can often be confidently attributed to seismic shaking as other potential mechanisms are unlikely there (Lu et al., 2017). In contrast, assigning a seismic trigger to SSDS becomes more complex for deltaic areas and near-coastal zones because wave action, surface currents, rapid sedimentation and groundwater flows can induce excess pore pressure and create similar structures (Owen & Moretti, 2011). Moreover, these areas typically exhibit a strong stratigraphic variability (Rodríguez-Pascua et al. 2000; Hubert-Ferrari et al., 2017), facilitating the spontaneous development of load structures. Intrusion-like liquefaction features can be triggered by strong seismic shaking but often fail to reach the contemporaneous sediment–water interface and thus their age cannot be accurately determined.

A wide range of processes, such as heavy precipitation, freeze–thaw cycles or seismic shaking, can result in terrestrial mass movements. When these occur in the lake catchment rather than at its shoreline, they can merely result in a signal of catchment response in the lake sequence and the synchronicity criterion can thus not be directly applied. Attributing a seismic origin to such changes in lacustrine sedimentation thus requires identifying preceding coseismic structures or deposits in the sequence. This is often unproblematic in basins with small catchments as it has been inferred that significant catchment response occurs at relatively high shaking intensities and thus generally also results in coseismic imprints in the studied lake basin (Sect. 4.2; Howarth et al., 2014; Oswald et al., 2021b).

More information on the strategies used to single out sedimentary shaking evidence in lacustrine studies can be found in different publications (Archer et al., 2019; Avşar et al., 2016; Beck, 2009; Brooks, 2018; Kremer et al., 2017; Sabatier et al., 2022; Strasser et al., 2013). Given the wide range of lake systems and possible earthquake imprints, it is crucial to develop a site-specific approach, especially for small and/or shallow lakes, non-laminated sediment sequences and closed basins with large lake level fluctuations (e.g. Archer et al., 2019; Hubert-Ferrari et al., 2012; Lu et al., 2017; Monecke et al., 2018; Polonia et al., 2021). In some lake systems, earthquake shaking and associated processes may severely disrupt lake ecosystems (Brancelj et al., 2002; Wojewódka-Przybył et al., 2022) and thus abrupt changes in biotic communities could potentially be used as paleoseismic proxy.

7.3 Dating Paleoseismic Events in a Continuous Record

7.3.1 Age Control

In lacustrine paleoseismology, event ages are extracted from continuous accumulation histories built by using absolute ages such as calibrated ^{14}C ages from organic macroremains (Howarth et al., 2013) as input data for Bayesian age-depth modeling techniques (e.g. Bacon, Blaauw & Christen, 2011; Oxcal P_sequence, Bronk Ramsey, 2009). Sedimentation rates can be further constrained by $^{210}Pb_{xs}$ accumulation models for the last 150 years and complemented by anthropogenic peaks in ^{137}Cs and ^{14}C present since the 1960's, and by the shape of the post-bomb decay curve of ^{14}C. Identification of regional tephra layers that were independently dated in other terrestrial or lacustrine archives can provide additional age constraints (Moernaut et al., 2018; Wils et al., 2018). The continuous and rather constant rate of lacustrine sedimentation allows for very high resolution paleoseismic age control, even exhibiting a narrower uncertainty range compared to the calibrated ^{14}C ages that underlie the model. This is possible through the use of numerous ^{14}C dates (Fig. 7.4a; Howarth et al., 2016). The typically rather rigid sedimentation rate also implies that, for closely-spaced event layers in the sequence, the uncertainty on interevent time (relative age) is likely much smaller compared to the uncertainty range of the absolute ages of the two considered successive events (Moernaut, 2020). This allows for a precise estimation of recurrence parameters such as the average recurrence interval or the coefficient of variation of the interevent times. A statistical distribution of these parameters can be obtained by extracting the interevent times from each iteration that builds the final age-depth model (Kempf & Moernaut, 2021). In addition to generally low event numbers, terrestrial paleoseismic archives mostly lack such continuous and constant sedimentation rates, making their recurrence statistics less constrained.

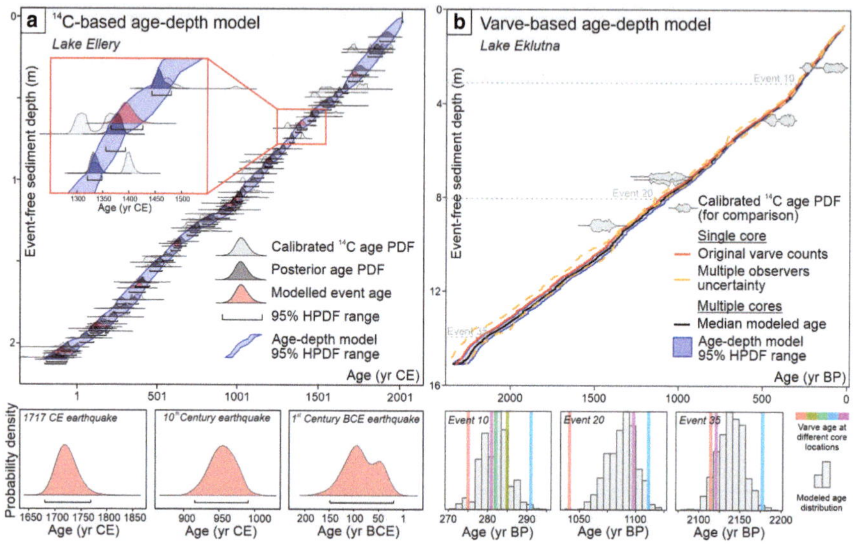

Fig. 7.4 Two examples of different approaches to obtain precise age-depth models and dates for lacustrine paleo-earthquake imprints. **a** Bayesian age-depth model based on numerous ^{14}C dates obtained on organic macro-remains in a sediment core from Lake Ellery (Howarth et al., 2016). The posterior age probability density function (PDF) is narrower than the calibrated ^{14}C age PDF due to the constraints given by the continuous lacustrine sedimentation process. Event deposits (i.e. instantaneous deposits) are removed from the sedimentary sequence before modeling. HPDF: highest probability density functions. Three representative modeled event age PDFs are given with indication of their 95% HPDF. **b** Bayesian age-depth model based on varve counting that also incorporates information regarding age uncertainty from multi-observer counts and age distributions of marker layers from multiple cross-correlated cores in Lake Eklutna (Fortin et al., 2019). The modeled age distribution and varve age obtained from different cores is given for three representative event layers. Note the increase in age uncertainty with depth, which is typical for varve-counted age models

In lakes where seasonal forcing results in annually-laminated (varved) records, lamination counting leads to a superior age control, especially when integrating these with independent absolute age information into a Bayesian framework (Vandergoes et al., 2018). Varve age precision can be further improved by a combination of multiple varve-counted cores in a lake, integrating data from multiple observers and using lake-wide marker layers as chronostratigraphic horizons (Fig. 7.4b; Fortin et al., 2019). Varve chronology often represents minimum ages for paleoseismic events as some varve years might be "missing" due to unclear lamination boundaries, subtle erosion by sediment gravity flows and deformation by coring or earthquake shaking. In contrast, river floods may result in sub-laminations, which can lead to age overestimation when these are not adequately detected. Therefore, complementary dating techniques are advisable, firstly to prove that laminations are formed on an annual basis, and secondly to mitigate the effect of possible under-or overestimation. Chronological precision of varved lake records is estimated by comparing repeated counting, ideally by multiple observers, and is typically in the range of 0.5–4% (Ojala et al., 2012). This implies that the uncertainty on varve chronology increases with age and, beyond a certain age, may thus underperform the more standard ^{14}C based chronologies.

7.3.2 Record Continuity

Continuous sedimentation supplies subaqueous slopes with fresh soft sediments that influence the overall static and dynamic slope stability. Earthquake shaking can remove some potentially unstable sequences, whereas other slopes may remain stable. These remnant slope sequences may get closer to near-critical stability due to removal of lateral and frontal buttresses facilitating failure during the next large earthquake. Accordingly, in areas where large earthquake frequency is low and sedimentation rates are sufficiently high, every large earthquake is capable of triggering multiple subaqueous landslides that result in MTDs (e.g. Strasser et al., 2011). However, this strategy is not applicable to high-seismicity settings where MTD stratigraphy typically underrepresents large earthquake recurrence (Moernaut et al., 2019), unless sedimentation rates are very high as is the case in proglacial lakes (Praet et al., 2017). These requisites for successful MTD paleoseismology are apparently not applicable to seismo-turbidite records as these are inferred to form continuous paleoseismic archives in a range of high-seismicity environments (see compilation in Moernaut, 2020). This can be explained by the recently-discovered process of surficial remobilization, involving the stripping of only a thin veneer (few cm) of surficial slope sediments over large slope areas (Moernaut et al., 2017a). Accordingly, slopes do not need to immediately recharge with fresh sediment in order to produce a turbidity current during a next earthquake. Long-term stable and continuous seismo-turbidite records are possible as long as the overall net slope accumulation is positive over several millennia and the dominant underlying sedimentary process involves the remobilization of surficial slope sediments. This process also implies that earthquakes in close succession can each generate seismo-turbidites, as exemplified by the observation of short interevent times between seismo-turbidites in a wide range of settings: ~18 yr in Alpine Fault lakes (Howarth et al., 2016), ~13–14 yr in Chilean lakes (Moernaut et al., 2014; Van Daele et al., 2019), ~10 yr in Alaskan proglacial lakes (Praet, 2020), ~13 yr in Late Glacial eastern Canadian lakes (Brooks & Adams, 2020), 1 to 8 yr in (European) Alpine lakes (Rapuc et al., 2018; Sabatier et al., 2017; Simonneau et al., 2013), and ~1 yr in Lake Tutira (New Zealand; Gomez et al., 2015). The theoretical resolution for distinguishing separate turbidites is in the order of days to months, i.e. the time required for deposition of sufficient fine-grained material that allows distinction of individual turbidity currents (Van Daele et al., 2017). However, a recent study in a Sumatran lake demonstrated that even earthquakes separated by only a few hours can result in individual turbidity currents identifiable in the sedimentary record. This insight was obtained through microfabric analysis of multipulsed turbidites providing information on depositional and flow dynamics (Wils et al., 2021b). Although this suggests that there is virtually no limit to triggering and identifying lacustrine seismo-turbidites, erosion and overprinting can affect the record continuity. Erosion is especially significant for sandy turbidites (Hizzett et al., 2018), but quasi-negligible for the more frequently studied muddy turbidites in distal depocenters (Van Daele et al., 2017). For MTD-stratigraphies, renewed landslide activity can obliterate older evidence or complicate identification of seismic-stratigraphic levels related to MTDs (Daxer et al., 2020; Strupler et al., 2018; Wils et al., 2018). Overprinting of SSDS mainly affects the registration of short interevent times and thus needs to be considered when interpreting recurrence statistics (Agnon et al., 2006; Migowski et al., 2004; Molenaar et al., 2021). Additional complexity in the interpretation of SSDS can be caused by sub-surface deformation, i.e. localized deformation of intrastratal horizons that are significantly older than the seismic shaking event (Alsop et al., 2022). In any of these cases, integration of different types of lacustrine paleoseismic evidence, basin-wide stratigraphic mapping and comparison of closely-located basins may identify and effectively fill such gaps in the record continuity.

Overall, high sedimentation rates relative to the average recurrence interval of strong shaking have a positive effect on the above-mentioned aspects regarding age control and record continuity. However, these factors lead to a short record time span and only few recorded events, unless deep drilling efforts are undertaken at the site (such as within ICDP projects). Low event numbers in a record have negative implications for long-term regional site correlations and a statistically robust analysis of recurrence parameters. Therefore, the ideal research design targets both high and low sedimentation rate lacustrine sites within a study area.

7.4 Quantifying Shaking Strength

7.4.1 Methods for Linking Seismic Ground Motion and Sedimentary Imprint

A paleoseismic record only becomes meaningful when some quantitative information on the recorded paleo-earthquakes is obtained. For lake paleoseismology, the primary parameter is the "strength" of the seismic ground motion that has induced one or more types of earthquake imprint in a lacustrine record. Several calibration methods have been developed that aim at (i) constraining a threshold in shaking strength that must be exceeded to produce a specific earthquake imprint at a given site and (ii) establishing scaling relationships between the strength of shaking and the size or type of sedimentary imprint. Comparison of such quantitative calibration techniques revealed relatively comparable relationships for different geodynamic settings (Sect. 7.4.2).

The most common calibration technique involves comparing the presence or absence of sedimentary imprints in the uppermost accurately-dated lacustrine sequence with the timing and shaking strength of recent and historical earthquakes in the region. For recent events, quantitative constraints are provided by instrumental ground motion data, while historical earthquakes rely on documentation of the more subjective macroseismic intensity data. Different seismic intensity scales are in use for the historical calibration of lacustrine earthquake imprints (i.e. Modified Mercalli Intensity MMI-56; Medvedev-Sponheuer-Karnik MSK-64; European Macroseismic Scale EMS-98), but as their equivalence is roughly one to one (Li et al., 2021; Musson et al., 2010) we use the general term "(seismic) intensity" in this book chapter. With respect to these calibration techniques, most studies can be grouped into two approaches (Fig. 7.5a, b):

(i) Estimating the local seismic intensity at the lake site for well-documented historical and recent events, also called the EQRT approach ("EarthQuake-Recording Threshold", Moernaut et al., 2014). These intensity values can be obtained from damage assessments for communities near the lake, by interpolation between inferred intensity values in the region or by application of

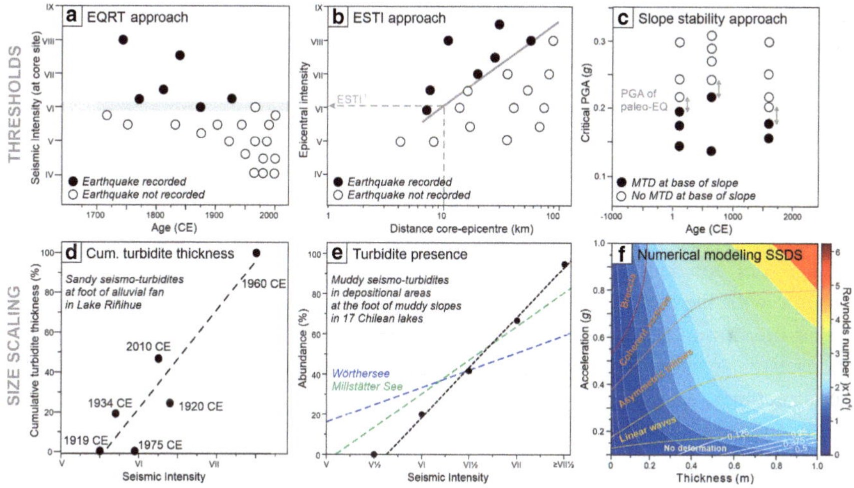

Fig. 7.5 **a–c** Three different methods to constrain the ground motion threshold or sensitivity of a lake site to record earthquake shaking in its sedimentary archive (modified after Moernaut, 2020). **a** and **b** refer to a single core site, whereas C concerns an entire lake basin. The data are simulated. EQRT: EarthQuake-Recording Threshold (Moernaut et al., 2014), ESTI: Earthquake-Sensitivity Threshold Index (Wilhelm et al., 2016), slope stability approach (Strasser et al., 2011). The grey lines delimit the parameter space where a sedimentary imprint is recorded or not recorded. **d–e** Size scaling relationships between sedimentary imprint and ground motion as derived from natural case studies (**d**, **e**: Chilean lakes: Moernaut et al., 2014; Van Daele et al., 2015) and numerical simulations (**f** Lu et al., 2020). In **d** and **e**, a generally linear relationship was proposed between seismic intensity and cumulative turbidite thickness or the presence of muddy turbidites in different depositional areas, respectively. Seismo-turbidites in Wörthersee and Millstätter See (Austria) also show size-scaling relationships (**e** Daxer et al., 2022b). **f** shows that increasing PGA levels lead to a more developed SSDS type, i.e. from linear waves to intra-clast breccia. Other influencing factors are the layer thickness, Reynolds number and Richardson number

empirical intensity attenuation relationships that calculate the intensity at the lake site based on earthquake magnitude and the distance to the rupture area. The EQRT is the intensity value (or range) that delimits data with positive (earthquake imprint present) and negative (no earthquake imprint) evidence.

(ii) Comparing the epicentral seismic intensity and lake-epicenter distance for different historical events, also called the ESTI approach ("Earthquake-Sensitivity Threshold Index", Wilhelm et al., 2016). A threshold line with a fixed slope value is drawn between data points with positive and negative evidence. The ESTI index is the inverse of the intercept value of the threshold line with the intensity axis, considering a distance of 10 km from the lake. A higher ESTI index indicates a higher sensitivity of the lake site to record earthquake shaking. In several studies, the earthquake magnitude is used instead of the epicentral seismic intensity (e.g. Avşar et al., 2016; Migowski et al., 2004)

It is important to note that the EQRT and ESTI values are core site-specific and more (lower EQRT, higher ESTI) or less sensitive locations can thus exist within a single lake basin (Sect. 7.4.2). A more detailed comparison of these calibration techniques is presented in Moernaut (2020).

Alternative calibration approaches involve calculating the required ground motions for inducing subaqueous slope instability or the occurrence of plastic in-situ deformation structures. When geometrical and geotechnical data for the subaqueous slope sequences is available, it is possible to calculate the required ground motion (i.e. PGA—peak ground acceleration) to induce slope failure by applying slope stability equations that allow incorporation of seismic loading. These calculations can be applied on different slope areas of a lake basin and this for different episodes over the entire studied depositional history. Comparing the outcome with the mapped MTD event stratigraphy in the basin then allows calculation of the minimum/maximum ground motion values at the lake site for different paleoseismic events (Fig. 7.5c; Strasser et al., 2011). For ductile in-situ deformation structures, numerical simulations have been performed using the concept of Kelvin–Helmholtz instability, formed by transient shear stress between different near-surface sediment layers (Wetzler et al., 2010). By implementing rheological data into the models, the required PGA for different stages and thicknesses of soft-sediment deformation can be calculated and compared with the identified structures in the sedimentary sequences. This method allowed a multi-scale paleoshaking reconstruction for the Dead Sea basin over the past 220 kyr (Lu et al., 2020). These two alternative approaches have each been applied in a single basin only (Lake Lucerne and Dead Sea basin).

7.4.2 Obtained Relationships: Thresholds and Size Scaling

Seismo-turbidites form the most-used subaqueous earthquake imprint. Turbidites at the foot of steep and coarse-grained alluvial fans and delta slopes are documented from intensities as low as ~V¼-V½ in Chilean and Alaskan lakes (Moernaut et al.,

2014; Van Daele et al., 2020), while intensity values of ~ VI-VII are required to trigger turbidity currents on muddy hemipelagic slopes in a wide suite of lake settings, such as Chile (Moernaut et al., 2014), Alaska (Van Daele et al., 2020), Japan (Lamair, 2018), New Zealand (Gomez et al. 2015; Howarth et al., 2016), European Alps (Daxer et al., 2022b; Rapuc et al., 2018; Sabatier et al., 2017). This threshold range of ~ VI-VII is similar to values obtained for MTD records in Swiss Alpine foreland lakes, where coeval translational slope failures were attributed to historical earthquakes (Sect. 7.6; Kremer et al., 2017). The threshold value for ductile in-situ deformation structures is proposed to be slightly lower, for example ~ V to VI½ at the Dead Sea region (Agnon et al., 2006; Lu et al., 2020; Migowski et al., 2004), ~ V for a lake in SW Turkey (Avşar et al., 2016) and ~VI½ for lakes in central Switzerland (Monecke et al., 2004) and on lacustrine slopes in Chile (Molenaar et al., 2021). This variability indicates a site-specific control on the formation of in-situ SSDS, exerted by sediment-physical parameters, stratification, slope gradient and earthquake type (Molenaar et al., 2022, 2024).

In comparison to these in-lake processes, relatively higher seismic intensities are proposed for extensive terrestrial mass wasting, inducing a significant catchment response and a long-term postseismic imprint in the lake sequence. This idea is based on the observation that (i) only few events recorded by subaqueous processes also show a catchment response signature (Avşar et al., 2016; Howarth et al., 2014; Oswald et al., 2021b), (ii) extensive onshore landsliding during recent earthquakes only occurred in areas of high seismic intensity, such as ≥ IX in southern New Zealand (Hancox et al., 2002) and in China (Xu et al., 2013) and (iii) significant post-seismic increase in suspended sediment yield is associated with regions of extensive terrestrial landsliding along the fault rupture (Hovius et al., 2011; Wang et al., 2015).

Besides the identification of intensity thresholds, a few multi-core studies also propose that the "size" of the sedimentary imprint may scale to the strength of the seismic shaking (Fig. 7.5d–f). For example, in a Chilean lake, a roughly linear relationship was suggested between seismic intensity and the cumulative thickness (summed thickness over multiple cores) of clastic turbidites in front of an alluvial fan (Moernaut et al., 2014). This idea was further refined by comparing the seismic intensities of the 1960 (M_W 9.5) and 2010 (M_W 8.8) south-central Chilean earthquakes with their sedimentary imprint in 17 lakes using 107 cores (Van Daele et al., 2015). When only considering suitable basins surrounded by sedimentary slopes, it was found that the number of muddy seismo-turbidites linearly increases with seismic intensity, starting with no muddy seismo-turbidites for intensities lower than V¾ and reaching 100% when intensities are VII½ or higher. Such scaling relationships in the Chilean lakes are possible because the mapped turbidites are mainly generated through the process of surficial remobilization, for which it is proposed that stronger shaking is capable of remolding slightly deeper stratigraphic levels, leading to larger turbidites (Moernaut et al., 2017a; Molenaar et al., 2021). Linear scaling relationships between turbidite abundance and seismic intensity of historical earthquakes were also documented in two Austrian lakes: Wörthersee and Millstätter See (Daxer et al., 2022b). The obtained trendlines are less steep than in the Chilean lake studies, indicating a wider range of seismic intensities that can be distinguished by mapping the size of the sedimentary imprints.

For in-situ SSDS, scaling relationships between the deformation type or thickness and seismic shaking have been proposed in some outcrop-based studies (Hibsch et al., 1997; Rodríguez-Pascua et al. 2000), however, without the means of robustly testing and quantifying such relationships. The numerical simulations for the Dead Sea margin sediments (Sect. 7.4.1; Fig. 7.5f) show a progressive development with increasing PGA: from disturbed laminations over folds (asymmetric billows and coherent vortices) to finally intraclast breccia at PGAs of ca. 0.14, 0.20 and 0.58 g, respectively, considering a 5 cm thick unit (Lu et al., 2020; Wetzler et al., 2010). Comparable PGA values were obtained by historical event calibration for in-situ SSDS on Chilean lake slopes (PGA ca. 0.17, 0.23, 0.57 g), even though very different lithologies and earthquake types are involved in the process (Molenaar et al., 2021). An overview of existing methods to obtain information on past earthquakes based on SSDS characteristics (e.g. type, thickness, spatial distribution) can be found in Zhong et al. (2022).

For subaqueous landslides, the volume of a single MTD is not directly related to the strength of shaking. This is because the thickness and area of the critically unstable slope is controlled by its failure history and how fast it was recharged with fresh slope sediments, in addition to several preconditioning factors such as local morphology, excess pore pressure distribution, spatial lithological changes, etc. (Strasser et al., 2007). Comparing the number of MTDs in a lake for a given earthquake and their total transported volumes can partly reduce these effects (Praet et al., 2017) and rough scaling relationships between seismic intensity and MTD number or volume can thus be established (Daxer et al., 2022b). Yet, whether quantitative paleoseismology based on MTD records is feasible still needs to be assessed by investigating more areas with a rich historical earthquake record and multiple basins that contain MTDs.

7.4.3 Controlling Factors on the Lacustrine Paleoseismograph

In most lacustrine paleoseismology studies, thresholds and scaling relationships are obtained from relatively recent earthquakes and sedimentary sequences and subsequently applied on the paleo-record. This disregards the potentially dynamic nature of lake systems on longer time scales during which the sensitivity of the lacustrine seismometer can be affected by significant changes in sedimentation rate and sediment type on the slopes and basin floor. In this way, the obtained paleo-earthquake record may form an under- or overestimation during specific periods. Such in-lake sediment-dynamic changes are caused by enhanced sediment availability in the lake catchment (e.g. landslide deposits, volcanic deposits, deglaciation, deforestation, land-use), increased fluvial transport capacity (e.g. extreme precipitation events, snow melt), or significant lake-level changes influencing the rate or spatial extent of slope charging (Lu et al., 2021b; Tournier et al., 2023). For example, periods of overall higher sedimentation rates in a Chilean lake and several Alpine lakes coincided with a much higher frequency of seismo-turbidites (Bertrand et al., 2008; Rapuc et al., 2018, 2022). A dominant role of sedimentation rate is further supported by a comparative study of the record sensitivity (ESTI) on a range of Alpine lakes with different basin floor sedimentation rates (Wilhelm et al., 2016). The mechanisms behind the effects of sedimentation rate and type on the lacustrine seismometer are still debated and are likely manifold: (i) higher sedimentation rates provide a faster recharging of slopes, potentially bringing slopes earlier to a near-critical state ((Strasser et al., 2011), (ii) higher sedimentation rates of fine-grained sediments may lead to excess pore pressure and underconsolidated sequences, thus requiring less strong shaking to fail (Moernaut et al., 2017b), (iii) changes in sediment cohesion may affect earthquake-triggered erosion of surficial sediments (Moernaut et al., 2017a), (iv) changes in sediment permeability may facilitate the development of liquefaction structures (Obermeier, 1996) and the initiation of slip planes ("weak layers") along which translational slope failures or ductile in-situ SSDS develop (Alsop et al., 2020; Moernaut et al., 2019), (v) higher sedimentation rates lead to relatively younger sequences. These have experienced a lower number of earthquakes that can have dynamically compacted the sequence and are thus less affected by seismic strengthening (e.g. Sawyer & Devore, 2015). As the current recommendation is to be very cautious in making paleoseismic interpretations when a significant change in sedimentation rate or type is encountered over a longer period (Rapuc et al., 2018, 2022), it is crucial to advance our understanding of these sediment-dynamic aspects for a full exploitation of lake paleoseismic records.

As seismic intensity only describes seismic ground motion in a semi-quantitative manner, there is an increasing tendency to consider other parameters for the thresholds and scaling relationships. These include peak ground acceleration (PGA; Avşar et al., 2016), peak ground velocity (PGV; Howarth et al., 2021b) or bracketed duration. For example in Chilean lakes, the amount of sediment erosion by surficial remobilization correlates better to the calculated bracketed duration of historical earthquakes than to their PGA or seismic intensity (Molenaar et al., 2021). An alternative parameter could be the Arias intensity (integration of the squared acceleration values over time), for which a good correlation to the number and size of terrestrial mass movements has been found (Wang et al., 2010). Besides statistical support, the selection of relevant ground motion parameters should rely on physics-based understanding of the underlying sedimentary process. For example, it is proposed that surficial remobilization is driven by the velocity contrast between the seismic motion of the topmost sediments and the inert water column (Molenaar et al., 2021), whereas ductile in-situ SSDS may be formed by the acceleration-driven relative motion between layers of different density (Wetzler et al., 2010). Significant challenges remain to include site-specific amplification effects for characterization of seismic ground motions at the lake or ocean bottom (Gomberg, 2018). These include topographic effects due to basin geometry and are related to the mechanical properties of the entire stratigraphic succession, often involving thick sequences of poorly consolidated sediments.

7.5 Constraining the Paleoseismic Source Using Multiple Subaqueous Basins

A regional correlation of paleoseismic evidence between different records is required to estimate the possible epicenter location, rupture length and magnitude of past earthquakes. This principally relies on comparison of event age probability density functions (PDFs) derived from individual age-depth models, potentially supported by statistical tests (Howarth et al., 2021a). Given that event age uncertainty (95% confidence range) is typically several decades or more, synchronicity of event deposits between sites cannot be proven by age information alone. This issue is overcome by incorporating other stratigraphic evidence that provides information on seismic intensity (Sect. 7.4.2) and by comparing the number of earthquake imprints between multiple sites within the timeframe set by the uncertainty on the earthquake ages that potentially correlate. For example, if a lake shows sedimentary evidence of strong shaking (e.g. intensity > VII) in a certain period, then sedimentary evidence with a significantly overlapping age PDF in a neighboring lake is most likely related to the same earthquake event (see case studies of Sect. 7.6, 7.7 and 7.8). If the sedimentary evidence represented a different large earthquake, each of the lakes should exhibit two closely-spaced imprints. However, when distances are large and the imprints reflect low seismic intensity (e.g. VI), several scenarios are possible and additional sites and ground-motion modeling are needed to evaluate these (Sect. 7.6).

Two ground motion modeling methods are currently in use to estimate the location, magnitude and size of paleoearthquakes based on the earthquake-induced imprints in multiple subaqueous basins (Fig. 7.6). The grid-search approach (Bakun & Wentworth, 1997) calculates a range of moment magnitudes over a grid of trial source locations using backward application of an empirical intensity prediction equation (IPE), in which the threshold intensities for positive and/or negative shaking evidence for a specific earthquake are used as input intensities. This allows constraining an area of possible epicenter locations and a range of magnitudes that could explain the distribution of lacustrine earthquake imprints. The larger the number of lakes with positive and negative evidence, the more constrained the potential source area is. Especially negative evidence plays an important role in constraining the reconstructed area of possible epicenters. The used threshold intensity mainly affects the resulting range of potential magnitudes.

In contrast to the grid-search method, the probabilistic modeling approach developed by Vanneste et al. (2018) simultaneously allows for the actual intensity to exceed or to be lower than the assumed positive or negative threshold values, respectively, and takes into account the uncertainties of IPEs. It is a forward modeling approach that handles positive and negative evidence in the same way by considering the (non-)exceedance probability given by probabilistic seismic hazard assessment (Cornell, 1968; McGuire, 1976). Moreover, it uses a simplified model of known active faults, segmented to allow for all geometrically-possible combinations of rupture lengths and linkages (UCERF3, Field et al., 2014). The

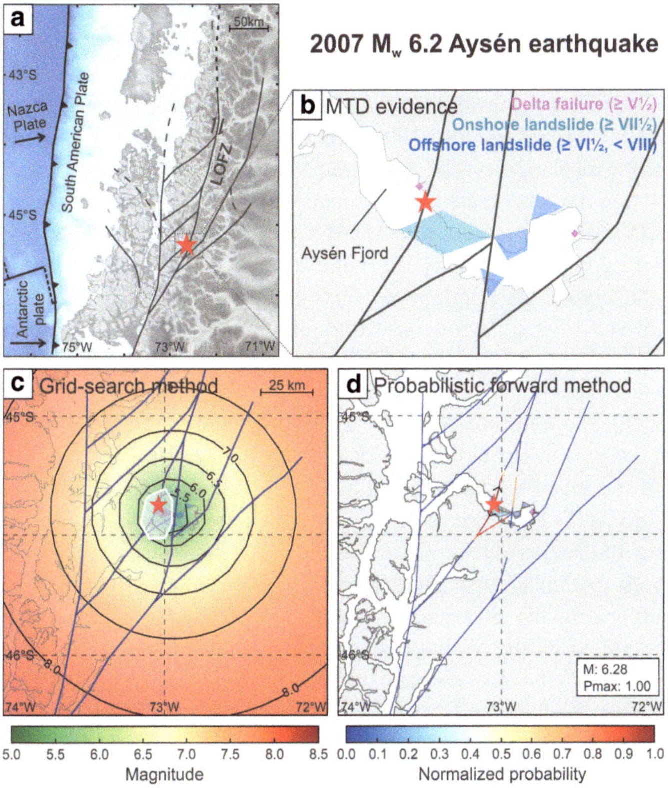

Fig. 7.6 Application of the grid-search method and probabilistic forward method to reconstruct earthquake source parameters as applied to mass-wasting evidence in Aysén Fjord (Chile) for the 2007 M_W 6.2 earthquake (epicenter indicated by red star). **a** Location of Aysén Fjord, crossed by several branches of the Liquiñe-Ofqui Fault Zone (LOFZ) that accommodates the trench-parallel component of oblique subduction; **b** Potential source areas for mass-transport deposits (MTD) in the fjord, color-coded according to the type of MTDs generated by the 2007 earthquake. This distribution is used as input for the grid-search method and probabilistic forward method, and includes a combination of positive and negative evidence: delta failures ($\geq$ V½, pink), onshore landslides ($\geq$ VII½, turquoise) and offshore landslides in absence of onshore landslides ($\geq$ VI½, < VIII, blue). **c** Results of the grid-search method showing all possible combinations of magnitude and location that could generate the imprints observed in B. White line and enclosed hatched area mark the most likely epicentral region. **d** map showing the probability that an earthquake with magnitude 6.28 (i.e. the magnitude for which the highest probability is obtained) on each of the LOFZ fault sections caused the spatial pattern of positive and negative evidence for MTDs in the fjord presented in B. Arrows indicate limits of rupture trace with highest probability. Modified from Vanneste et al. (2018)

probability for a rupture on each of these fault sections to have caused a given combination of positive and negative paleoshaking evidence is subsequently calculated. This method not only indicates the most likely fault segment, but also provides a magnitude estimate by considering the rupture area and a magnitude-scaling relationship (Wells & Coppersmith, 1994). In this way, the probabilistic method can also generate information on the maximum distance and associated magnitude of earthquakes that have generated the mapped distribution of sedimentary imprints.

Both methods rely on good age control and detailed basin-wide mapping for earthquake-triggered sedimentary imprints, so that both positive and negative evidence for the considered earthquake timing can be compiled. Especially when the lakes are very distant, erroneous combinations of paleoseismic data are more frequent and may result in extremely high magnitude estimates (Kremer et al., 2017). Another important requirement is well-constrained intensity thresholds for the used earthquake imprints (Sect. 7.4.2). In addition, the outcome of the forward probabilistic approach strongly depends on the quality of the regional active fault model. Both the grid-search and probabilistic modeling approach were compared and applied to MTD evidence in Aysén Fjord, southern Chile, resulting from various shaking intensities during crustal earthquakes (Vanneste et al., 2018). The results of both methods are in good agreement, although the probabilistic method tends to outperform the grid-search method and is generally more constraining. This is exemplified by the results obtained for a recent M_W 6.2 crustal earthquake in 2007, for which the probabilistic method provides a better magnitude range estimate (6.05–6.5) compared to the grid-search method (5.5–6.0) and indicates the actual fault segment that ruptured rather than a (relatively small) area covering multiple fault strands, as is the case for the grid-search method.

7.6 Intraplate Settings: Many Source Fault Candidates and Irregular Recurrence Patterns

Intraplate settings are generally characterized by low deformation rates and, consequently, long interseismic periods that exceed the time period covered by instrumental and historical earthquake data. Moreover, instrumental earthquake records often do not allow identification of active faults as seismicity is distributed over wide regions that contain many faults. Thus, the knowledge of active faults and their hazard potential is often limited for intraplate settings. Given the typically episodic and migrating nature of earthquakes there, a regional approach in which the frequency of large earthquake occurrence in a certain area is determined may be more relevant for hazard analysis than studying every potentially active fault section. Lacustrine paleoseismological studies in intraplate settings have been carried out around the world (Fig. 7.1 and Supplementary Table 1) and include the eastern USA (Monecke et al., 2018), eastern Canada (Brooks, 2018; Doughty et al., 2014), China (Fan et al., 2020, 2022, 2023; Jiang et al., 2017; Liu et al., 2022; Shi et al., 2022; Wang et al., 2021; Wei et al., 2021), Kyrgyzstan (Lauterbach et al., 2019), Norway (Bellwald et al., 2019), Sweden (Mörner, 2005), Finland (Ojala et al., 2019), the French Massif Central (Chapron et al., 2021; Chassiot et al., 2016) and the European Alps (Banjan et al., 2023; Chapron et al., 1999; Kremer et al., 2017; Monecke et al., 2004; Oswald et al., 2021b; Strasser et al., 2013; Wilhelm et al., 2016). Most of these studies are performed in lakes of glacial origin, limiting the maximum time span of their sedimentary record (e.g. max. ~18–19 kyr in the European Alps; Ivy-Ochs et al., 2004). Within these lakes, the general earthquake imprints are MTDs, turbidites and SSDS. In particular cases, active fault structures can be identified so that both on-fault and off-fault paleoseismic evidence is recorded within a single stratigraphic framework (Gastineau et al., 2021; Oswald et al., 2021a). This allows for a direct integration of information on surface rupture dimensions and seismic shaking strength (Oswald et al., 2021a). Many of the lacustrine paleoseismic studies involve comparison of several lacustrine (and terrestrial) paleoseismological records within a certain area (Fig. 7.7), in this way revealing different phases of large earthquake activity (Bellwald et al., 2019; Brooks & Adams, 2020; Kremer et al., 2017; Mörner, 2011; Ojala et al., 2019; Oswald et al., 2022). In the Swiss Alps, over 20 years of limnogeological research with a paleoseismological focus on more than 10 lakes generated an extensive dataset that formed the base for a regional multi-lake approach (Fig. 7.8b, c) to reconstruct the potential epicentral areas and magnitude ranges of paleo-earthquakes using the above-mentioned grid-search approach (Sect. 7.5).

7.6.1 Regional Lacustrine Paleoseismology in Central Switzerland

In the studied Swiss lakes, the earthquake imprints are essentially formed by MTDs and SSDS. Their interpretation was based on a comprehensive understanding of the sedimentary processes in each lake basin and a thorough analysis of all potential triggers. The most important support for an earthquake trigger is the existence of multiple synchronous MTDs in one or several basins (i.e. synchronicity criterion (Schnellmann et al., 2002). MTDs that are on the same seismic-stratigraphic level or are dated within the same time window were considered the product of synchronous slope instabilities. Age-depth models were based on dating with ^{14}C, ^{137}Cs, ^{210}Pb and supported by historically-documented events. In these records, average age uncertainty is around 250 years for sediments dated with ^{14}C, whereas age control for the last 100–150 yr is much more precise when ^{137}Cs and ^{210}Pb data are available. A prominent example of the lacustrine paleoseismological approach relates to the historical 1601 CE M_W 5.9 Unterwalden earthquake (location on Fig. 7.8a) which caused multiple subaqueous (and subaerial) MTDs as well as SSDS in Lake Lucerne, Lake Lungern, Lake Baldegg, Lake Seelisberg and Lake Aegeri (Fig. 7.8b, c; Kremer et al., 2017; Monecke et al., 2006; Schnellmann et al., 2006). For example in Lake Lucerne, at least 14 individual MTDs related to this historical earthquake have been mapped in different subbasins (Hilbe & Anselmetti, 2014; Monecke et al., 2006). The grid-search approach was tested on

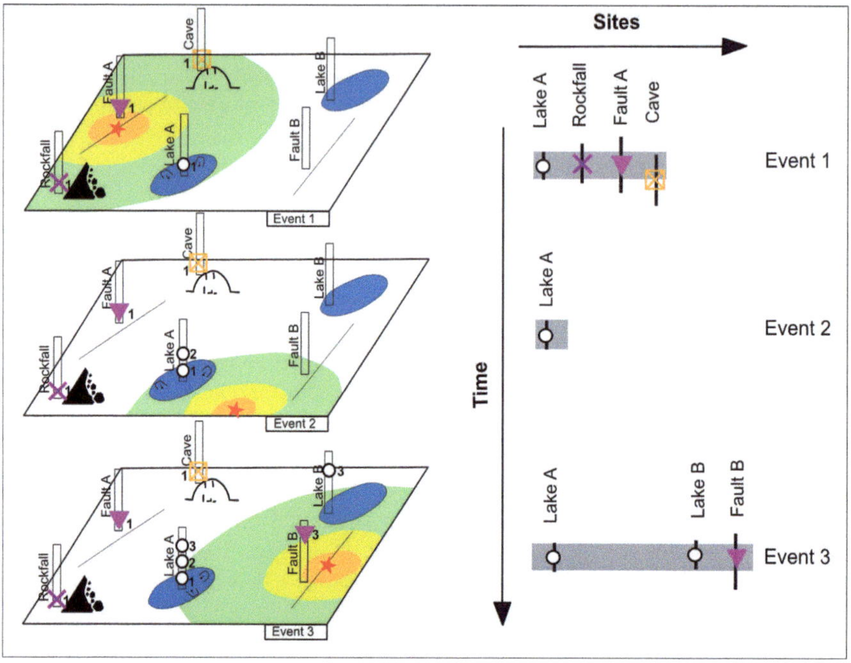

Fig. 7.7 Cartoon describing the concept of integrative regional paleoseismology. Spatial distribution of earthquake-related processes such as mass movements (terrestrial and sublacustrine), destruction in caves and on-fault surface ruptures for three paleo-events (left; modified from Becker et al., 2005), compared to their corresponding paleoseismic evidence including dating uncertainty (right). Spatiotemporal correlation of evidence forms the basis for paleo-earthquake scenarios. Red star: earthquake epicenter; colored zones: decreasing seismic intensity with distance from the rupture area

this historical case study (Fig. 7.8a) using a total of five lakes with positive evidence and four lakes with negative evidence. The resulting epicentral area matches with the location of the earthquake epicenter determined from numerous macroseismic intensity data points (EMS-98 scale) inferred from historical documents (Fäh et al., 2011). The intensity threshold for lacustrine paleoseismic evidence was modified until the obtained magnitude was similar to the actual magnitude of the earthquake obtained from historical seismology, resulting in a threshold intensity of VI 2/10 (Kremer et al., 2017). The same threshold value was used for applying the grid-search approach on well-dated MTD records in eleven lakes over the past 12 kyr, allowing potential areas and magnitude ranges of paleoearthquakes to be estimated. This revealed several phases of enhanced MTD occurrence in multiple lakes (at ~ 12 kyr BP, 9.7 kyr BP; 6.5 kyr BP and during the last 4 kyr Becker et al., 2005; Monecke et al., 2006; Strasser et al., 2013; Kremer et al., 2017). Because of the age uncertainty range of ~250 years, it remains unclear whether these phases were each caused by a single large earthquake or by a series of smaller earthquakes closely-spaced in time but large enough to locally generate MTDs. Therefore, the

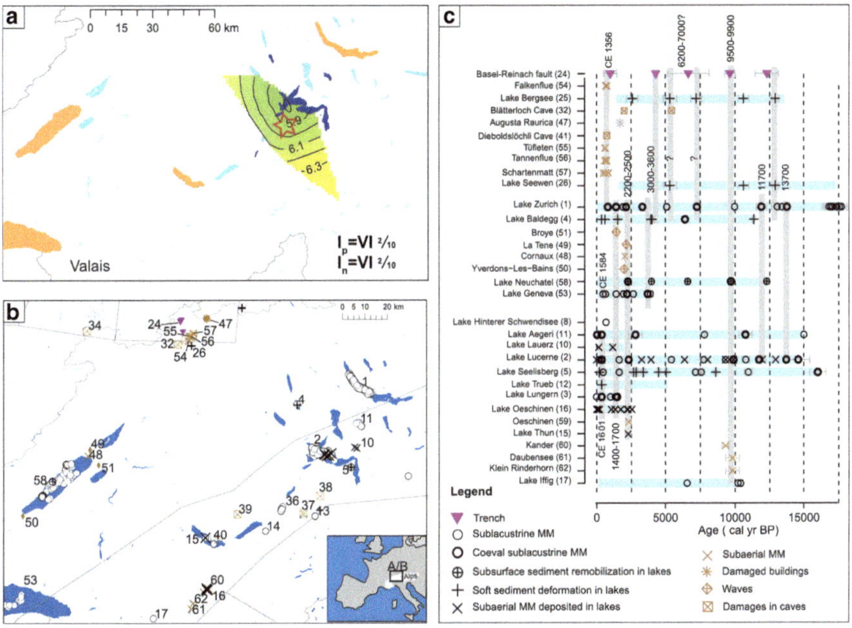

Fig. 7.8 Lacustrine and integrative paleoseismology in the intraplate case study of Central Switzerland. **a** Reconstruction of the possible epicentral area and magnitude range of the historical 1601 CE earthquake (M_W 5.9; epicenter marked by the red star) based on the grid-search approach on earthquake imprints in lake sediments (Kremer et al., 2017). Earthquake-triggered mass movements and deformation are recorded in lakes marked in dark blue while lakes that do not record any evidence for the considered period are marked in orange. **b** Spatial distribution of potential paleoseismological evidence mentioned in literature and summarized in a database (Kremer et al., 2020). **c** Temporal distribution of the potential paleoseismological evidence at each site (sites are numbered and shown in **b**. Only evidence based on reliable age models is shown. The length of the blue horizontal bars represents the length of the continuous lacustrine records. The grey vertical bars show potential coeval evidence (within dating error range) and may represent single past earthquakes or periods of enhanced seismicity

reconstruction of epicenters and magnitudes of paleoearthquakes is based on different alternative scenarios, especially when distance between the lakes is large (up to several 100 km). In Kremer et al. (2017), the phases of widespread MTD occurrence were firstly treated as the product of single large events to test if there are any plausible scenarios that could explain the distribution of positive and negative evidence. This approach suggests that larger earthquakes with magnitude exceeding 6.4 could have occurred at least four times in the past 12 kyr in the Valais region, which is one of the seismically most active regions in Switzerland.

7.6.2 Integrative Paleoseismology

The reconstruction of epicenters and magnitudes of paleoearthquakes based on the grid-search approach relies on how well temporal correlation of earthquake imprints between individual sites can be constrained. In order to improve the confidence level on this correlation, one approach is to integrate the lacustrine results with other paleoseismological archives. Although these other evidence types also suffer from (often larger) dating uncertainties, they may provide support for one or the other reconstruction scenario. For the case study in Switzerland, a database of potential primary and secondary evidence was compiled taking into account data from sedimentological (e.g. lakes), geomorphological (e.g. fault displacement, terrestrial mass movements; e.g. Kremer et al., 2022), archeological (e.g. evidence of tsunami waves) and speleological (e.g. broken speleothems and other damages in caves) research (Kremer et al., 2020). This new compilation corroborates the proposed periods of enhanced seismicity from lacustrine paleoseismology and provides a more comprehensive perspective on the associated secondary hazards related to such rare high-impact events. One example of the value of integrative paleoseismology relates to the 9.5–9.9 kyr BP phase (Fig. 7.8c). Within this phase, primary on-fault evidence suggests a seismic source in the Basel region (Ferry et al., 2005) whereas the epicenter reconstruction based on the lacustrine datasets suggests a seismic source in the Valais region (Kremer et al., 2017). Hence, the spatial distribution of paleoseismic evidence at 9.5–9.9 kyr BP suggests the occurrence of at least two earthquakes, i.e. a phase of enhanced seismic activity rather than a single large event. This hypothesis is furthermore supported by an increased occurrence of subaqueous and terrestrial MTDs over large part of the Alpine region in this period: (i) in French alpine lakes Le Bourget, Paladru and Blanc Huez, MTDs and turbidites are dated to 9.55 ± 0.15 kyr, 9.01 ± 0.6 kyr and 9.68 ± 0.14 kyr BP, respectively (Chapron et al., 2016; Simonneau et al., 2014); (ii) in an Austrian alpine lake (Lake Piburgersee), well-developed SSDS indicating severe seismic shaking at 9.76–10.13 kyr BP (Oswald et al., 2021b); (iii) the two largest Holocene mass movements of the Alps, the Flims (Switzerland) and Köfels (Austria) rockslides, are dated at 9.48–9.43 kyr BP (Deplazes et al., 2007) and 9.527–9.498 kyr BP (Nicolussi et al., 2015), respectively; (iv) two rock avalanches that are discussed within a paleoseismological context occurred in the Bernese Alps at 9.8 ± 0.5 kyr BP (Grämiger et al., 2016).

Lacustrine and integrative paleoseismological research in Canada (Brooks & Adams, 2020), Sweden (Mörner, 2005), Finland (Ojala et al., 2019) and in Norwegian lakes and fjords (Bellwald et al., 2019) also show evidence for multiple large earthquakes in the Late Glacial and Early Holocene, similar to the Swiss case study. Enhanced seismicity in these periods were interpreted to result from removal of the ice overburden and fast isostatic uplift after the last glaciation (Banjan et al., 2023; Brooks, 2018; Mörner, 2011). It is still debated whether this phenomenon might still be modulating the current seismic hazard in formerly-glaciated intraplate regions (Brandes et al., 2015).

7.7 Transform Settings: The Role of Segment Boundaries

Lacustrine paleoseismology has been applied to understand the sedimentary signatures of earthquake shaking and to develop earthquake records for a growing number of Earth's transform faults. These include the Alpine Fault in New Zealand (Howarth et al., 2012, 2014, 2016, 2021a), the North and East Anatolian faults in Turkey (Avşar et al., 2014a, b; Boës et al., 2010; Gastineau et al., 2021; Hubert-Ferrari et al., 2012, 2020), the Dead Sea Transform in the middle east (Alsop et al., 2022; Begin et al., 2005; Kagan et al., 2018; Ken-Tor et al., 2001; Lu et al., 2020; Marco et al., 1996; Migowski et al., 2004), the San Andreas Fault in California (Goldfinger, 2021), the Polochic Fault in Guatemala (Brocard et al., 2016), the Boconó Fault in Venezuela (Beck et al., 2019), the Magallanes-Fagnano Transform in Tierra del Fuego (Waldmann et al., 2011) and the Palu-Koro Fault in Sulawesi Island (Tournier et al., 2023). To date, most studies have been limited to single lakes with a main focus on understanding the sedimentary signatures of strong ground motions using temporal correlation with historic earthquakes (Avşar et al., 2014b; Boës et al., 2010; Ken-Tor et al., 2001). Sedimentary signatures range from SSDS (Lu et al., 2020) to MTDs (Waldmann et al., 2011), seismo-turbidites (Hubert-Ferrari et al., 2020), organic rich intercalations (Avşar et al., 2014b), catchment responses (Avşar et al., 2014a; Howarth et al., 2014) or a combination of the above (Howarth et al., 2021a) depending on the specific lake setting.

A more limited set of lacustrine paleoseismic studies have established recurrence statistics for large earthquakes on the main and subsidiary strands of transform faults (Avşar et al., 2014a; Hubert-Ferrari et al., 2020; Lu et al., 2020; Waldmann et al., 2011). For example, clustered earthquake behavior for the Dead Sea Transform was inferred using SSDS identified in a 220 kyr drill core. SSDS were linked to $M_W > 7$ earthquakes using numerical modeling of deformation to relate the degree of deformation to ground motions and finally earthquake magnitude (Sect. 7.4.2; Lu et al., 2020). An important limitation of single-lake studies is that it can be difficult to conclusively link inferred ground motions at the site to rupture on one specific fault or fault section. Consequently, single lake studies often do not contribute to spatiotemporal reconstructions of earthquake behavior for individual faults unless they can be correlated with independent paleoseismic information (Beck et al., 2019).

Multi-lake quantitative lacustrine paleoseismic studies provide a regional distribution of strong ground motions for a given earthquake and thus do have the potential to better constrain the specific earthquake source. Such studies have contributed substantially to understanding the spatiotemporal behavior of transform faults (e.g. Howarth et al., 2021a). In the following sections, quantitative lacustrine paleoseismic records of Alpine Fault earthquakes in New Zealand, are reviewed to illustrate how such studies are conducted and demonstrate their utility for understanding earthquake behavior at transform plate boundary faults.

7.7.1 Developing a Lacustrine Paleoseismic Record for the Alpine Fault

The Alpine Fault is an 850 km-long transform fault that forms the boundary between the Pacific and Australian plates in New Zealand's South Island (Fig. 7.9a). It is an example of a structurally isolated, mature transform fault because the southern two thirds of the fault have strike-slip rates of ~30 $mm.yr^{-1}$ that accommodate the majority of the plate motion in this region (DeMets et al., 2010; Norris & Cooper, 2007). Segmentation of the fault into five sections is defined by variation in slip rate, structure and geometry (Barth et al., 2013). Lacustrine paleoseismology has been applied along the Central section and integrated with paleoseismic records from near-fault wetlands along the South Westland section. These combined records cover the last ~4 kyr and have been integral in the development of one of the most spatially and temporally precise rupture reconstructions yet developed for a single fault (Fig. 7.9f; Howarth et al., 2021a).

In four lakes (Kaniere, Mapourika, Paringa and Ellery) proximal to the Alpine Fault, the sedimentary signature of large earthquakes was determined by examining the lake response to the last earthquake. The most recent earthquake occurred in 1717 CE, had an estimated magnitude of M_W 8.1 and ruptured > 380 km of the fault (Fig. 7.9c; De Pascale & Langridge, 2012; Wells et al., 1999). It was identified in the lakes using temporal correlation. High-precision ages for earthquake signatures were obtained using Bayesian age-depth modeling of numerous levels (many tens per lake) of ^{14}C-dated terrestrial macrofossils, providing earthquake ages with decadal precision (Fig. 7.4a). The 1717 CE earthquake formed a distinctive sequence of coseismic and postseismic deposits, henceforth termed Earthquake Event Sequences (EES; Fig. 7.9b). The sequence starts with lake-wide subaqueous mass-wasting triggered by shaking that was recognized in cores as decimeter-thick turbidites. This was followed by years to decades of accelerated fluvial influx and deposition of terrigenous sediments, which were derived from earthquake-induced landslide deposits in the lake catchment (Sect. 7.2: catchment response). Far field earthquakes that produce lower shaking intensities at the lake sites were recorded by thin turbidites formed by subaqueous mass-wasting alone (Fig. 7.9b; Howarth et al., 2014, 2016, 2018).

Cores from the four lakes contain between 6 and 15 EES over the past 1.6–4.0 kyr (Fig. 7.9f). Establishing whether or not the lakes faithfully record Alpine Fault earthquakes requires that (i) non-seismic trigger mechanisms can be discounted and (ii) non-Alpine Fault seismic sources can be eliminated. EES were inferred to be synchronous between lakes because there is no statistical difference in EES ages between sites and the temporally correlated EES contain only a single turbidite derived from subaqueous mass-wasting, which precludes multiple events closely spaced in time. A series of earthquakes several hours to years apart would be recorded as separate subaqueous mass-wasting derived turbidites, each with a clay cap (Van Daele et al., 2017) and potentially a few silt laminations between them, a signature which was not observed (Howarth et al., 2018). The sequence of subaqueous mass-wasting followed by increased fluvial sediment flux eliminates

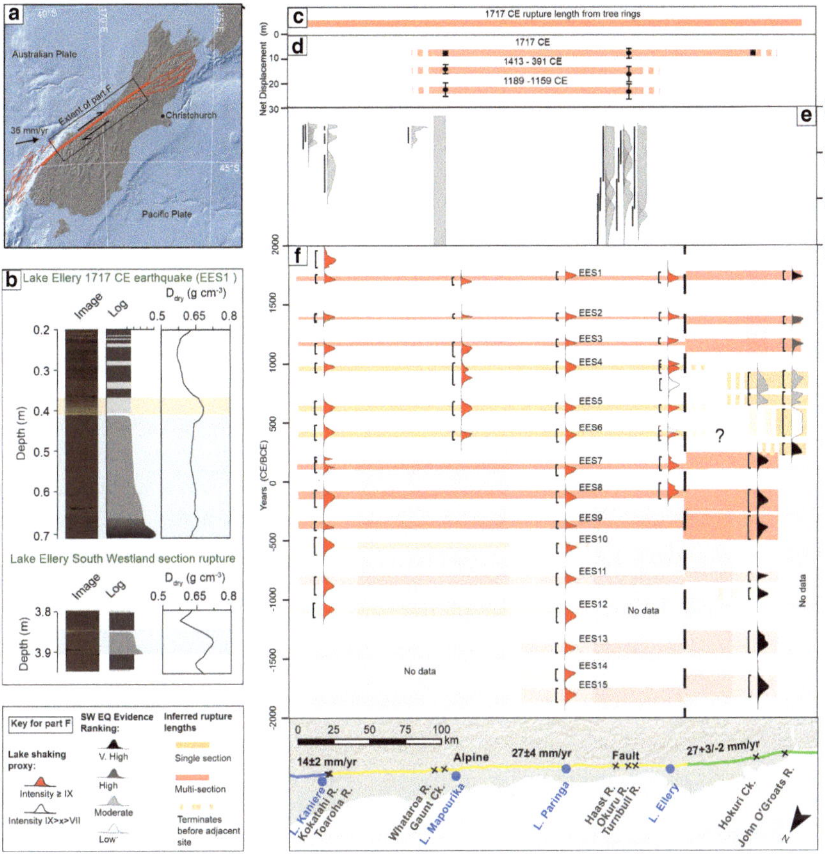

Fig. 7.9 Spatiotemporal reconstruction of earthquakes on the Alpine Fault, New Zealand, using lacustrine paleoseismology (modified after Howarth et al., 2021a). **a** The Alpine Fault in the context of the boundary between the Pacific and Australian plates in the South Island of New Zealand. **b** Core image, graphic log and density of terrestrial sediment for earthquake event sequences (EES) and subaqueous mass-wasting derived turbidites in Lake Ellery that record seismic intensity (MMI) ≥ IX and IX > x > VII shaking, respectively. **c** Rupture extent of the 1717 CE great earthquake, which validates the rupture length derived from the lake records. **d** Single event displacement locations (black crosses) and amounts (shown cumulatively) for the last three Alpine Fault earthquakes. **e** Earthquake ages from trench sites along the Alpine Fault's Central section. **f** Timing of intensity ≥ IX shaking events (red probability density functions, PDFs) recorded at each lake site and surface rupturing earthquakes on the SW section recorded in fault-proximal wetland stratigraphies (black to grey PDFs). Intensity IX > x > VII shaking events from Lake Ellery that correlate temporally with rupture of the SW section are shown as white PDFs. Horizontal lines represent rupture extents inferred on the basis of earthquake age correlations between sites and details of the paleoseismic event signatures, where red horizontal lines show ruptures that propagated through the Central-South Westland (C-SW) section boundary and yellow horizontal lines those that did not. The thickness of the inferred rupture extents depicts the 95% highest PDF range for earthquake ages and shading depicts correlation confidence. Black vertical dashed line shows the location of C-SW section boundary

non-seismic drivers, such as spontaneous failure of subaqueous slopes, lake level variations or hydrometeorological events. Moreover, non-seismic processes fail to explain the regional synchronicity of EES in up to four lakes over 240 km distance. Hence, large earthquakes were interpreted as the only process capable of generating EES in two or more of the lakes (Howarth et al., 2014, 2016).

To produce a lacustrine paleoseismic record specific to the Alpine Fault, it is also necessary to eliminate triggering by other seismogenic sources. The EQRT approach was used to assign a seismic intensity threshold (based on Modified Mercalli Intensity: MMI) to earthquake signatures in the lakes using modeled isoseismals for the 1717 CE and other historic earthquakes (Howarth et al., 2016). The analysis demonstrated that EES are produced at seismic intensity (MMI) $\geq$ IX. Conversely, turbidites that archive only subaqueous mass-wasting result from lower shaking intensities of > VI but < IX depending on the lake (Howarth et al., 2021a). Forward modeling of isoseismals for 110 active or potentially active faults surrounding the Alpine Fault (Cox et al., 2012) was then used to demonstrate that even synchronous multi-fault rupture was not capable of generating the distribution of high intensity shaking inferred from the lake records. Also, a M_W 7.2 earthquake located directly under any of the lake sites failed to produce the observed multi-lake sedimentary imprint. Hence, EES correlated between multiple lakes provide a paleoseismic record of ruptures along the Alpine Fault´s Central section.

The lake-derived paleoseismic record for the Alpine Fault provides a catalogue of rupture extent along the Central section for the past 2.5 kyr and recurrence information for this section over the last 4 kyr (Fig. 7.9f). As intensity IX shaking correlates spatially with the extent of fault surface rupture, it can be used to map earthquake rupture length when correlated temporally between sites (Howarth et al., 2016). On this basis, minimum Central section rupture extents were mapped, showing that the entire length of the Central section ruptured in the last eight events. Further back in time, the record is less complete because of the limited temporal extent of some lake cores (Fig. 7.9f). However, the lake record does precisely record the timing of the last 15 earthquakes, providing an average recurrence interval (~249 yr) on the Central section. The Central section has one of the lowest CoVs (0.22) yet documented for any fault section for which a sufficient number of paleoseismic events is available to allow robust recurrence analysis (Howarth et al., 2021a).

7.7.2 Comparison of Lacustrine and On-Fault Paleoseismic Records for the Alpine Fault

Integration of other forms of paleoseismic evidence can be used to strengthen spatiotemporal paleoearthquake reconstructions for transform faults derived from lakes. For example, rupture length reconstructions from tree ring damage and fault trenching for the 1717 CE M_W 8.1 Alpine Fault earthquake match the lake-derived rupture length reconstructions (Fig. 7.9c). The large right-lateral single-event displacements (>7 m) of geomorphic features (terrace risers and stream channels) and the timing of surface rupture along the Central section suggest the last three large earthquakes spanned a large part of the section (Fig. 7.9d, f; Berryman et al., 2012a, b; De Pascale et al., 2014) and thus further corroborate the lake reconstruction (Howarth et al., 2021a).

Comparisons between the on-fault paleoseismic evidence and the lake-based reconstruction for the last three events on the Alpine Fault also highlight some of the strengths of lacustrine paleoseismic data. Earthquake ages derived from on-fault trenching typically span multiple centuries due to limitations in the availability of high-quality target fractions for radiocarbon dating and the stratigraphic context of the trenches (Fig. 9e; Berryman et al., 2012a, b). The lake-based reconstructions provide an order of magnitude better event age precision, which aids the development of unique along-strike temporal correlations that are not possible with the fault trench data on its own (Fig. 7.9f). Trench investigations on the Alpine Fault rarely record more than the last three events due to high rates of landscape evolution (De Pascale & Langridge, 2012; Langridge et al., 2018, 2021). By contrast, the lakes used in the paleoseismic reconstruction record many times this number of earthquakes. The total record length is currently limited by the available coring technology. As these lakes formed ~17 kyr BP when glaciers retreated at the end of the last glacial maximum (Barrell et al., 2011), there is potential to generate spatiotemporal earthquake reconstructions that span many tens of events and thus robustly constrain the earthquake behavior for the Alpine Fault (Howarth et al., 2018).

7.7.3 Rupture Terminations and Spatiotemporal Rupture Behavior

The spatiotemporal rupture behavior of the Alpine Fault's Central (C) and South Westland (SW) sections was resolved by combining lacustrine paleoseismology with paleoearthquake evidence from the SW section derived from two wetland sites located on the fault that provide precise ages for surface ruptures (M_W > 7) of this section (Berryman et al., 2012b; Cochran et al., 2017). The wetland stratigraphies record Alpine Fault earthquakes as abrupt transitions between peats and silts because ~7.5 m of dextral and ~1 m of west-side up displacements during earthquakes impounded rivers draining across the fault, causing ponding of water against the fault scarp. These near-fault swamp records are thus unique to the Alpine Fault and further information on how they were generated can be found elsewhere (Berryman et al., 2012b; Cochran et al., 2017; Howarth et al., 2018).

Statistical comparison of earthquake ages between the sections, details of the shaking signatures preserved in the lakes and published event signature confidence ranks for the SW section were combined to identify both synchronous and independent ruptures of the Alpine Fault's C and SW sections. The location of Lake Ellery near the C-SW section boundary makes the site uniquely placed to record sedimentary evidence of through-going or terminating ruptures. Three through-going ruptures between 1717 and 1100 CE were inferred because there was no significant difference between earthquake ages for the sections, the occurrence of EES containing single turbidites at Lake Ellery implying a single rupture, and strong event confidence rankings from the wetlands implying large surface ruptures (Fig. 7.9f). It was inferred that seven ruptures terminated at or near the section boundary between ~1100 CE and ~300 CE. Significant differences in earthquake ages between the sections, earthquake signatures indicative of lower shaking intensities than for through-going ruptures at Lake Ellery, and low event confidence rankings in the wetlands all imply smaller terminating ruptures. For example, shaking events of seismic intensity < IX and > VII at Lake Ellery record the termination of the ~800 CE and ~500 CE SW section ruptures. Between ~300 CE and ~400 BCE three ruptures likely propagated through the C-SW section boundary as there was no significant difference between earthquake ages and EES with single turbidites occurred at Lake Ellery.

The pattern of alternating modes (termed mode-switching) of through-going and terminating ruptures observed in the paleoseismic data indicates that the section boundary plays an important role in controlling the rupture behavior of the Alpine Fault (Fig. 7.9f). Whether or not ruptures terminate at this boundary influences rupture length and produces two modes of earthquake behavior typified by periods of major (M_W 7–8) and great (M_W > 8) earthquakes. The spatial and temporal completeness of the Alpine Fault paleoseismic record make it sufficiently precise to validate physics-based models of the earthquake cycle. Earthquake catalogues from physics-based simulations of multiple Alpine Fault earthquake cycles show close resemblance to the paleoseismic observations when parameterized with realistic fault geometry (Howarth et al., 2021a). In the simulations switching between rupture modes occurs because heterogeneous stress states evolve over multiple seismic cycles in response to along-strike differences in geometry, demonstrating that minor changes in strike and dip between the sections control the mode-switching behavior. These geometric complexities exert an important control on rupture behavior that is not currently accounted for in fault source models generated for seismic hazard analysis.

7.8 Subduction Settings: Giant Earthquakes and Multiple Earthquake Sources

Subduction zones constitute some of the most complex tectonic settings, affected by high-magnitude tsunamigenic interplate earthquakes on the subduction megathrust, as well as intraplate events in the downgoing oceanic slab (intraslab earthquakes) and in the shallow continental crust (crustal earthquakes). In areas characterized by oblique subduction, large transform fault systems develop, accommodating the trench-parallel component of subduction. In this way, the seismic hazards at subduction zones combine those that exist in intraplate and transform settings, augmented by the occurrence of extensive megathrust earthquakes. Disentangling earthquake sources thus forms a key issue specific to subduction zone paleoseismology and constitutes a prerequisite in order to confidently infer source parameters. Lacustrine paleoseismic studies have proven particularly useful in recording seismic shaking that results from different earthquake source mechanisms. This was demonstrated by traces of both historical megathrust and intraplate earthquakes found in lake sediments worldwide, from proglacial Lake Eklutna in Alaska (Van Daele et al., 2020), over two tropical lakes (Maninjau, Singkarak) in West Sumatra (Wils et al., 2021a), to Lake Tutira in New Zealand (Gomez et al., 2015), Lake Biwa in Japan (Inouchi et al., 1996) and several subaquatic basins in southern Chile (Van Daele et al., 2019; Wils et al., 2018).

The source mechanism for past earthquakes can be constrained by considering the difference in spatial extent between inter- and intraplate earthquakes. The extensive rupture area of subduction earthquakes often leads to the presence of synchronous paleoshaking evidence over wide areas, as is the case for the 2010 M_W 8.8 Maule earthquake in south-central Chile, which was traced in the sediments of several lakes along a trench-parallel transect of 770 km long (Van Daele et al., 2015). The sedimentary imprints show a clear tapering southward beyond the rupture termination following the decreasing seismic intensity with distance from the rupture area. The southernmost evidence for the 2010 earthquake is only found in shaking-sensitive depositional areas, such as in front of alluvial fans (Sect. 7.4.2). Accordingly, by combining the presence/absence and type of simultaneous paleoshaking evidence in a series of lacustrine earthquake records, it is possible to infer the along-strike extent of a megathrust rupture. Ideally, such paleoseismic sites are distributed along-strike of the subduction zone and comprise a combination of multi-threshold study sites (Wils et al., 2020), allowing quantitative approximations by applying ground-motion attenuation relations. Due to the limited area that is generally affected by intraplate earthquakes, the lack of contemporaneous paleoshaking evidence for a specific event at distant locations may indicate a local intraplate source mechanism. Moreover, correlation with other types of paleoseismic archives may further constrain earthquake sources. For example, the presence of contemporaneous tsunami deposits or coseismic coastal elevation changes (Garrett et al., 2016; Kempf et al., 2020; Shennan et al., 2016; Walton et al., 2021) rule out intraslab events or in-land crustal earthquakes (Fig. 7.10), while trenching records or terrestrial fault scarps provide unambiguous evidence for a crustal earthquake (Leithold et al., 2019). In this particular case, similar approaches for retrieving rupture locations and/or magnitude as those elaborated in the previous sections on intraplate (Sect. 7.6) and transform (Sect. 7.7) settings can be applied, as demonstrated by Vanneste et al. (2018) for shallow crustal earthquakes recorded in Aysén Fjord, located along the Chilean subduction zone (Sect. 7.5).

At subduction zones, most lake paleoseismic studies cover only a single basin, but some multi-lake records are under development in Cascadia (Morey et al., 2013), central Japan (Lamair, 2018) and south-central Alaska (Praet, 2020). Southern Chile is currently the only subduction setting where multiple long-term lacustrine paleoseismic records have been established, and where these have been correlated and combined with other paleoseismic evidence in order to reveal past megathrust rupture modes (Sect. 7.8.2; Wils et al., 2020). Distinguishing between megathrust earthquakes and intraplate earthquakes using a multi-lake approach can

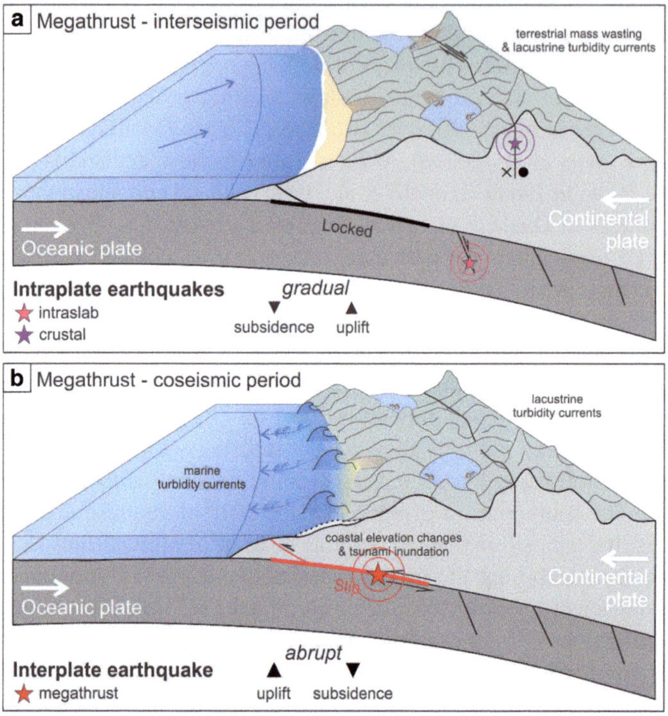

Fig. 7.10 Schematic visualization of processes related to earthquakes in subduction zones. **a** During the interseismic period of the megathrust fault, the plate interface is locked and gradual deformation occurs in the continental plate, often resulting in slow coastal uplift. Intraslab as well as crustal earthquakes can occur, potentially triggering terrestrial mass wasting and lacustrine turbidity currents. **b** Rupture of the interplate megathrust fault (coseismic period) generates abrupt uplift of the near-trench portion of the continental plate and more distant subsidence, triggering a tsunami. The coseismic elevation changes and/or associated tsunami can be recorded in coastal archives such as coral microatolls, marshes and lakes. Megathrust earthquakes also trigger turbidity currents in lakes as well as in the marine realm

be challenging as a result of, for example, age uncertainties and alternative seismic scenarios that fit the paleoseismic data. This requires further characterization of the effects of different types of earthquakes on individual lacustrine shaking records by considering the characteristics of seismic ground motions at the lake sites, potentially affecting the type and size of seismically-generated deposits.

Each type of earthquake imprint typically requires a site-dependent minimal seismic intensity, which tends to increase from moderate-threshold delta failures over failures of hemipelagic subaqueous slopes to widespread onshore landslides and rock falls that only occur in case of very strong to severe seismic shaking (Sect. 4.2). As a result, intense shaking produced by large intraplate earthquakes near the lake site may result in SSDS, subaqueous mass wasting and onshore slope failures in its catchment, while far-field megathrust earthquakes produce lower shaking intensities at the study site and thus a signal for catchment response is missing. This difference in imprint is further supported by the distinct frequency content of ground motions produced by inter- and intraplate earthquakes. Intraslab earthquakes typically produce higher-frequency accelerations compared to interplate (megathrust) earthquakes, which likely results from relatively larger stress drops due to increased frictional strength at depth and/or the directivity effect of up-dip propagating earthquake sources (Leyton et al., 2009). The high-frequency response spectra of bedrock thus facilitate rock slope failure during intraslab earthquakes, while sedimentary slopes amplify lower-frequency accelerations so that subaqueous mass wasting is favored during megathrust earthquakes (Fig. 7.10). This concept has been particularly useful for distinguishing between turbidites related to inter- and intraplate earthquakes in Lo Encañado Lake in the central Chilean Andes by tracking the distinct sedimentary signature of failures of hemipelagic sediments compared to subaerial slopes (Van Daele et al., 2019). Although the specific setting of the lake and its catchment largely influences the characteristics of seismically-generated deposits, distinct sedimentological imprints for inter- and intraplate earthquakes have also been found in a fjord in southern Chile (Wils et al., 2020), proglacial lakes in Alaska (Praet et al., 2022) and tropical lakes along the Sumatra Fault (Wils et al., 2021a).

7.8.1 Subaqueous Paleoseismology in Southern Chile

Several subaqueous records of past shaking with variable recording thresholds were investigated along the Valdivia master segment of the Chilean subduction zone. In 1960 CE, this segment hosted the largest instrumentally recorded earthquake worldwide with M_W 9.5, a rupture length of about 1000 km and patches of up to 40 m coseismic slip (Moreno et al., 2009). Its lacustrine imprint was tracked in 13 lakes in the northern half of its rupture area (Van Daele et al., 2015), from which lakes Riñihue and Calafquén (Fig. 7.11) were selected for establishing long and continuous seismo-turbidite records related to past megathrust earthquakes (Moernaut et al., 2014, 2018). This selection was based on the inferred one-to-one correlation between large historical megathrust earthquakes in 1575, 1737, 1837 and 1960 CE, and extensive, thin muddy turbidites in distal sub-basins. Temporal correlation was made using a precise chronology combining varve counting, short-lived radionuclides (^{210}Pb and ^{137}Cs) and identification of sedimentary traces for historically-documented volcanic eruptions (Van Daele et al., 2014b). Chronology of the long cores covering the last 4.7 kyr and 3.5 kyr was established by ^{14}C dating of organic macro-remains and (offset-corrected) bulk organic fraction and identification of regionally-documented tephra layers. For the studied distal sub-basins, non-seismic triggers were ruled out because these basins are protected from major turbidity currents related to river floods and volcanic lahars, and most of the slope sequences are deep enough to be unaffected by surface currents, wave action and other coastal processes. The historical calibration revealed a different EQRT for different sites in these sub-basins, which depends on the length of the source

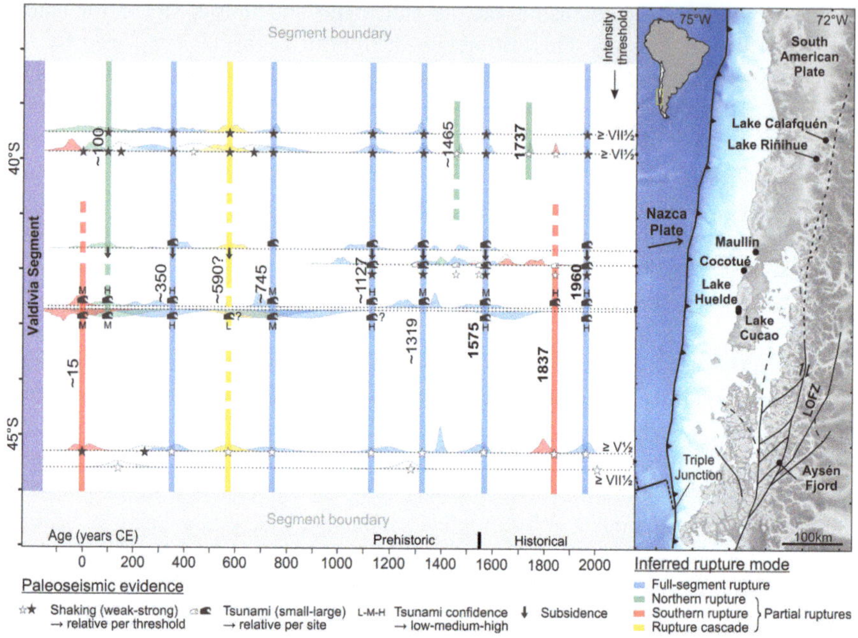

Fig. 7.11 Proposed correlation of paleoseismic evidence (shaking, tsunami inundation and/or coastal subsidence) along the Valdivia segment of the Chilean subduction zone. Inferred full-segment ruptures are indicated in blue, partial ruptures in green (northern), red (southern) or yellow (cascade). For each event registered at a paleoseismic site, the age range is presented as a color-coded probability-density function. Grey age ranges constitute local shaking events and can represent either small megathrust earthquakes or intraplate events. Modified from Wils et al. (2020)

slope, the travel distance of the turbidity current to the site and the compositional nature of the slope sediments (Moernaut et al., 2014; Molenaar et al., 2021). As a consequence, the long coring site in Lake Calafquén (CAL1) only records events with high shaking intensities (≥ VII½), while the site in Lake Riñihue (RIN2) also records events with intensities down to VI½. Given the short distance between the lakes compared to the extent of large megathrust ruptures, one can regard the CAL1 and RIN2 records as one multi-threshold paleoseismic record. This shows an average recurrence rate of ~290 yr for the strongest shaking events (≥ VII½: such as in 1960 CE). Lowering the threshold to ≥ VI½ produces an average of ca. 140 yr. Importantly, the records reveal that the strongest shaking events occurred in a more periodic manner (CoV 0.32) than when also including the intensity VI½-VII events (CoV 0.5), which implies that seismic cycle concepts are justified when describing the recurrence pattern of the largest megathrust earthquakes in this region, i.e. giant M_W 9-type events (Moernaut et al., 2018). Even though local intraplate events in the last 100 yr only induced small sandy turbidity currents on alluvial fan slopes in these lakes, it cannot be fully excluded that some prehistorical intraplate earthquakes were strong enough to be recorded as muddy turbidites at the long core sites in distal basins. Therefore, a regional paleoseismic approach was adopted for discrimination of seismic sources and evaluation of recurrence parameters for megathrust events.

Another multi-threshold paleoshaking record is constructed from the sedimentary infill of Aysén Fjord in the southern part of the Valdivia segment. Interest in this particular area spiked after a seismic swarm along the Liquiñe-Ofqui Fault Zone in 2007 CE, of which the highest-magnitude earthquake (M_W 6.2) had its epicenter in the fjord (Naranjo et al., 2009). It triggered multiple landslides and rock slope failures along the fjord shoreline, the deposits of which can be identified as several MTDs and turbidites on the basin floor (Van Daele et al., 2013). These turbidites show an amalgamated nature, shaped by individual synchronous turbidity currents arriving at various core sites from different directions (Van Daele et al., 2014a). Similar MTDs have been identified on nine older seismic-stratigraphic levels throughout the fjord, of which the chronology builds on numerous radiocarbon dates in combination with ^{210}Pb activities for the most recent sediments. By application of the synchronicity criterion, these mass-wasting events have been attributed to (pre-)Holocene earthquake shaking (Wils et al., 2018). The forward probabilistic modeling approach (Sect. 7.5) revealed that seven of these events are the result of crustal earthquake activity with magnitudes generally between 5 and 7 along one of the Liquiñe-Ofqui fault strands in the vicinity of the fjord, while inconclusive results were obtained for the two remaining MTD-generating events (Vanneste et al., 2018). A common feature of these two events is the lack of onshore-originating MTDs (Wils et al., 2018). Such large rock slope failures and landslides become more frequent as of intensities of VII-VIII (Keefer, 1984; Serva et al., 2016), indicating that ground motions were less strong during both of these events (Vanneste et al., 2018). Considering the large distance between the subduction zone and Aysén Fjord, megathrust earthquakes typically result in lower shaking intensities in the fjord area as demonstrated by a value of only ~VI during the 1960 CE earthquake. Accordingly, both MTD levels are considered to result from megathrust earthquakes. The same reasoning applies to 22 turbidites characterized by end-member modelling of grain-size data from a sediment core retrieved in the inner fjord. These occurred in the last 9 kyr and were not related to any mapped MTDs (Wils et al., 2020). Analysis of their compositional trends supports a seismic origin, resulting from low-intensity (≥ V½, < VI½) failures of deltaic slopes, such as those present at the Aysén River mouth, although non-seismic delta instabilities could not be fully ruled out.

7.8.2 Variable Rupture Mode Inferred from Integrated Subduction Zone Paleoseismology

Apart from temporal correlation of sedimentary paleoshaking evidences along the Valdivia segment, the proposed seismic source mechanism for each of the identified deposits can be further supported by integration with other types of paleoseismic archives. This is illustrated by, for example, the only dated prehistoric megaturbidite identified in Reloncaví Fjord (St-Onge et al., 2012). For this event, a seismic origin is inferred based on visual similarities to three voluminous megaturbidites that were attributed to historical megathrust earthquakes (1960, 1837, 1575 CE) using ^{210}Pb dating and two radiocarbon ages, in addition to age overlap with coseismic subsidence and tsunami inundation in a coastal marsh (Cisternas et al., 2005). Many of these terrestrial coastal records exist along the Chilean subduction zone (e.g. Cisternas et al., 2018; Dura et al., 2017; Ely et al., 2014; Garrett et al., 2015; Hocking et al., 2021; Nelson et al., 2009), but generally do not cover more than 1 kyr in this region, except for the Maullín site (Fig. 7.11; Cisternas et al., 2005). A longer record span similar to the paleoshaking records has been obtained by studying tsunami inundation in coastal lakes in the region (Kempf et al., 2017, 2020). Combination of these different paleoseismic archives led to a spatiotemporal reconstruction of megathrust ruptures in southern Chile (Wils et al., 2020). The proposed scenarios are most reliable for the last two millennia, a period for which paleoseismic sites are available along the entire length of the Valdivia segment, and include an alternation of full and partial ruptures of the master segment as well as one rupture cascade. Only a representative selection of paleoseismic/tsunami records containing all known events in this time period is presented here (Fig. 7.11).

Events for which paleoshaking evidence could be correlated to the presence of tsunami deposits and/or coastal elevation changes were interpreted as megathrust earthquakes with a high confidence level. This implies that all of the recorded low-intensity earthquakes in Aysén Fjord can be considered as megathrust earthquakes, as is true for the high-intensity shaking events registered in Lake Calafquén. On the other hand, not all of the low-intensity events recorded in Lake Riñihue correspond to a tsunami deposit and these might thus represent intraplate earthquakes or local single-asperity megathrust earthquakes on the deeper section of the seismogenic zone. The latter is exemplified by the 1737 CE megathrust earthquake which led to relatively small lacustrine seismo-turbidites (Moernaut et al., 2014) and no conclusive traces for tsunami inundation or coastal elevation changes have been found yet. Paleo-events for which segment-wide shaking evidence coincides with the presence of tsunami deposits and/or coastal elevation changes in the central part of the Valdivia segment were interpreted as megathrust ruptures over the complete length of this master segment. A total of six of these full-segment ruptures have been recorded, including the 1960 and 1575 CE historical earthquakes, resulting in an average recurrence time of ~ 320 yr. One of these occurred around ~ 745 CE and shows an aberrant coastal elevation pattern. In contrast to all other proposed full-segment ruptures, no evidence of coseismic coastal subsidence has been documented yet. This implies different rupture characteristics, e.g. a locally deeper or very wide slip distribution (e.g. Cisternas et al., 2017), or rupture along a splay fault instead of the up-dip zone of the megathrust (e.g. Melnick et al., 2012). Such a different rupture pattern may have disrupted the megathrust earthquake cycle initiating a ca. 400-yr long quiet period during which no imprints are recorded at any of the studied sites. This quiescence is about twice the length of any other recorded megathrust interseismic period.

While shaking evidence for the event around ~590 CE can be found along the entire Valdivia segment and also coseismic subsidence has been registered, no conclusive tsunami deposits associated with this earthquake have been identified up to now. Full ruptures unaccompanied by widespread tsunamis are unlikely, especially considering that deeper parts of the megathrust have a very patchy asperity distribution (Lay et al., 2012). This hampers the possibility for long and deep ruptures that fail to include the shallower seismogenic sections and thus do not trigger a significant tsunami. Accordingly, this event was interpreted as a rupture cascade in which several asperities within the Valdivia segment ruptured within a short period of maximum a few decades. Combinations of partial ruptures can also occur over longer timescales, where a partial rupture of the northern part of the Valdivia segment is followed by a rupture of the southern part or vice-versa. Each of those partial ruptures can affect a large portion of the megathrust, some of them releasing a significant portion of the accumulated slip deficit along part of the Valdivia segment (Moernaut et al., 2014). Moreover, they are often tsunamigenic as exemplified by the 1837 CE event that induced a trans-Pacific tsunami. The recurrence patterns of partial ruptures are, however, highly variable and much less periodic compared to full-segment ruptures, the latter forming the dominant rupture style in the Valdivia master segment (Wils et al., 2020).

7.9 Conclusions and Future Perspectives

The herein outlined principles of lacustrine paleoseismology and its regional application in different settings worldwide (Fig. 7.1, Sects. 7.6, 7.7 and 7.8) show the importance of lake sedimentary records for paleoseismological research. This book chapter provides a state-of-the-art summary and guidance for future advancements in this young and quickly-developing discipline. In this final section, we summarize the main advantages of lacustrine paleoseismology and identify promising approaches to push the current methodological frontiers and advance our understanding of past large earthquakes and their recurrence patterns.

(i) Lake sedimentary records can provide excellent age control due to continuous sedimentation (Sect. 7.3), which strongly facilitates the **assessment of synchronicity** between individual lacustrine sites. The combined use in a Bayesian framework of annual lamination counting, numerous absolute dates and marker layers in multiple cores in a lake may significantly reduce event age ranges and provide more reliable quantification of their uncertainties. Moreover, proposed inter-lake correlations can be enhanced by robust statistical testing. However, the success of such tests requires a comparable dataset quality and a standardized approach for constructing age-depth models.

(ii) Although some event age uncertainty will remain, the systematic evaluation of event deposit synchronicity allows **rupture scenarios** to be built using different inverse and forward ground motion modeling techniques (Sect. 7.5).

Incorporating reconstructed seismic intensities from the sedimentary earthquake imprints allows different rupture scenarios to be ranked, for which the most plausible scenarios can be tested against quantitative information from other paleoseismic evidence and results from physics-based earthquake recurrence simulations (Howarth et al., 2021a). Due to the high spatial and temporal precision of lake based paleoseismic reconstructions, it is anticipated that these will form an increasingly important part of multi-archive rupture scenarios that go beyond characterizing recurrence behavior at a single point.

(iii) A key challenge in paleoseismology is to evaluate **fault segmentation models** of subduction megathrusts and transform faults, and assess how persistent proposed segment boundaries are (Philibosian & Meltzner, 2020). Primary paleoseismic techniques are generally not capable of distinguishing between a single passing rupture and a rupture cascade of adjacent sections, whereas the high-resolution continuous nature of lacustrine shaking records provides a way to test these rupture scenarios (Sect. 7.7; Howarth et al., 2021a). Especially lakes located between asperities or at section boundaries form strategic sites in which passing ruptures would leave a single turbidite whereas rupture cascades can produce two turbidites in very close stratigraphic succession. Moreover, the relative size of turbidites generated through surficial remobilization may scale to the bracketed duration of strong shaking (Molenaar et al., 2021). This opens perspectives for distinguishing single- from multi-asperity megathrust ruptures because ground motion for the latter is of much longer duration but produces rather similar PGA and/or intensity values at a site.

(iv) In contrast to methods developed for rupture characterization in intraplate and transform settings, no direct mathematical relationships linking paleoshaking evidence to rupture parameters have been established for **megathrust earthquakes**. This is mainly due to the large intrinsic variability of these ruptures, including a range of different possible slip distributions in both along-strike and down-dip direction. As a result, it remains unclear how the spatial distribution of paleoseismic evidence relates to the actual rupture depth and extent, and whether the persistence of megathrust rupture boundaries over long timescales can be evaluated with paleo-shaking records. The way forward is (at least) four-fold: (i) systematic mapping of the imprint of recent megathrust earthquakes along lake transects to compare rupture terminations with the distance over which lacustrine evidence tapers out (e.g. Van Daele et al., 2015); (ii) developing a new approach for forward probabilistic ground-motion modeling for subduction megathrust ruptures, capable of incorporating along-strike and down-dip rupture variability, and rupture directivity effects; (iii) integrating calibrated paleoshaking records with coastal records of tsunami inundation and coseismic elevation changes, the patterns of which provide insights in the down-dip location of the megathrust rupture (Ely et al., 2014); (iv) evaluate whether lake sedimentary imprints can reflect the different frequency content of ground motion emanated by deep versus shallow megathrust ruptures, as has been proposed recently for Chilean lakes

(Molenaar et al., 2021). There, high frequency ground motion and high PGA originating from deep ruptures is interpreted to have preferentially induced SSDS over seismo-turbidites.

(v) The **direction of rupture propagation** causes asymmetry in the distribution of strong ground motions that exerts substantial influence on seismic hazard but is often unknown or poorly constrained (e.g. Bradley et al., 2017). At subduction zones, the effect of rupture directivity can be compounded by the presence of low-velocity sedimentary prisms that amplify (low-frequency) ground motions at large distances from source (e.g. Wallace et al., 2017). To date there are few approaches for reconstructing the rupture direction and asymmetry in strong ground motions during past earthquakes (Kearse et al., 2019). The ability of lacustrine paleoseismic records to record evidence of shaking across a range of shaking intensities may provide an avenue to do so, particularly when direct paleoseismic evidence independently constrains rupture extent. In such cases, quantitative lacustrine paleoseismology applied to a network of lakes located around rupture tips can allow asymmetric patterns of ground motions to be reconstructed. A recent example is provided by the M_W 7.8 Kaikōura earthquake where the spatial distribution of marine turbidites faithfully records the asymmetric distribution of strong ground motions hundreds of km northeast of the rupture tip (Howarth et al., 2021b).

Within the broader perspective of seismic hazard analysis, the contribution of lacustrine paleoseismic data can be highly informative for different aspects of the seismic hazard modeling process:

(i) Results from lake paleoseismic records can inform **source fault models.** For example, integrated lacustrine datasets with other paleoseismic archives along the Alpine Fault (Sect. 7.7) and in Southern Chile (Sect. 7.8) have elucidated past rupture location and extent over several millennia and this for a transform plate boundary fault and subduction megathrust, respectively. These records revealed the occurrence of passing ruptures and unitary segment ruptures, generating new insights in the role of presumed segment boundaries and in the relative contribution of differently-sized ruptures in spending accumulated slip deficit. These studies support the idea that simple quasi-stationary segmentation models are obsolete. In intraplate regions where the faults that can cause large earthquakes are poorly constrained, a regional lacustrine paleoseismic approach can delimit the areas ("area sources") or faults where important ruptures have occurred in the past and where on-fault investigation should be planned to better constrain coseismic slip distribution (and thus magnitude) of rare high-magnitude events (Sect. 7.6).

(ii) Given the long and continuous nature of lake records and the possibility to obtain multi-threshold paleoseismic records, quantitative lacustrine paleoseismology can provide data points for constraining and testing **frequency-magnitude relations** for specific faults or for regional seismicity, i.e. the high-magnitude part in a Gutenberg-Richter relationship (e.g. Becker et al., 2005). For the Chilean case study (Sect. 7.8), the recurrence rates obtained from lake studies (Moernaut et al., 2018) have been used to validate a hybrid frequency-magnitude relationship that combines instrumental seismicity with geodetic data on plate convergence and plate locking (Pagani et al., 2020). Moreover, lake records at subduction zones may provide a unique perspective on the recurrence parameters of large intraslab events in a region, the potential hazard of which is exemplified by recent intraslab events (e.g. 2017 M_W 7.1 in Mexico, 2018 M_W 7.1 in Alaska) and devastating historical events (e.g. 1939 M_S 7.8 Chillán event, the deadliest historical earthquake in Chile). In intraplate settings, lake paleoseismology can additionally be used to better constrain the maximum credible earthquake (Oswald et al., 2022), i.e. where the frequency-magnitude relationship should taper off. Given low deformation rates, instrumental and historical records likely did not record such extreme events.

(iii) Finally, lake records are useful for **testing hazard curves** that underlie seismic hazard maps (Daxer et al., 2022a). This involves direct comparison of the extracted hazard curve near the lake site and the recurrence rate of strong seismic shaking derived from the lacustrine paleoseismic archive. This comparison forms a means for testing seismic hazard maps, which constitutes the final step of seismic hazard analysis.

Acknowledgements We gratefully acknowledge all authors that are dedicated to the advancement of lacustrine paleoseismology. Special thanks to Ulaş Avşar, Christoph Daxer, Ariana Molenaar, Yin Lu, Nore Praet and Maarten Van Daele for contributing data to the figures. We also thank the anonymous colleagues for reviewing an earlier version of this work. This research was funded in whole, or in part, by the Austrian Science Fund (FWF) [https://doi.org/10.55776/P34504]. For the purpose of open access, the author has applied a CC BY public copyright licence to any Author Accepted Manuscript version arising from this submission.

References

Abaimov, S. G., Turcotte, D. L., Shcherbakov, R., Rundle, J. B., Yakovlev, G., Goltz, C., & Newman, W. I. (2008). Earthquakes: Recurrence and interoccurrence times. *Pure and Applied Geophysics, 165*, 777–795. https://doi.org/10.1007/s00024-008-0331-y

Adams, J. (1990). Paleoseismicity of the cascadia subduction zone: Evidence from turbidites off the oregon-Washington margin. *Tectonics, 9*, 569–583. https://doi.org/10.1029/TC009i004p00569

Agnon, A., Migowski, C., & Marco, S. (2006). Intraclast breccias in laminated sequences reviewed: Recorders of paleo-earthquakes. *Special Papers Geological Society of America, 401*, 195–214. https://doi.org/10.1130/2006.2401(13)

Alsop, G. I., Marco, S., & Levi, T. (2022). Recognising surface versus sub-surface deformation of soft-sediments: Consequences and considerations for palaeoseismic studies. *Journal of Structural Geology, 154*, 104493. https://doi.org/10.1016/j.jsg.2021.104493

Alsop, G. I., Weinberger, R., Marco, S., & Levi, T. (2020). Bed-parallel slip: Identifying missing displacement in mass transport deposits. *Journal of Structural Geology, 131*, 103952. https://doi.org/10.1016/j.jsg.2019.103952

Archer, C., Noble, P., Rosen, M. R., Sagnotti, L., Florindo, F., Mensing, S., Piovesan, G., & Michetti, A. M. (2019). Lakes as paleoseismic records in a seismically-active, low-relief area (Rieti Basin, central Italy). *Quaternary Science Reviews, 211*, 186–207. https://doi.org/10.1016/j.quascirev.2019.03.004

Arnaud, F., Lignier, V., Revel, M., Desmet, M., Beck, C., Pourchet, M., Charlet, F., Trentesaux, A., & Tribovillard, N. (2002). Flood and earthquake disturbance of 210Pb geochronology (Lake Anterne, NW Alps). *Terra Nova, 14*, 225–232. https://doi.org/10.1046/j.1365-3121.2002.00413.x

Atwater, B. F., Carson, B., Griggs, G. B., Johnson, P. P., & Salmi, M. S. (2014). Rethinking turbidite paleoseismology along the Cascadia subduction zone. *Geology, 42*, 827–830. https://doi.org/10.1130/G35902.1

Avşar, U., Hubert-Ferrari, A., De Batist, M., & Fagel, N. (2014a). A 3400 year lacustrine paleoseismic record from the North Anatolian Fault, Turkey: Implications for bimodal recurrence behavior. *Geophysical Research Letters, 41*, 377–384. https://doi.org/10.1002/2013GL058221

Avşar, U., Hubert-Ferrari, A., De Batist, M., Lepoint, G., Schmidt, S., & Fagel, N. (2014b). Seismically-triggered organic-rich layers in recent sediments from Göllüköy Lake (North Anatolian Fault, Turkey). *Quaternary Science Reviews, 103*, 67–80. https://doi.org/10.1016/j.quascirev.2014.08.020

Avşar, U., Jónsson, S., Avşar, Ö., & Schmidt, S. (2016). Earthquake-induced soft-sediment deformations and seismically amplified erosion rates recorded in varved sediments of Köyceğiz Lake (SW Turkey). *Journal of Geophysical Research: Solid Earth, 121*, 4767–4779. https://doi.org/10.1002/2016JB012820

Bakun, W. H., & Wentworth, C. M. (1997). Estimating earthquake location and magnitude from seismic intensity datae. *Bulletin of the Seismological Society of America, 87*, 1502–1521.

Banjan, M., Christian, C., Pierre, S., Hervé, J., Manon, B., Francois, D., Anne-Lise, D., Jean-Philippe, J., Bernard, F., Emmanuel, M., Findling, N., Philippe, A., Julien, D., Vincent, B., Sylvain, C., & Erwan, M. (2023). Did the Younger Dryas to Holocene climate transition favour high seismicity rates in the north-western Alps? *Sedimentology, 70*, 538–568. https://doi.org/10.1111/sed.13050

Barrell, D. J. A., Andersen, B. G., & Denton, G. H. (2011). *Glacial geomorphology of the central South Island, New Zealand*. GNS Science. GNS Science Monograph, vol. 27, p. 81.

Barth, N. C., Boulton, C., Carpenter, B. M., Batt, G. E., & Toy, V. G. (2013). Slip localization on the Southern Alpine Fault, New Zealand. *Tectonics, 32*, 620–640. https://doi.org/10.1002/tect.20041

Beck, C. (2009). Late Quaternary lacustrine paleo-seismic archives in north-western Alps: Examples of earthquake-origin assessment of sedimentary disturbances. *Earth-Science Reviews, 96*, 327–344. https://doi.org/10.1016/j.earscirev.2009.07.005

Beck, C., Carrillo, E., Audemard, F., van Welden, A., & Disnar, J. R. (2019). Tentative integration of paleoseismic data from lake sediments and from nearby trenches: The central section of the Boconó Fault (Northern Venezuela). *Journal of South American Earth Sciences, 92*, 646–657. https://doi.org/10.1016/j.jsames.2019.03.028

Becker, A., Ferry, M., Monecke, K., Schnellmann, M., & Giardini, D. (2005). Multiarchive paleoseismic record of late Pleistocene and Holocene strong earthquakes in Switzerland. *Tectonophysics, 400*, 153–177. https://doi.org/10.1016/j.tecto.2005.03.001

Begin, Z. B., Steinberg, D. M., Ichinose, G. A., & Marco, S. (2005). A 40,000 year unchanging seismic regime in the dead sea rift. *Geology, 33*, 257–260. https://doi.org/10.1130/G21115.1

Bellwald, B., Hjelstuen, B. O., Sejrup, H. P., Stokowy, T., & Kuvås, J. (2019). Holocene mass movements in west and mid-Norwegian fjords and lakes. *Marine Geology, 407*, 192–212. https://doi.org/10.1016/j.margeo.2018.11.007

Bernhardt, A., Melnick, D., Hebbeln, D., Lückge, A., & Strecker, M. R. (2015). Turbidite paleoseismology along the active continental margin of Chile—Feasible or not? *Quaternary Science Reviews, 120*, 71–92. https://doi.org/10.1016/j.quascirev.2015.04.001

Berryman, K., Cooper, A., Norris, R., Villamor, P., Sutherland, R., Wright, T., Schermer, E., Langridge, R., & Biasi, G. (2012a). Late Holocene rupture history of the alpine fault in South Westland, New Zealand. *Bulletin of the Seismological Society of America, 102*, 620–638. https://doi.org/10.1785/0120110177

Berryman, K. R., Cochran, U. A., Clark, K. J., Biasi, G. P., Langridge, R. M., & Villamor, P. (2012b). Major earthquakes occur regularly on an isolated plate boundary fault. *Science, 80*(336), 1690–1693. https://doi.org/10.1126/science.1218959

Bertrand, S., Charlet, F., Chapron, E., Fagel, N., & De Batist, M. (2008). Reconstruction of the Holocene seismotectonic activity of the Southern Andes from seismites recorded in Lago Icalma, Chile, 39°S. *Palaeogeography, Palaeoclimatology, Palaeoecology, 259*, 301–322. https://doi.org/10.1016/j.palaeo.2007.10.013

Bilek, S. L., & Lay, T. (2018). Subduction zone megathrust earthquakes. *Geosphere, 14*, 1468–1500. https://doi.org/10.1130/GES01608.1

Blaauw, M., & Christen, J. A. (2011). Flexible paleoclimate age-depth models using an autoregressive gamma process. *Bayesian Analysis, 6*, 457–474. https://doi.org/10.1214/11-BA618

Blais-Stevens, A., Rogers, G. C., & Clague, J. J. (2011). A revised earthquake chronology for the last 4,000 years inferred from varve-bounded debris-flow deposits beneath an inlet near Victoria, British Columbia. *Bulletin of the Seismological Society of America, 101*, 1–12. https://doi.org/10.1785/0120090360

Boës, X., Moran, S. B., King, J., Cağatay, M. N., & Hubert-Ferrari, A. (2010). Records of large earthquakes in lake sediments along the North Anatolian Fault, Turkey. *Journal of Paleolimnology, 43*, 901–920. https://doi.org/10.1007/s10933-009-9376-x

Bondevik, S., Gjevik, B., & Sørensen, M. B. (2013). Norwegian seiches from the giant 2011 Tohoku earthquake. *Geophysical Research Letters, 40*, 3374–3378. https://doi.org/10.1002/grl.50639

Bradley, B. A., Bae, S. E., Polak, V., Lee, R. L., Thomson, E. M., & Tarbali, K. (2017). Ground motion simulations of great earthquakes on the Alpine Fault: Effect of hypocentre location and comparison with empirical modelling. *New Zealand Journal of Geology and Geophysics, 60*, 188–198. https://doi.org/10.1080/00288306.2017.1297313

Brancelj, A., Šiško, M., Muri, G., Appleby, P., Lami, A., Shilland, E., Rose, N. L., Kamenik, C., Brooks, S. J., & Dearing, J. A. (2002). Lake Jezero v Ledvici (NW Slovenia)—changes in sediment records over the last two centuries. *Journal of Paleolimnology, 28*, 47–58. https://doi.org/10.1023/A:1020367818144

Brandes, C., Steffen, H., Steffen, R., & Wu, P. (2015). Intraplate seismicity in northern Central Europe is induced by the last glaciation. *Geology, 43*, 611–614. https://doi.org/10.1130/G36710.1

Brocard, G., Anselmetti, F. S., & Teyssier, C. (2016). Guatemala paleoseismicity: From late classic maya collapse to recent fault creep. *Science and Reports, 6*, 36976. https://doi.org/10.1038/srep36976

Bronk Ramsey, C. (2009). Bayesian analysis of radiocarbon dates. *Radiocarbon, 51*, 337–360. https://doi.org/10.1017/S0033822200033865

Brooks, G. R. (2018). Deglacial record of palaeoearthquakes interpreted from mass transport deposits at three lakes near Rouyn-Noranda, north-western Quebec, Canada. *Sedimentology, 65*, 2439–2467. https://doi.org/10.1111/sed.12473

Brooks, G. R., & Adams, J. (2020). A review of evidence of glacially-induced faulting and seismic shaking in eastern Canada. *Quaternary Science Reviews, 228*. https://doi.org/10.1016/j.quascirev.2019.106070

Bussmann, F., & Anselmetti, F. S. (2010). Rossberg landslide history and flood chronology as recorded in Lake Lauerz sediments (Central Switzerland). *Swiss Journal of Geosciences, 103*, 43–59. https://doi.org/10.1007/s00015-010-0001-9

Chapron, E., Beck, C., Pourchet, M., & Deconinck, J. F. (1999). 1822 earthquake-triggered homogenite in Lake Le Bourget (NW Alps). *Terra Nova, 11,* 86–92. https://doi.org/10.1046/j.1365-3121.1999.00230.x

Chapron, E., Foucher, A., Chassiot, L., Fleurdeus, W., Arricau, V., Perdereaux, L., Gay-Ovejero, I., Lavrieux, M., Motelica-Heino, M., & Salvador-Blanes, S. (2021). Evaluating Holocene natural hazards in the French Massif Central from a regional lake sediment approach. *Quaternary International.* https://doi.org/10.1016/j.quaint.2021.05.018

Chapron, E., Simonneau, A., Ledoux, G., Arnaud, F., Lajeunesse, P., & Albéric, P. (2016). French Alpine foreland Holocene paleoseismicity revealed by coeval mass wasting deposits in glacial Lakes. In G. Lamarche, J. Mountjoy, S. Bull, T. Hubble, S. Krastel, E. Lane, A. Micallef, L. Moscardelli, C. Mueller, I. Pecher, & S. Woelz (Eds.), *Submarine Mass Movements and their Consequences* (pp. 341–349). Springer International Publishing, Cham.

Chassiot, L., Chapron, E., Di Giovanni, C., Lajeunesse, P., Tachikawa, K., Garcia, M., & Bard, E. (2016). Historical seismicity of the Mont Dore volcanic province (Auvergne, France) unraveled by a regional lacustrine investigation: New insights about lake sensitivity to earthquakes. *Sedimentary Geology, 339*, 134–150. https://doi.org/10.1016/j.sedgeo.2016.04.007

Cisternas, M., Atwater, B. F., Torrejón, F., Sawai, Y., Machuca, G., Lagos, M., Eipert, A., Youlton, C., Salgado, I., Kamataki, T., Shishikura, M., Rajendran, C. P., Malik, J. K., Rizal, Y., & Husni, M. (2005). Predecessors of the giant 1960 Chile earthquake. *Nature, 437*, 404–407. https://doi.org/10.1038/nature03943

Cisternas, M., Carvajal, M., Wesson, R., Ely, L. L., & Gorigoitia, N. (2017). Exploring the historical earthquakes preceding the giant 1960 Chile earthquake in a time-dependent seismogenic zone. *Bulletin of the Seismological Society of America, 107*, 2664–2675. https://doi.org/10.1785/0120170103

Cisternas, M., Garrett, E., Wesson, R., Dura, T., & Ely, L. L. (2018). Unusual geologic evidence of coeval seismic shaking and tsunamis shows variability in earthquake size and recurrence in the area of the giant 1960 Chile earthquake. *Marine Geology, 396*, 54–66. https://doi.org/10.1016/j.margeo.2018.01.005

Clare, M. A., Hughes Clarke, J. E., Talling, P. J., Cartigny, M. J. B., & Pratomo, D. G. (2016). Preconditioning and triggering of offshore slope failures and turbidity currents revealed by most detailed monitoring yet at a fjord-head delta. *Earth and Planetary Science Letters, 450*, 208–220. https://doi.org/10.1016/j.epsl.2016.06.021

Clare, M. A., Talling, P. J., & Hunt, J. E. (2015). Implications of reduced turbidity current and landslide activity for the initial eocene thermal maximum—evidence from two distal, deep-water sites. *Earth and Planetary Science Letters, 420*, 102–115. https://doi.org/10.1016/j.epsl.2015.03.022

Cochran, U. A., Clark, K. J., Howarth, J. D., Biasi, G. P., Langridge, R. M., Villamor, P., Berryman, K. R., & Vandergoes, M. J. (2017). A plate boundary earthquake record from a wetland adjacent to the Alpine fault in New Zealand refines hazard estimates. *Earth and Planetary Science Letters, 464*, 175–188. https://doi.org/10.1016/j.epsl.2017.02.026

Cornell, C. A. (1968). Engineering seismic risk analysis. *Bulletin of the Seismological Society of America, 58*, 1583–1606.

Cox, S. C., Stirling, M. W., Herman, F., Gerstenberger, M., & Ristau, J. (2012). Potentially active faults in the rapidly eroding landscape adjacent to the Alpine Fault, central Southern Alps, New Zealand. *Tectonics, 31.* https://doi.org/10.1029/2011TC003038

Daxer, C., Huang, J. J. S., Weginger, S., Hilbe, M., Strasser, M., & Moernaut, J. (2022a). Validation of seismic hazard curves using a calibrated 14 ka lacustrine record in the Eastern Alps, Austria. *Science and Reports, 12*, 1–14. https://doi.org/10.1038/s41598-022-24487-w

Daxer, C., Ortler, M., Fabbri, S. C., Hilbe, M., Hajdas, I., Dubois, N., Piechl, T., Hammerl, C., Strasser, M., & Moernaut, J. (2022b). High-resolution calibration of seismically-induced lacustrine deposits with historical earthquake data in the Eastern Alps (Carinthia, Austria). *Quaternary Science Reviews, 284*, 107497. https://doi.org/10.1016/j.quascirev.2022.107497

Daxer, C., Sammartini, M., Molenaar, A., Piechl, T., Strasser, M., & Moernaut, J. (2020). Morphology and spatio-temporal distribution of lacustrine mass-transport deposits in Wörthersee,

Eastern Alps, Austria. *Special Publication of the Geological Society of London, 500*, 235–254. https://doi.org/10.1144/SP500-2019-179
Daxer, C., Wils, K., Ramisch, A., Strasser, M., & Moernaut, J. (2024). Contrasting sedimentary and long-lasting geochemical imprints of seismic shaking in a small, groundwater-fed lake basin (Klopeiner See, Eastern European Alps). *Sedimentologika, 2,* 1–27 . https://doi.org/10.57035/journals/sdk.2024.e21.1296
de la Fuente, A., Meruane, C., Contreras, M., Ulloa, H., & Niño, Y. (2010). Strong vertical mixing of deep water of a stratified reservoir during the Maule earthquake, central Chile (Mw 8.8). *Geophysical Research Letters, 37*, 1–5. https://doi.org/10.1029/2010GL045798
De Pascale, G. P., & Langridge, R. M. (2012). New on-fault evidence for a great earthquake in A.D. 1717, central Alpine fault. *New Zealand. Geology, 40*, 791–794. https://doi.org/10.1130/G33363.1
De Pascale, G. P., Quigley, M. C., & Davies, T. R. H. (2014). Lidar reveals uniform Alpine fault offsets and bimodal plate boundary rupture behavior, New Zealand. *Geology, 42*, 411–414. https://doi.org/10.1130/G35100.1
DeMets, C., Gordon, R. G., & Argus, D. F. (2010). Geologically current plate motions. *Geophysical Journal International, 181*, 1–80. https://doi.org/10.1111/j.1365-246X.2009.04491.x
Deplazes, G., Anselmetti, F. S., & Hajdas, I. (2007). Lake sediments deposited on the Flims rockslide mass: The key to date the largest mass movement of the Alps. *Terra Nova, 19*, 252–258. https://doi.org/10.1111/j.1365-3121.2007.00743.x
Doig, R. (1986). A method for determining the frequency of large-magnitude earthquakes using lake sediments. *Canadian Journal of Earth Sciences, 23*, 930–937. https://doi.org/10.1139/e86-094
Doughty, M., Eyles, N., Eyles, C. H., Wallace, K., & Boyce, J. I. (2014). Lake sediments as natural seismographs: Earthquake-related deformations (seismites) in central Canadian lakes. *Sedimentary Geology, 313*, 45–67. https://doi.org/10.1016/j.sedgeo.2014.09.001
Drab, L., Hubert Ferrari, A., Schmidt, S., & Martinez, P. (2012). The earthquake sedimentary record in the western part of the Sea of Marmara, Turkey. *Natural Hazards and Earth Systems Sciences, 12*, 1235–1254. https://doi.org/10.5194/nhess-12-1235-2012
Dura, T., Engelhart, S. E., Vacchi, M., Horton, B. P., Kopp, R. E., Peltier, W. R., & Bradley, S. (2016). The role of Holocene relative sea-level change in preserving records of subduction zone earthquakes. *Current Climate Change Reports, 2*, 86–100. https://doi.org/10.1007/s40641-016-0041-y
Dura, T., Horton, B. P., Cisternas, M., Ely, L. L., Hong, I., Nelson, A. R., Wesson, R. L., Pilarczyk, J. E., Parnell, A. C., & Nikitina, D. (2017). Subduction zone slip variability during the last millennium, south-central Chile. *Quaternary Science Reviews, 175*, 112–137. https://doi.org/10.1016/j.quascirev.2017.08.023
Ely, L. L., Cisternas, M., Wesson, R. L., & Dura, T. (2014). Five centuries of tsunamis and land-level changes in the overlapping rupture area of the 1960 and 2010 Chilean earthquakes. *Geology, 42*, 995–998. https://doi.org/10.1130/G35830.1
Fäh, D., Giardini, D., Kästli, P., Deichmann, N., Gisler, M., Schwarz-Zanetti, G., Alvarez-Rubio, S., Sellami, S., Edwards, B., Allmann, B., Bethmann, F., Wössner, J., Gassner-Stamm, G., Fritsche, S., & Eberhard, D. (2011). ECOS-09 Earthquake Catalogue of Switzerland Release 2011. Report and Database.
Fan, J., Jiang, H., Shi, W., Guo, Q., Zhang, S., Wei, X., Xu, H., Zhong, N., Huang, S., Chang, X., & Xiao, J. (2020). A 450-year lacustrine record of recurrent seismic activities around the Fuyun fault, Altay Mountains, Northwest China. *Quaternary International, 558*, 75–88. https://doi.org/10.1016/j.quaint.2020.08.051
Fan, J., Xu, H., Shi, W., Guo, Q., Zhang, S., Wei, X., Cai, M., Huang, S., Wang, J., & Xiao, J. (2022). A ~28-kyr continuous lacustrine paleoseismic record of the intraplate, slow-slipping Fuyun Fault in northwest China. *Frontiers in Earth Science, 10*, 1–16. https://doi.org/10.3389/feart.2022.828801
Fan, J., Zhai, D., Xu, H., Wei, X., Jin, C., Jiang, H., Shi, W., & Liu, X. (2023). Distinct lake sedimentary imprints of earthquakes, floods and human activities in the Xiaojiang Fault zone:

Towards a quantitative paleoseismograph in the southeastern Tibetan Plateau. *Science of the Total Environment, 868*, 161662. https://doi.org/10.1016/j.scitotenv.2023.161662

Ferry, M., Meghraoui, M., Delouis, B., & Giardini, D. (2005). Evidence for Holocene palaeoseismicity along the Basel-Reinach active normal fault (Switzerland): A seismic source for the 1356 earthquake in the Upper Rhine graben. *Geophysical Journal International, 160*, 554–572. https://doi.org/10.1111/j.1365-246X.2005.02404.x

Field, E. H., Arrowsmith, R. J., Biasi, G. P., Bird, P., Dawson, T. E., Felzer, K. R., Jackson, D. D., Johnson, K. M., Jordan, T. H., Madden, C., Michael, A. J., Milner, K. R., Page, M. T., Parsons, T., Powers, P. M., Shaw, B. E., Thatcher, W. R., Weldon, R. J., & Zeng, Y. (2014). Uniform California earthquake rupture forecast, version 3 (UCERF3)–the time-independent model. *Bulletin of the Seismological Society of America, 104*, 1122–1180. https://doi.org/10.1785/0120130164

Fortin, D., Praet, N., McKay, N. P., Kaufman, D. S., Jensen, B. J. L., Haeussler, P. J., Buchanan, C., & De Batist, M. (2019). New approach to assessing age uncertainties—the 2300-year varve chronology from Eklutna Lake, Alaska (USA). *Quaternary Science Reviews, 203*, 90–101. https://doi.org/10.1016/j.quascirev.2018.10.018

Franco, A., Moernaut, J., Schneider-Muntau, B., Strasser, M., & Gems, B. (2020). The 1958 Lituya Bay tsunami—pre-event bathymetry reconstruction and 3D numerical modelling utilising the computational fluid dynamics software flow-3D. *Natural Hazards and Earth Systems Sciences, 20*, 2255–2279. https://doi.org/10.5194/nhess-20-2255-2020

Garrett, E., Fujiwara, O., Garrett, P., Heyvaert, V. M. A., Shishikura, M., Yokoyama, Y., Hubert-Ferrari, A., Brückner, H., Nakamura, A., & De Batist, M. (2016). A systematic review of geological evidence for Holocene earthquakes and tsunamis along the Nankai-Suruga Trough, Japan. *Earth-Science Rev, 159*, 337–357. https://doi.org/10.1016/j.earscirev.2016.06.011

Garrett, E., Shennan, I., Woodroffe, S. A., Cisternas, M., Hocking, E. P., & Gulliver, P. (2015). Reconstructing paleoseismic deformation, 2: 1000 years of great earthquakes at Chucalén, south central Chile. *Quaternary Science Reviews, 113*, 112–122. https://doi.org/10.1016/j.quascirev.2014.10.010

Gastineau, R., de Sigoyer, J., Sabatier, P., Fabbri, S. C., Anselmetti, F. S., Develle, A. L., Şahin, M., Gündüz, S., Niessen, F., & Gebhardt, A. C. (2021). Active subaquatic fault segments in Lake Iznik along the middle strand of the North Anatolian Fault, NW Turkey. *Tectonics, 40*, 1–22. https://doi.org/10.1029/2020TC006404

Gastineau, R., Sabatier, P., Fabbri, S. C., Anselmetti, F. S., Roeser, P., Findling, N., Şahin, M., Gündüz, S., Arnaud, F., Franz, S. O., Ünsal, N. D., & de Sigoyer, J. (2023). Lateral variations in the signature of earthquake-generated deposits in Lake Iznik, NW Turkey. *The Depositional Record, 2*. https://doi.org/10.1002/dep2.232

Girardclos, S., Schmidt, O. T., Sturm, M., Ariztegui, D., Pugin, A., & Anselmetti, F. S. (2007). The 1996 AD delta collapse and large turbidite in Lake Brienz. *Marine Geology, 241*, 137–154. https://doi.org/10.1016/j.margeo.2007.03.011

Glur, L., Wirth, S. B., Büntgen, U., Gilli, A., Haug, G. H., Schär, C., Beer, J., & Anselmetti, F. S. (2013). Frequent floods in the European Alps coincide with cooler periods of the past 2500 years. *Science and Reports, 3*. https://doi.org/10.1038/srep02770

Goldfinger, C. (2011). Submarine paleoseismology based on turbidite records. *Annual Review of Marine Science, 3*, 35–66. https://doi.org/10.1146/annurev-marine-120709-142852

Goldfinger, C. (2021). Paleoseismic record of Peninsula Segment earthquakes on the San Andreas Fault near San Francisco, CA and possible NSAF linkage to Cascadia—NEHRP Final Technical Report G18AP00058. https://ir.library.oregonstate.edu/concern/technical_reports/ms35th18w

Goldfinger, C., Nelson, C. H., & Johnson, J. E. (2003). Holocene earthquake records from the Cascadia subduction zone and northern San Andreas fault based on precise dating of offshore turbidites. *Annual Review of Earth and Planetary Sciences, 31*, 555–577. https://doi.org/10.1146/annurev.earth.31.100901.141246

Goldfinger, C., Nelson, C. H., Morey, A. E., Johnson, J. E., Patton, J. R., Karabanov, E., Gutiérrez-Pastor, J., Eriksson, A. T., Gràcia, E., Dunhill, G., Enkin, R. J., Dallimore, A., & Vallier, T. (2012). Turbidite event history-methods and implications for holocene paleoseismicity of the

Cascadia subduction zone. US Geological Survey Prof Pap 1661–F 170 p. . https://doi.org/10.3133/PP1661F
Goldfinger, C., Patton, J. R., Van Daele, M., Moernaut, J., Nelson, C. H., de Batist, M., & Morey, A. E. (2014). Can turbidites be used to reconstruct a paleoearthquake record for the central Sumatran margin?: Comment. *Geology, 42*, 344–344. https://doi.org/10.1130/G35558C.1
Gomberg, J. (2018). Cascadia onshore-offshore site response, submarine sediment mobilization, and earthquake recurrence. *Journal of Geophysical Research: Solid Earth, 123*, 1381–1404. https://doi.org/10.1002/2017JB014985
Gomez, B., Corral, Á., Orpin, A. R., Page, M. J., Pouderoux, H., & Upton, P. (2015). Lake Tutira paleoseismic record confirms random, moderate to major and/or great Hawke's Bay (New Zealand) earthquakes. *Geology, 43*, 103–106. https://doi.org/10.1130/G36006.1
Grämiger, L. M., Moore, J. R., Vockenhuber, C., Aaron, J., Hajdas, I., & Ivy-Ochs, S. (2016). Two early Holocene rock avalanches in the Bernese Alps (Rinderhorn, Switzerland). *Geomorphology, 268*, 207–221. https://doi.org/10.1016/j.geomorph.2016.06.008
Hancox, G. T., Perrin, N. D., & Dellow, G. D. (2002). Recent studies of historical earthquake-induced landsliding, ground damage, and MM intensity in New Zealand. *Bulletin of the New Zealand Society Earthquake Engineering, 35*, 59–95. https://doi.org/10.5459/bnzsee.35.2.59-95
Hibsch, C., Alvarado, A., Yepes, H., Perez, V. H., & Sébrier, M. (1997). Holocene liquefaction and soft-sediment deformation in Quito (Ecuador): A paleoseismic history recorded in lacustrine sediments. *Journal of Geodynamics, 24*, 259–280. https://doi.org/10.1016/S0264-3707(97)00010-0
Hilbe, M., & Anselmetti, F. S. (2014). Mass Movement-Induced Tsunami Hazard on Perialpine Lake Lucerne (Switzerland): Scenarios and Numerical Experiments. *Pure and Applied Geophysics, 172*, 545–568. https://doi.org/10.1007/s00024-014-0907-7
Hilbe, M., Anselmetti, F. S., Eilertsen, R. S., Hansen, L., & Wildi, W. (2011). Subaqueous morphology of Lake Lucerne (Central Switzerland): Implications for mass movements and glacial history. *Swiss Journal of Geosciences, 104*, 425–443. https://doi.org/10.1007/s00015-011-0083-z
Hizzett, J. L., Hughes Clarke, J. E., Sumner, E. J., Cartigny, M. J. B., Talling, P. J., & Clare, M. A. (2018). Which triggers produce the most erosive, frequent, and longest runout turbidity currents on deltas? *Geophysical Research Letters, 45*, 855–863. https://doi.org/10.1002/2017GL075751
Hocking, E. P., Garrett, E., Aedo, D., Carvajal, M., & Melnick, D. (2021). Geological evidence of an unreported historical Chilean tsunami reveals more frequent inundation. *Communications Earth Environironment, 2*, 245. https://doi.org/10.1038/s43247-021-00319-z
Hovius, N., Meunier, P., Lin, C. W., Chen, H., Chen, Y. G., Dadson, S., Horng, M. J., & Lines, M. (2011). Prolonged seismically induced erosion and the mass balance of a large earthquake. *Earth and Planetary Science Letters, 304*, 347–355. https://doi.org/10.1016/j.epsl.2011.02.005
Howarth, J. D., Barth, N. C., Fitzsimons, S. J., Richards-Dinger, K., Clark, K. J., Biasi, G. P., Cochran, U. A., Langridge, R. M., Berryman, K. R., & Sutherland, R. (2021a). Spatiotemporal clustering of great earthquakes on a transform fault controlled by geometry. *Nature Geoscience, 1–7*. https://doi.org/10.1038/s41561-021-00721-4
Howarth, J. D., Cochran, U. A., Langridge, R. M., Clark, K., Fitzsimons, S. J., Berryman, K., Villamor, P., & Strong, D. T. (2018). Past large earthquakes on the Alpine Fault: Paleoseismological progress and future directions. *New Zealand Journal of Geology and Geophysics, 61*, 309–328.
Howarth, J. D., Fitzsimons, S. J., Jacobsen, G. E., Vandergoes, M. J., & Norris, R. J. (2013). Identifying a reliable target fraction for radiocarbon dating sedimentary records from lakes. *Quaternary Geochronology, 17*, 68–80. https://doi.org/10.1016/j.quageo.2013.02.001
Howarth, J. D., Fitzsimons, S. J., Norris, R. J., & Jacobsen, G. E. (2012). Lake sediments record cycles of sediment flux driven by large earthquakes on the Alpine fault, New Zealand. *Geology, 40*, 1091–1094. https://doi.org/10.1130/G33486.1

Howarth, J. D., Fitzsimons, S. J., Norris, R. J., & Jacobsen, G. E. (2014). Lake sediments record high intensity shaking that provides insight into the location and rupture length of large earthquakes on the Alpine Fault, New Zealand. *Earth and Planetary Science Letters, 403*, 340–351. https://doi.org/10.1016/j.epsl.2014.07.008

Howarth, J. D., Fitzsimons, S. J., Norris, R. J., Langridge, R., & Vandergoes, M. J. (2016). A 2000 yr rupture history for the Alpine fault derived from Lake Ellery, South Island, New Zealand. *Bulletin Geological Society of America, 128*, 627–643. https://doi.org/10.1130/B31300.1

Howarth, J. D., Orpin, A. R., Kaneko, Y., Strachan, L. J., Nodder, S. D., Mountjoy, J. J., Barnes, P. M., Bostock, H. C., Holden, C., Jones, K., & Cağatay, M. N. (2021b). Calibrating the marine turbidite palaeoseismometer using the 2016 Kaikōura earthquake. *Nature Geoscience, 14*, 161–167. https://doi.org/10.1038/s41561-021-00692-6

Hubert-Ferrari, A., Avsar, U., El, O. M., Lepoint, G., Martinez, P., & Fagel, N. (2012). Paleoseismic record obtained by coring a sag-pond along the north anatolian fault (Turkey). *Annales Geophysicae, 55*, 929–953. https://doi.org/10.4401/ag-5460

Hubert-Ferrari, A., El-Ouahabi, M., Garcia-Moreno, D., Avşar, U., Altınok, S., Schmidt, S., Fagel, N., & Çağatay, M. N. (2017). Earthquake imprints on a lacustrine deltaic system: The Kürk Delta along the East Anatolian Fault (Turkey). *Sedimentology, 64*, 1322–1353. https://doi.org/10.1111/sed.12355

Hubert-Ferrari, A., Lamair, L., Hage, S., Schmidt, S., Çağatay, M. N., & Avşar, U. (2020). A 3800 yr paleoseismic record (Lake Hazar sediments, eastern Turkey): Implications for the East Anatolian Fault seismic cycle. *Earth and Planetary Science Letters, 538*, 116152. https://doi.org/10.1016/j.epsl.2020.116152

Ikehara, K., Kanamatsu, T., Nagahashi, Y., Strasser, M., Fink, H., Usami, K., Irino, T., & Wefer, G. (2016). Documenting large earthquakes similar to the 2011 Tohoku-oki earthquake from sediments deposited in the Japan Trench over the past 1500 years. *Earth and Planetary Science Letters, 445*, 48–56. https://doi.org/10.1016/j.epsl.2016.04.009

Inouchi, Y., Kinugasa, Y., Kumon, F., Nakano, S., Yasumatsu, S., & Shiki, T. (1996). Turbidites as records of intense palaeoearthquakes in Lake Biwa, Japan. *Sedimentary Geology, 104*, 117–125. https://doi.org/10.1016/0037-0738(95)00124-7

Ivy-Ochs, S., Schäfer, J., Kubik, P. W., Synal, H.-A., & Schlüchter, C. (2004). Timing of deglaciation on the northern Alpine foreland (Switzerland). *Eclogae Geologicae Helvetiae, 97*, 47–55. https://doi.org/10.1007/s00015-004-1110-0

Jiang, H., Zhong, N., Li, Y., Ma, X., Xu, H., Shi, W., Zhang, S., & Nie, G. (2017). A continuous 13.3-ka record of seismogenic dust events in lacustrine sediments in the eastern Tibetan Plateau. *Science and Reports, 7*, 4–13. https://doi.org/10.1038/s41598-017-16027-8

Kagan, E., Stein, M., & Marco, S. (2018). Integrated paleoseismic chronology of the Last Glacial Lake Lisan: From lake margin seismites to deep-lake mass transport deposits. *Journal of Geophysical Research: Solid Earth, 123*, 2806–2824. https://doi.org/10.1002/2017JB014117

Karlin, R. E., Holmes, M., Abella, S. E. B., & Sylwester, R. (2004). Holocene landslides and a 3500-year record of Pacific Northwest earthquakes from sediments in Lake Washington. *Bulletin Geological Society of America, 116*, 94–108. https://doi.org/10.1130/B25158.1

Kearse, J., Kaneko, Y., Little, T., & Van Dissen, R. (2019). Curved slickenlines preserve direction of rupture propagation. *Geology, 47*, 838–842. https://doi.org/10.1130/G46563.1

Keefer, D. K. (1984). Landslides caused by earthquakes. *GSA Bulletin, 95*, 406–421. https://doi.org/10.1130/0016-7606(1984)95<406:LCBE>2.0.CO;2

Kempf, P., & Moernaut, J. (2021). Age uncertainty in recurrence analysis of paleoseismic records. *Journal of Geophysics Research: Solid Earth*. https://doi.org/10.1029/2021JB021996

Kempf, P., Moernaut, J., & De, B. M. (2019). Bimodal recurrence pattern of tsunamis in South-Central Chile: A statistical exploration of paleotsunami data. *Seismological Research Letters, 90*, 194–202. https://doi.org/10.1785/0220180204

Kempf, P., Moernaut, J., Van Daele, M., Pino, M., Urrutia, R., & De Batist, M. (2020). Paleotsunami record of the past 4300 years in the complex coastal lake system of Lake Cucao, Chiloé Island, south central Chile. *Sedimentary Geology, 401*, 105644. https://doi.org/10.1016/j.sedgeo.2020.105644

Kempf, P., Moernaut, J., Van Daele, M., Vandoorne, W., Pino, M., Urrutia, R., & De Batist, M. (2017). Coastal lake sediments reveal 5500 years of tsunami history in south central Chile. *Quaternary Science Reviews, 161*, 99–116. https://doi.org/10.1016/j.quascirev.2017.02.018

Ken-Tor, R., Agnon, A., Enzel, Y., Stein, M., Marco, S., & Negendank, J. F. W. (2001). High-resolution geological record of historic earthquakes in the Dead Sea basin. *Journal of Geophysics Research: Solid Earth, 106*, 2221–2234. https://doi.org/10.1029/2000JB900313

Kioka, A., Schwestermann, T., Moernaut, J., Ikehara, K., Kanamatsu, T., Eglinton, T. I., & Strasser, M. (2019). Event stratigraphy in a hadal oceanic trench: The Japan trench as sedimentary archive recording recurrent giant subduction zone earthquakes and their role in organic carbon export to the deep sea. *Frontiers in Earth Science, 7*, 1–24. https://doi.org/10.3389/feart.2019.00319

Kluger, M. O., Lowe, D. J., Moon, V. G., Chaneva, J., Johnston, R., Villamor, P., Ilanko, T., Melchert, R. A., Orense, R. P., Loame, R. C., & Ross, N. (2023). Seismically-induced downsagging structures in tephra layers (tephra-seismites) preserved in lakes since 17.5 cal ka, Hamilton lowlands, New Zealand. *Sedimentary Geology, 445*, 106327. https://doi.org/10.1016/j.sedgeo.2022.106327

Kremer, K., Fabbri, S. C., Evers, F. M., Schweizer, N., & Wirth, S. B. (2022). Traces of a prehistoric and potentially tsunamigenic mass movement in the sediments of Lake Thun (Switzerland). *Swiss Journal of Geosciences, 115*, 13. https://doi.org/10.1186/s00015-022-00405-0

Kremer, K., Gassner-Stamm, G., Grolimund, R., Wirth, S. B., Strasser, M., & Fäh, D. (2020). A database of potential paleoseismic evidence in Switzerland. *Journal of Seismology, 24*, 247–262. https://doi.org/10.1007/s10950-020-09908-5

Kremer, K., Hilbe, M., Simpson, G., Decrouy, L., Wildi, W., & Girardclos, S. (2015). Reconstructing 4000 years of mass movement and tsunami history in a deep peri-Alpine lake (Lake Geneva, France-Switzerland). *Sedimentology, 62*, 1305–1327. https://doi.org/10.1111/sed.12190

Kremer, K., Simpson, G., & Girardclos, S. (2012). Giant lake geneva tsunami in AD 563. *Nature Geoscience, 5*, 756–757. https://doi.org/10.1038/ngeo1618

Kremer, K., Wirth, S. B., Reusch, A., Fäh, D., Bellwald, B., Anselmetti, F. S., Girardclos, S., & Strasser, M. (2017). Lake-sediment based paleoseismology: Limitations and perspectives from the Swiss Alps. *Quaternary Science Reviews, 168*, 1–18. https://doi.org/10.1016/j.quascirev.2017.04.026

Kulkarni, R., Wong, I., Zachariasen, J., Goldfinger, C., & Lawrence, M. (2013). Statistical analyses of great earthquake recurrence along the Cascadia subduction zone. *Bulletin of the Seismological Society of America, 103*, 3205–3221. https://doi.org/10.1785/0120120105

Lamair, L. (2018). *Holocene history of natural hazards in central Japan (Fuji Five Lakes): Imprints of earthquakes, typhoons and volcanic events in lacustrine sediments* [PhD Thesis]. University of Liège

Langridge, R. M., Howarth, J. D., Cox, S. C., Palmer, J. G., & Sutherland, R. (2018). Frontal fault location and most recent earthquake timing for the Alpine Fault at Whataroa, Westland, New Zealand. *New Zealand Journal of Geology and Geophysics, 61*, 329–340. https://doi.org/10.1080/00288306.2018.1509878

Langridge, R. M., Villamor, P., Howarth, J. D., Ries, W. F., Clark, K. J., & Litchfield, N. J. (2021). Reconciling an Early Nineteenth-Century Rupture of the Alpine Fault at a Section End, Toaroha River, Westland, New Zealand. *Bulletin of the Seismological Society of America, 111*, 514–540. https://doi.org/10.1785/0120200116

Larsen, D. J., Crump, S. E., Abbott, M. B., Harbert, W., Blumm, A., Wattrus, N. J., & Hebberger, J. J. (2019). Paleoseismic evidence for climatic and magmatic controls on the Teton Fault, WY. *Geophysical Research Letters, 46*, 13036–13043. https://doi.org/10.1029/2019GL085475

Lauterbach, S., Mingram, J., Schettler, G., & Orunbaev, S. (2019). Two twentieth-century M LH = 7.5 earthquakes recorded in annually laminated lake sediments from Sary Chelek, western Tian Shan, Kyrgyzstan. *Quaternary Research, 92*, 288–303. https://doi.org/10.1017/qua.2019.21

Lay, T., Kanamori, H., Ammon, C. J., Koper, K. D., Hutko, A. R., Ye, L., Yue, H., & Rushing, T. M. (2012). Depth-varying rupture properties of subduction zone megathrust faults. *Journal of Geophysical Research: Solid Earth, 117*, 4311. https://doi.org/10.1029/2011JB009133

Lefevre, M., Souloumiac, P., Cubas, N., & Klinger, Y. (2020). Experimental evidence for crustal control over seismic fault segmentation. *Geology, 48*, 844–848. https://doi.org/10.1130/G47115.1

Leithold, E. L., Wegmann, K. W., Bohnenstiehl, D. R., Joyner, C. N., & Pollen, A. F. (2019). Repeated megaturbidite deposition in Lake Crescent, Washington, USA, triggered by Holocene ruptures of the Lake Creek-Boundary Creek fault system. *Bulletin Geological Society of America, 131*, 2039–2055. https://doi.org/10.1130/B35076.1

Leyton, F., Ruiz, J., Campos, J., & Kausel, E. (2009). Intraplate and interplate earthquakes in Chilean subduction zone: A theoretical and observational comparison. *Physics of the Earth and Planetary Interiors, 175*, 37–46. https://doi.org/10.1016/j.pepi.2008.03.017

Li, S., Chen, Y., & Yu, T. (2021). Comparison of macroseismic-intensity scales by considering empirical observations of structural seismic damage. *Earthquake Spectra, 37*, 449–485. https://doi.org/10.1177/8755293020944174

Liu, L., Yang, J., Liu, X., Mao, X., & Qin, R. (2022). Historic earthquakes for the Xianshuihe Fault derived from Lake Mugeco in the southeastern margin of the Tibetan Plateau during the past 300 years. *Frontiers in Earth Science, 10*, 1–10. https://doi.org/10.3389/feart.2022.859471

Lu, Y., Moernaut, J., Bookman, R., Waldmann, N., Wetzler, N., Agnon, A., Marco, S., Alsop, G. I., Strasser, M., & Hubert-Ferrari, A. (2021a). A new approach to constrain the seismic origin for prehistoric turbidites as applied to the Dead Sea Basin. *Geophysical Research Letters, 48*, 1–10. https://doi.org/10.1029/2020GL090947

Lu, Y., Moernaut, J., Waldmann, N., Bookman, R., Ian Alsop, G., Hubert-Ferrari, A., Strasser, M., Agnon, A., & Marco, S. (2021b). Orbital- and millennial-scale changes in lake-levels facilitate earthquake-triggered mass failures in the Dead Sea Basin. *Geophysical Research Letters, 48*. https://doi.org/10.1029/2021GL093391

Lu, Y., Waldmann, N., Alsop, G. I., & Marco, S. (2017). Interpreting soft sediment deformation and mass transport deposits as seismites in the Dead Sea depocenter. *Journal of Geophysics Research: Solid Earth, 122*, 8305–8325. https://doi.org/10.1002/2017JB014342

Lu, Y., Wetzler, N., Waldmann, N., Agnon, A., Biasi, G. P., & Marco, S. (2020). A 220,000-year-long continuous large earthquake record on a slow-slipping plate boundary. *Science Advances, 6*, 4170–4197. https://doi.org/10.1126/sciadv.aba4170

Marco, S., Stein, M., Agnon, A., & Ron, H. (1996). Long-term earthquake clustering: A 50,000-year paleoseismic record in the Dead Sea Graben. *Journal of Geophysics Research B Solid Earth, 101*, 6179–6191. https://doi.org/10.1029/95jb01587

McCalpin, J. P. (2009). Chapter 9 Application of Paleoseismic Data to Seismic Hazard Assessment and Neotectonic Research. *Paleoseismology* (pp. 1–106). International Geophysics. Academic Press.

McGuire, R. K. (1976). FORTRAN computer program for seismic risk analysis. *USGS Open-File Report, 76–67*. https://doi.org/10.3133/ofr7667

McHugh, C. M., Seeber, L., Rasbury, T., Strasser, M., Kioka, A., Kanamatsu, T., Ikehara, K., & Usami, K. (2020). Isotopic and sedimentary signature of megathrust ruptures along the Japan subduction margin. *Marine Geology, 428*, 106283. https://doi.org/10.1016/j.margeo.2020.106283

Melnick, D., Moreno, M., Motagh, M., Cisternas, M., & Wesson, R. L. (2012). Splay fault slip during the Mw 8.8 2010 Maule Chile earthquake. *Geology, 40*, 251–254. https://doi.org/10.1130/G32712.1

Migowski, C., Agnon, A., Bookman, R., Negendank, J. F. W., & Stein, M. (2004). Recurrence pattern of Holocene earthquakes along the Dead Sea transform revealed by varve-counting and radiocarbon dating of lacustrine sediments. *Earth and Planetary Science Letters, 222*, 301–314. https://doi.org/10.1016/j.epsl.2004.02.015

Miller, D. J. (1960). The Alaska earthquake of July 10, 1958: Giant wave in Lituya Bay. *Bulletin of the Seismological Society of America, 50*, 253–266.

Moernaut, J. (2020). Time-dependent recurrence of strong earthquake shaking near plate boundaries: A lake sediment perspective. *Earth-Science Reviews, 210*, 103344. https://doi.org/10.1016/j.earscirev.2020.103344

Moernaut, J., De Batist, M., Heirman, K., Van Daele, M., Pino, M., Brümmer, R., & Urrutia, R. (2009). Fluidization of buried mass-wasting deposits in lake sediments and its relevance for paleoseismology: Results from a reflection seismic study of lakes Villarrica and Calafquén (South-Central Chile). *Sedimentary Geology, 213*, 121–135. https://doi.org/10.1016/j.sedgeo.2008.12.002

Moernaut, J., Van Daele, M., Fontijn, K., Heirman, K., Kempf, P., Pino, M., Valdebenito, G., Urrutia, R., Strasser, M., & De Batist, M. (2018). Larger earthquakes recur more periodically: New insights in the megathrust earthquake cycle from lacustrine turbidite records in south-central Chile. *Earth and Planetary Science Letters, 481*, 9–19. https://doi.org/10.1016/j.epsl.2017.10.016

Moernaut, J., Van Daele, M., Heirman, K., Fontijn, K., Strasser, M., Pino, M., Urrutia, R., & De Batist, M. (2014). Lacustrine turbidites as a tool for quantitative earthquake reconstruction: New evidence for a variable rupture mode in south central Chile. *Journal of Geophysics Research Solid Earth, 119*, 1607–1633. https://doi.org/10.1002/2013JB010738

Moernaut, J., Van Daele, M., Heirman, K., Wiemer, G., Molenaar, A., Vandorpe, T., Melnick, D., Hajdas, I., Pino, M., Urrutia, R., & De Batist, M. (2019). The subaqueous landslide cycle in south-central Chilean lakes: The role of tephra, slope gradient and repeated seismic shaking. *Sedimentary Geology, 381*, 84–105. https://doi.org/10.1016/j.sedgeo.2019.01.002

Moernaut, J., Van Daele, M., Strasser, M., Clare, M. A., Heirman, K., Viel, M., Cardenas, J., Kilian, R., Ladrón de Guevara, B., Pino, M., Urrutia, R., & De Batist, M. (2017a). Lacustrine turbidites produced by surficial slope sediment remobilization: A mechanism for continuous and sensitive turbidite paleoseismic records. *Marine Geology, 384*, 159–176. https://doi.org/10.1016/j.margeo.2015.10.009

Moernaut, J., Wiemer, G., Reusch, A., Stark, N., De Batist, M., Urrutia, R., Ladrón de Guevara, B., Kopf, A., & Strasser, M. (2017b). The influence of overpressure and focused fluid flow on subaquatic slope stability in a formerly glaciated basin: Lake Villarrica (South-Central Chile). *Marine Geology, 383*, 35–54. https://doi.org/10.1016/j.margeo.2016.11.012

Molenaar, A., Van Daele, M., Huang, J.-J.S., Strasser, M., De Batist, M., Pino, M., Urrutia, R., & Moernaut, J. (2022). Disentangling factors controlling earthquake-triggered soft-sediment deformation in lakes. *Sedimentary Geology, 438*, 106200. https://doi.org/10.1016/J.SEDGEO.2022.106200

Molenaar, A., Van Daele, M., Vandorpe, T., Degenhart, G., De Batist, M., Urrutia, R., Pino, M., Strasser, M., & Moernaut, J. (2021). What controls the remobilization and deformation of surficial sediment by seismic shaking? Linking lacustrine slope stratigraphy to great earthquakes in South-Central Chile. *Sedimentology, 12856*. https://doi.org/10.1111/sed.12856

Molenaar, A., Wils, K., Van Daele, M., Daxer, C., Dubois, N., Grießer, A., Oswald, P., Ramisch, A., Strasser, M., & Moernaut, J. (2024). Shaken and stirred: A comparative study of earthquake-triggered soft-sediment deformation structures in lake sediments. *Geochemical Geophysics Geosystem, 25*, 1–18. https://doi.org/10.1029/2023GC011402

Monecke, K., Anselmetti, F. S., Becker, A., Schnellmann, M., Sturm, M., & Giardini, D. (2006). Earthquake-induced deformation structures in lake deposits: A late pleistocene to holocene paleoseismic record for Central Switzerland. *Eclogae Geologicae Helvetiae, 99*, 343–362. https://doi.org/10.1007/s00015-006-1193-x

Monecke, K., Anselmetti, F. S., Becker, A., Sturm, M., & Giardini, D. (2004). The record of historic earthquakes in lake sediments of Central Switzerland. *Tectonophysics, 394*, 21–40. https://doi.org/10.1016/j.tecto.2004.07.053

Monecke, K., Brabander, D. J., Howey, E., Janigian, G., McCarthy, F. G., Pentesco, J., Bradford Hubeny, J., Kielb, S., & Ebel, J. E. (2018). The 1755 cape ann earthquake recorded in lake sediments of eastern new England: An interdisciplinary paleoseismic approach. *Seismological Research Letters, 89*, 1212–1222. https://doi.org/10.1785/0220170220

Moreno, M. S., Bolte, J., Klotz, J., & Melnick, D. (2009). Impact of megathrust geometry on inversion of coseismic slip from geodetic data: Application to the 1960 Chile earthquake. *Geophysical Research Letters, 36*. https://doi.org/10.1029/2009GL039276

Morey, A. E., Goldfinger, C., Briles, C. E., Gavin, D. G., Colombaroli, D., & Kusler, J. E. (2013). Are great Cascadia earthquakes recorded in the sedimentary records from small forearc lakes? *Natural Hazards and Earth Systems Sciences, 13*, 2441–2463. https://doi.org/10.5194/nhess-13-2441-2013

Mörner, N. A. (2005). An interpretation and catalogue of paleoseismicity in Sweden. *Tectonophysics, 408*, 265–307. https://doi.org/10.1016/j.tecto.2005.05.039

Mörner, N. A. (2011). Paleoseismology: The application of multiple parameters in four case studies in Sweden. *Quaternary International, 242*, 65–75. https://doi.org/10.1016/j.quaint.2011.03.054

Musson, R. M. W., Grünthal, G., & Stucchi, M. (2010). The comparison of macroseismic intensity scales. *Journal of Seismology, 14*, 413–428. https://doi.org/10.1007/s10950-009-9172-0

Naranjo, J. A., Arenas, M., Clavero, J., & Muñoz, O. (2009). Mass movement-induced tsunamis: Main effects during the Patagonian Fjordland seismic crisis in Aisén (45°25'S), Chile. *Andean Geology, 36*, 137–145. https://doi.org/10.4067/S0718-71062009000100011

Nelson, A. R., Kashima, K., & Bradley, L. A. (2009). Fragmentary evidence of great-earthquake subsidence during holocene emergence, Valdivia estuary, South Central Chile. *Bulletin of the Seismological Society of America, 99*, 71–86. https://doi.org/10.1785/0120080103

Nicolussi, K., Spötl, C., Thurner, A., & Reimer, P. J. (2015). Precise radiocarbon dating of the giant Köfels landslide (Eastern Alps, Austria). *Geomorphology, 243*, 87–91. https://doi.org/10.1016/j.geomorph.2015.05.001

Norris, R. J., & Cooper, A. F. (2007). The Alpine Fault, New Zealand: Surface geology and field relationships. In: *A Continental Plate Boundary: Tectonics at South Island, New Zealand*. American Geophysical Union (AGU), pp. 157–175

Obermeier, S. F. (1996). Use of liquefaction-induced features for paleoseismic analysis-An overview of how seismic liquefaction features can be distinguished from other features and how their regional distribution and properties of source sediment can be used to infer the location.

Ojala, A. E. K., Francus, P., Zolitschka, B., Besonen, M., & Lamoureux, S. F. (2012). Characteristics of sedimentary varve chronologies—a review. *Quaternary Science Reviews, 43*, 45–60. https://doi.org/10.1016/j.quascirev.2012.04.006

Ojala, A. E. K., Mattila, J., Hämäläinen, J., & Sutinen, R. (2019). Lake sediment evidence of paleoseismicity: Timing and spatial occurrence of late- and postglacial earthquakes in Finland. *Tectonophysics, 771*. https://doi.org/10.1016/j.tecto.2019.228227

Oswald, P., Moernaut, J., Fabbri, S. C., De Batist, M., Hajdas, I., Ortner, H., Titzler, S., & Strasser, M. (2021a). Combined on-fault and off-fault paleoseismic evidence in the postglacial infill of the inner-Alpine lake Achensee (Austria, Eastern Alps). *Frontiers in Earth Science, 9*, 438. https://doi.org/10.3389/feart.2021.670952

Oswald, P., Strasser, M., Hammerl, C., & Moernaut, J. (2021b). Seismic control of large prehistoric rockslides in the Eastern Alps. *Nature Communications, 12*, 1059. https://doi.org/10.1038/s41467-021-21327-9

Oswald, P., Strasser, M., Skapski, J., & Moernaut, J. (2022). Magnitude and source area estimations of severe prehistoric earthquakes in the western Austrian Alps. *Natural Hazards and Earth Systems Sciences, 22*, 2057–2079. https://doi.org/10.5194/nhess-22-2057-2022

Owen, G., & Moretti, M. (2011). Identifying triggers for liquefaction-induced soft-sediment deformation in sands. *Sedimentary Geology, 235*, 141–147. https://doi.org/10.1016/j.sedgeo.2010.10.003

Pagani, M., Johnson, K., & Garcia Pelaez, J. (2020). Modelling subduction sources for probabilistic seismic hazard analysis. *Geological Society London*, Special Publication SP501-2019–120. https://doi.org/10.1144/sp501-2019-120

Park, H., Cox, D. T., Alam, M. S., & Barbosa, A. R. (2017). Probabilistic seismic and tsunami hazard analysis conditioned on a megathrust rupture of the Cascadia subduction zone. *Frontiers in Built Environment, 3*, 32. https://doi.org/10.3389/fbuil.2017.00032

Parsons, T. (2008). Monte Carlo method for determining earthquake recurrence parameters from short paleoseismic catalogs: Example calculations for California. *Journal of Geophysical Research: Solid Earth, 113*. https://doi.org/10.1029/2007JB004998

Patton, J. R., Goldfinger, C., Morey, A. E., Ikehara, K., Romsos, C., Stoner, J., Djadjadihardja, Y., Udrekh, A. S., Gaffar, E. Z., & Vizcaino, A. (2015). A 6600 year earthquake history in the region of the 2004 Sumatra- Andaman subduction zone earthquake. *Geosphere, 11*, 2067–2129. https://doi.org/10.1130/GES01066.1

Petersen, J., Wilhelm, B., Revel, M., Rolland, Y., Crouzet, C., Arnaud, F., Brisset, E., Chaumillon, E., & Magand, O. (2014a). Sediments of Lake Vens (SW European Alps, France) record large-magnitude earthquake events. *Journal of Paleolimnology, 51*, 343–355. https://doi.org/10.1007/s10933-013-9759-x

Petersen, M. D., Moschetti, M. P., Powers, P. M., Mueller, C. S., Haller, K. M., Frankel, A. D., Zeng, Y., Rezaeian, S., Harmsen, S. C., Boyd, O. S., Field, N., Chen, R., Rukstales, K. S., Luco, N., Wheeler, R. L., Williams, R. A., & Olsen, A. H. (2014b) Documentation for the 2014 Update of the United States National Seismic Hazard Maps. US Geology Survey Open-File Rep 243 p.

Philibosian, B., & Meltzner, A. J. (2020). Segmentation and supercycles: A catalog of earthquake rupture patterns from the Sumatran Sunda Megathrust and other well-studied faults worldwide. *Quaternary Science Reviews, 241*. https://doi.org/10.1016/j.quascirev.2020.106390

Polonia, A., Albertazzi, S., Bellucci, L. G., Bonetti, C., Bonetti, J., Giorgetti, G., Giuliani, S., Correa, M. L., Mayr, C., Peruzza, L., Stanghellini, G., & Gasperini, L. (2021). Decoding a complex record of anthropogenic and natural impacts in the Lake of Cavazzo sediments, NE Italy. *Science Total Environonment, 787*, 147659. https://doi.org/10.1016/j.scitotenv.2021.147659

Polonia, A., Panieri, G., Gasperini, L., Gasparotto, G., Bellucci, L. G., & Torelli, L. (2013). Turbidite paleoseismology in the Calabrian Arc subduction complex (Ionian Sea). *Geochemistry, Geophysics Geosystems, 14*, 112–140. https://doi.org/10.1029/2012GC004402

Pouderoux, H., Proust, J. N., & Lamarche, G. (2014). Submarine paleoseismology of the northern Hikurangi subduction margin of New Zealand as deduced from Turbidite record since 16ka. *Quaternary Science Reviews, 84*, 116–131. https://doi.org/10.1016/j.quascirev.2013.11.015

Praet, N. (2020). *Towards the construction of a multi-lake paleoseismic record in southcentral Alaska: A trembling tale of landslides and turbidites* [PhD Thesis]. Ghent University.

Praet, N., Moernaut, J., Van Daele, M., Boes, E., Haeussler, P. J., Strupler, M., Schmidt, S., Loso, M. G., & De Batist, M. (2017). Paleoseismic potential of sublacustrine landslide records in a high-seismicity setting (south-central Alaska). *Marine Geology, 384*, 103–119. https://doi.org/10.1016/j.margeo.2016.05.004

Praet, N., Van Daele, M., Collart, T., Moernaut, J., Vandekerkhove, E., Kempf, P., Haeussler, P. J., & De Batist, M. (2020). Turbidite stratigraphy in proglacial lakes: Deciphering trigger mechanisms using a statistical approach. *Sedimentology, 67*, 2332–2359. https://doi.org/10.1111/sed.12703

Praet, N., Van Daele, M., Moernaut, J., Mestdagh, T., Vandorpe, T., Jensen, B. J. L., Witter, R. C., Haeussler, P. J., & De Batist, M. (2022). Unravelling a 2300 year long sedimentary record of megathrust and intraslab earthquakes in proglacial Skilak Lake, south-central Alaska. *Sedimentology*. https://doi.org/10.1111/sed.12986

Rapuc, W., Arnaud, F., Sabatier, P., Anselmetti, F. S., Piccin, A., Peruzza, L., Bastien, A., Augustin, L., Régnier, E., Gaillardet, J., & Von Grafenstein, U. (2022). Instant sedimentation in a deep Alpine lake (Iseo, Italy) controlled by climate, human and geodynamic forcing. *Sedimentology*. https://doi.org/10.1111/sed.12972

Rapuc, W., Sabatier, P., Andrič, M., Crouzet, C., Arnaud, F., Chapron, E., Šmuc, A., Develle, A. L., Wilhelm, B., Demory, F., Reyss, J. L., Régnier, E., Daut, G., & Von Grafenstein, U. (2018). 6600 years of earthquake record in the Julian Alps (Lake Bohinj, Slovenia). *Sedimentology, 65*, 1777–1799. https://doi.org/10.1111/sed.12446

Ratzov, G., Cattaneo, A., Babonneau, N., Déverchère, J., Yelles, K., Bracene, R., & Courboulex, F. (2015). Holocene turbidites record earthquake supercycles at a slow-rate plate boundary. *Geology, 43*, 331–334. https://doi.org/10.1130/G36170.1

Rodríguez-Pascua MA, Calvo JP, De Vicente G, Gómez-Gras D,. (2000). Soft-sediment deformation structures interpreted as seismites in lacustrine sediments of the Prebetic Zone, SE

Spain, and their potential use as indicators of earthquake magnitudes during the Late Miocene. *Sedimentary Geology, 135*, 117–135. https://doi.org/10.1016/S0037-0738(00)00067-1

Sabatier, P., Moernaut, J., Bertrand, S., Van Daele, M., Kremer, K., Chaumillon, E., & Arnaud, F. (2022). A review of event deposits in lake sediments. *Quaternary, 5*, 1–47. https://doi.org/10.3390/quat5030034

Sabatier, P., Wilhelm, B., Ficetola, G. F., Moiroux, F., Poulenard, J., Develle, A.-L., Bichet, A., Chen, W., Pignol, C., Reyss, J.-L., Gielly, L., Bajard, M., Perrette, Y., Malet, E., Taberlet, P., & Arnaud, F. (2017). 6-kyr record of flood frequency and intensity in the western mediterranean alps—interplay of solar and temperature forcing. *Quaternary Science Reviews, 170*, 121–135. https://doi.org/10.1016/j.quascirev.2017.06.019

Salditch, L., Stein, S., Neely, J., Spencer, B. D., Brooks, E. M., Agnon, A., & Liu, M. (2020). Earthquake supercycles and long-term fault memory. *Tectonophysics, 774*, 228289. https://doi.org/10.1016/j.tecto.2019.228289

Sammartini, M., Moernaut, J., Anselmetti, F. S., Hilbe, M., Lindhorst, K., Praet, N., & Strasser, M. (2019). An atlas of mass-transport deposits in lakes. *Geophysical Monograph Series, 246*, 201–226. https://doi.org/10.1002/9781119500513.ch13

Sammartini, M., Moernaut, J., Kopf, A., Stegmann, S., Fabbri, S. C., Anselmetti, F. S., & Strasser, M. (2021). Propagation of frontally confined subaqueous landslides: Insights from combining geophysical, sedimentological, and geotechnical analysis. *Sedimentary Geology, 416*, 105877. https://doi.org/10.1016/j.sedgeo.2021.105877

Sawyer, D. E., & Devore, J. R. (2015). Elevated shear strength of sediments on active margins: Evidence for seismic strengthening. *Geophysical Research Letters, 42*, 10216–10221. https://doi.org/10.1002/2015GL066603

Schillereff, D. N., Chiverrell, R. C., Macdonald, N., & Hooke, J. M. (2014). Flood stratigraphies in lake sediments: A review. *Earth-Science Reviews, 135*, 17–37. https://doi.org/10.1016/j.earscirev.2014.03.011

Schnellmann, M., Anselmetti, F. S., Giardini, D., & McKenzie, J. A. (2005). Mass movement-induced fold-and-thrust belt structures in unconsolidated sediments in Lake Lucerne (Switzerland). *Sedimentology, 52*, 271–289. https://doi.org/10.1111/j.1365-3091.2004.00694.x

Schnellmann, M., Anselmetti, F. S., Giardini, D., & McKenzie, J. A. (2006). 15,000 Years of mass-movement history in Lake Lucerne: Implications for seismic and tsunami hazards. *Eclogae Geologicae Helvetiae, 99*, 409–428. https://doi.org/10.1007/s00015-006-1196-7

Schnellmann, M., Anselmetti, F. S., Giardini, D., McKenzie, J. A., & Ward, S. N. (2002). Prehistoric earthquake history revealed by lacustrine slump deposits. *Geology, 30*, 1131. https://doi.org/10.1130/0091-7613(2002)030%3c1131:PEHRBL%3e2.0.CO;2

Sergiou, S., Beckers, A., Geraga, M., Papatheodorou, G., Iliopoulos, I., & Papaefthymiou, H. (2017). Recent sedimentary processes in the western Gulf of Corinth, Greece: Seismic and aseismic turbidites. *Bulletin of the Geological Society of Greece, 50*, 383. https://doi.org/10.12681/bgsg.11739

Serva, L., Vittori, E., Comerci, V., Esposito, E., Guerrieri, L., Michetti, A. M., Mohammadioun, B., Mohammadioun, G. C., Porfido, S., & Tatevossian, R. E. (2016). Earthquake hazard and the environmental seismic intensity (ESI) scale. *Pure and Applied Geophysics, 173*, 1479–1515. https://doi.org/10.1007/s00024-015-1177-8

Shennan, I., Garrett, E., & Barlow, N. (2016). Detection limits of tidal-wetland sequences to identify variable rupture modes of megathrust earthquakes. *Quaternary Science Reviews, 150*, 1–30.

Shi, W., Jiang, H., Alsop, G. I., & Wu, G. (2022). A continuous 13.3-Ka paleoseismic record constrains major earthquake recurrence in the longmen shan collision zone. *Frontiers in Earth Science, 10*, 1–11. https://doi.org/10.3389/feart.2022.838299

Shrivastava, M. N., González, G., Moreno, M., Soto, H., Schurr, B., Salazar, P., & Báez, J. C. (2019). Earthquake segmentation in northern Chile correlates with curved plate geometry. *Science and Reports, 9*, 4403. https://doi.org/10.1038/s41598-019-40282-6

Simonneau, A., Chapron, E., Garçon, M., Winiarski, T., Graz, Y., Chauvel, C., Debret, M., Motelica-Heino, M., Desmet, M., & Di Giovanni, C. (2014). Tracking Holocene glacial and high-altitude alpine environments fluctuations from minerogenic and organic markers in proglacial lake sediments (Lake Blanc Huez, Western French Alps). *Quaternary Science Reviews, 89*, 27–43. https://doi.org/10.1016/j.quascirev.2014.02.008

Simonneau, A., Chapron, E., Vanniere, B., Wirth, S. B., Gilli, A., Di Giovanni, C., Anselmetti, F. S., Desmet, M., & Magny, M. (2013). Mass-movement and flood-induced deposits in Lake Ledro, southern Alps, Italy: Implications for holocene palaeohydrology and natural hazards. *Climate of the past, 9*, 825–840. https://doi.org/10.5194/cp-9-825-2013

Sims, J. D. (1975). Determining earthquake recurrence intervals from deformational structures in young lacustrine sediments. *Tectonophysics, 29*, 141–152.

St-Onge, G., Chapron, E., Mulsow, S., Salas, M., Viel, M., Debret, M., Foucher, A., Mulder, T., Winiarski, T., Desmet, M., Costa, P. J. M., Ghaleb, B., Jaouen, A., & Locat, J. (2012). Comparison of earthquake-triggered turbidites from the Saguenay (Eastern Canada) and Reloncavi (Chilean margin) Fjords: Implications for paleoseismicity and sedimentology. *Sedimentary Geology, 243–244*, 89–107. https://doi.org/10.1016/j.sedgeo.2011.11.003

St.-Onge, G., Mulder, T., Piper, D. J. W., Hillaire-Marcel, C., & Stoner, J. S. (2004). Earthquake and flood-induced turbidites in the saguenay fjord (Québec): A holocene paleoseismicity record. *Quaternary Science Reviews, 23*, 283–294. https://doi.org/10.1016/j.quascirev.2003.03.001

Stein, S., Liu, M., Camelbeeck, T., Merino, M., Landgraf, A., Hintersberger, E., & KüBler, S. (2017). Challenges in assessing seismic hazard in intraplate Europe. *Geological Society Special Publications, 432*, 13–28. https://doi.org/10.1144/SP432.7

Strasser, M., Anselmetti, F. S., Fäh, D., Giardini, D., & Schnellmann, M. (2006). Magnitudes and source areas of large prehistoric norhern Alpine earthquakes revealed by slope failures in lakes. *Geology, 34*, 1005–1008. https://doi.org/10.1130/G22784A.1

Strasser, M., Hilbe, M., & Anselmetti, F. S. (2011). Mapping basin-wide subaquatic slope failure susceptibility as a tool to assess regional seismic and tsunami hazards. *Marine Geophysical Researches, 32*, 331–347. https://doi.org/10.1007/s11001-010-9100-2

Strasser, M., Monecke, K., Schnellmann, M., & Anselmetti, F. S. (2013). Lake sediments as natural seismographs: A compiled record of late quaternary earthquakes in Central Switzerland and its implication for Alpine deformation. *Sedimentology, 60*, 319–341. https://doi.org/10.1111/sed.12003

Strasser, M., Stegmann, S., Bussmann, F., Anselmetti, F. S., Rick, B., & Kopf, A. (2007). Quantifying subaqueous slope stability during seismic shaking: Lake Lucerne as model for ocean margins. *Marine Geology, 240*, 77–97. https://doi.org/10.1016/j.margeo.2007.02.016

Strupler, M., Anselmetti, F. S., Hilbe, M., & Strasser, M. (2019). Quantitative characterization of subaqueous landslides in lake zurich (Switzerland) based on a high-resolution bathymetric dataset. *Geology Society of Special Publications, 477*, 399–412. https://doi.org/10.1144/SP477.7

Strupler, M., Danciu, L., Hilbe, M., Kremer, K., Anselmetti, F. S., Strasser, M., & Wiemer, S. (2018). A subaqueous hazard map for earthquake-triggered landslides in Lake Zurich, Switzerland. *Natural Hazards, 90*, 51–78. https://doi.org/10.1007/s11069-017-3032-y

Styron, R. (2019). The impact of earthquake cycle variability on neotectonic and paleoseismic slip rate estimates. *Solid Earth, 10*, 15–25. https://doi.org/10.5194/se-10-15-2019

Sumner, E. J., Siti, M. I., McNeill, L. C., Talling, P. J., Henstock, T. J., Wynn, R. B., Djajadihardja, Y. S., & Permana, H. (2013). Can turbidites be used to reconstruct a paleoearthquake record for the central Sumatran margin? *Geology, 41*, 763–766. https://doi.org/10.1130/G34298.1

Talling, P. J. (2021). Fidelity of turbidites as earthquake records. *Nature Geoscience, 14*, 113–116. https://doi.org/10.1038/s41561-021-00707-2

Tournier, N., Fabbri, S. C., Anselmetti, F. S., Cahyarini, S. Y., Bijaksana, S., Wattrus, N., Russell, J. M., & Vogel, H. (2023). Climate-controlled sensitivity of lake sediments to record earthquake-related mass wasting in tropical Lake Towuti during the past 40 kyr. *Quaternary Science Reviews, 305*, 108015. https://doi.org/10.1016/j.quascirev.2023.108015

Urlaub, M., Talling, P. J., Zervos, A., & Masson, D. (2015). What causes large submarine landslides on low gradient (<2°) continental slopes with slow (~0.15 m/kyr) sediment accumulation?

Journal of Geophysics Research: Solid Earth, 120, 6722–6739. https://doi.org/10.1002/2015JB012347

Van Daele, M., Araya-Cornejo, C., Pille, T., Vanneste, K., Moernaut, J., Schmidt, S., Kempf, P., Meyer, I., & Cisternas, M. (2019). Distinguishing intraplate from megathrust earthquakes using lacustrine turbidites. *Geology, 47*, 127–130. https://doi.org/10.1130/G45662.1

Van Daele, M., Cnudde, V., Duyck, P., Pino, M., Urrutia, R., & De Batist, M. (2014a). Multidirectional, synchronously-triggered seismo-turbidites and debrites revealed by X-ray computed tomography (CT). *Sedimentology, 61*, 861–880. https://doi.org/10.1111/sed.12070

Van Daele, M., Haeussler, P. J., Witter, R. C., Praet, N., & De Batist, M. (2020). The sedimentary record of the 2018 anchorage earthquake in Eklutna Lake, Alaska: Calibrating the lacustrine seismograph. *Seismological Research Letters, 91*, 126–141. https://doi.org/10.1785/0220190204

Van Daele, M., Meyer, I., Moernaut, J., De Decker, S., Verschuren, D., & De Batist, M. (2017). A revised classification and terminology for stacked and amalgamated turbidites in environments dominated by (hemi)pelagic sedimentation. *Sedimentary Geology, 357*, 72–82. https://doi.org/10.1016/j.sedgeo.2017.06.007

Van Daele, M., Moernaut, J., Doom, L., Boes, E., Fontijn, K., Heirman, K., Vandoorne, W., Hebbeln, D., Pino, M., Urrutia, R., Brümmer, R., & De Batist, M. (2015). A comparison of the sedimentary records of the 1960 and 2010 great Chilean earthquakes in 17 lakes: Implications for quantitative lacustrine palaeoseismology. *Sedimentology, 62*, 1466–1496. https://doi.org/10.1111/sed.12193

Van Daele, M., Moernaut, J., Silversmit, G., Schmidt, S., Fontijn, K., Heirman, K., Vandoorne, W., De Clercq, M., Van Acker, J., Wolff, C., Pino, M., Urrutia, R., Roberts, S. J., Vincze, L., & De Batist, M. (2014b). The 600 yr eruptive history of Villarrica Volcano (Chile) revealed by annually laminated lake sediments. *Bulletin Geological Society of America, 126*, 481–498. https://doi.org/10.1130/B30798.1

Van Daele, M., Versteeg, W., Pino, M., Urrutia, R., & De Batist, M. (2013). Widespread deformation of basin-plain sediments in aysén fjord (chile) due to impact by earthquake-triggered, onshore-generated mass movements. *Marine Geology, 337*, 67–79. https://doi.org/10.1016/j.margeo.2013.01.006

Vandekerkhove, E., Van Daele, M., Praet, N., Cnudde, V., Haeussler, P. J., & De Batist, M. (2020). Flood-triggered versus earthquake-triggered turbidites: A sedimentological study in clastic lake sediments (Eklutna Lake, Alaska). *Sedimentology, 67*, 364–389. https://doi.org/10.1111/sed.12646

Vandergoes, M. J., Howarth, J. D., Dunbar, G. B., Turnbull, J. C., Roop, H. A., Levy, R. H., Li, X., Prior, C., Norris, M., Keller, L. D., Baisden, W. T., Ditchburn, R., Fitzsimons, S. J., & Bronk Ramsey, C. (2018). Integrating chronological uncertainties for annually laminated lake sediments using layer counting, independent chronologies and Bayesian age modelling (Lake Ohau, South Island, New Zealand). *Quaternary Science Reviews, 188*, 104–120. https://doi.org/10.1016/j.quascirev.2018.03.015

Vanneste, K., Vleminckx, B., Stein, S., & Camelbeeck, T. (2016). Could Mmax be the same for all stable continental regions? *Seismological Research Letters, 87*, 1214–1223. https://doi.org/10.1785/0220150203

Vanneste, K., Wils, K., & Van Daele, M. (2018). Probabilistic evaluation of fault sources based on paleoseismic evidence from mass-transport deposits: The Example of Aysén Fjord, Chile. *Journal of Geophysics Research Solid Earth, 123*, 9842–9865. https://doi.org/10.1029/2018JB016289

Vermassen, F., Van Daele, M., Praet, N., Cnudde, V., Kissel, C., & Anselmetti, F. S. (2023). Unravelling megaturbidite deposition: Evidence for turbidite stacking/amalgamation and seiche influence during the 1601 CE earthquake at Lake Lucerne, Switzerland . *Sedimentology*. https://doi.org/10.1111/sed.13094

Waldmann, N., Anselmetti, F. S., Ariztegui, D., Austin, J. A., Pirouz, M., Moy, C. M., & Dunbar, R. (2011). Holocene mass-wasting events in Lago Fagnano, Tierra del Fuego (54°S): Implications for paleoseismicity of the Magallanes-Fagnano transform fault. *Basin Research, 23*, 171–190. https://doi.org/10.1111/j.1365-2117.2010.00489.x

Wallace, L. M., Kaneko, Y., Hreinsdóttir, S., Hamling, I., Peng, Z., Bartlow, N., D'Anastasio, E., & Fry, B. (2017). Large-scale dynamic triggering of shallow slow slip enhanced by overlying sedimentary wedge. *Nature Geoscience, 10*, 765–770. https://doi.org/10.1038/ngeo3021

Walton, M. A. L., Staisch, L. M., Dura, T., Pearl, J. K., Sherrod, B., Gomberg, J., Engelhart, S., Tréhu, A., Watt, J., Perkins, J., Witter, R. C., Bartlow, N., Goldfinger, C., Kelsey, H., Morey, A. E., Sahakian, V. J., Tobin, H., Wang, K., Wells, R., & Wirth, E. (2021). Toward an integrative geological and geophysical view of Cascadia subduction zone earthquakes. *Annual Review of Earth and Planetary Sciences, 49*. https://doi.org/10.1146/annurev-earth-071620-065605

Wang, J., He, Z., & Li, L. (2021). Palaeoseismic records in lacustrine sediments—a case study of the Daqingshan piedmont fault and Hasuhai Lake in Inner Mongolia, China. *Basin Research, 33*, 681–704. https://doi.org/10.1111/BRE.12490

Wang, J., Jin, Z., Hilton, R. G., Zhang, F., Densmore, A. L., Li, G., & Joshua West, A. (2015). Controls on fluvial evacuation of sediment from earthquake-triggered landslides. *Geology, 43*, 115–118. https://doi.org/10.1130/G36157.1

Wang, X., Nie, G., & Wang, D. (2010). Relationships between ground motion parameters and landslides induced by Wenchuan earthquake. *Earthquake Science, 23*, 233–242. https://doi.org/10.1007/s11589-010-0719-5

Wei, X., Jiang, H., Xu, H., Fan, J., Shi, W., Guo, Q., & Zhang, S. (2021). Response of sedimentary and pollen records to the 1933 Diexi earthquake on the eastern Tibetan Plateau. *Ecological Indicators, 129*, 107887. https://doi.org/10.1016/j.ecolind.2021.107887

Wells, A., Yetton, M. D., Duncan, R. P., & Stewart, G. H. (1999). Prehistoric dates of the most recent Alpine fault earthquakes, New Zealand. *Geology, 27*, 995–998. https://doi.org/10.1130/0091-7613(1999)027%3c0995:PDOTMR%3e2.3.CO;2

Wells, D. L., & Coppersmith, K. J. (1994). New empirical relationships among magnitude, rupture length, rupture width, rupture area, and surface displacement

Wetzler, N., Marco, S., & Heifetz, E. (2010). Quantitative analysis of seismogenic shear-induced turbulence in lake sediments. *Geology, 38*, 303–306. https://doi.org/10.1130/G30685.1

Wilhelm, B., Nomade, J., Crouzet, C., Litty, C., Sabatier, P., Belle, S., Rolland, Y., Revel, M., Courboulex, F., Arnaud, F., & Anselmetti, F. S. (2016). Quantified sensitivity of small lake sediments to record historic earthquakes: Implications for paleoseismology. *Journal of Geophysics Research F Earth Surface, 121*, 2–16. https://doi.org/10.1002/2015JF003644

Wilhelm, B., Sabatier, P., & Arnaud, F. (2015). Is a regional flood signal reproducible from lake sediments? *Sedimentology, 62*, 1103–1117. https://doi.org/10.1111/sed.12180

Wils, K., Daryono, M. R., Praet, N., Santoso, A. B., Dianto, A., Schmidt, S., Vervoort, M., Huang, J.-J.S., Kusmanto, E., Suandhi, P., Natawidjaja, D. H., & De Batist, M. (2021a). The sediments of Lake Singkarak and Lake Maninjau in West Sumatra reveal their earthquake, volcanic and rainfall history. *Sedimentary Geology, 416*, 105863. https://doi.org/10.1016/j.sedgeo.2021.105863

Wils, K., Deprez, M., Kissel, C., Vervoort, M., Van Daele, M., Daryono, M. R., Cnudde, V., Natawidjaja, D. H., & De Batist, M. (2021b). Earthquake doublet revealed by multiple pulses in lacustrine seismo-turbidites. *Geology, 49*, 1301–1306. https://doi.org/10.1130/G48940.1

Wils, K., Van Daele, M., Kissel, C., Moernaut, J., Schmidt, S., Siani, G., & Lastras, G. (2020). Seismo-turbidites in Aysén Fjord (southern Chile) reveal a complex pattern of rupture modes along the 1960 megathrust earthquake segment. *Journal of Geophysics Research: Solid Earth, 125*, e2020JB019405. https://doi.org/10.1029/2020JB019405

Wils, K., Van Daele, M., Lastras, G., Kissel, C., Lamy, F., & Siani, G. (2018). Holocene event record of Aysén Fjord (Chilean Patagonia): An interplay of volcanic eruptions and crustal and megathrust earthquakes. *Journal of Geophysics Research: Solid Earth, 123*, 324–343. https://doi.org/10.1002/2017JB014573

Wirth, S. B., Glur, L., Gilli, A., & Anselmetti, F. S. (2013). Holocene flood frequency across the Central Alps-solar forcing and evidence for variations in North Atlantic atmospheric circulation. *Quaternary Science Reviews, 80*, 112–128. https://doi.org/10.1016/j.quascirev.2013.09.002

Wojewódka-Przybył, M., Krahn, K. J., Hamerlík, L., Macario-González, L., Cohuo, S., Charqueño-Celis, F., Cisneros, A., Hoelzmann, P., Yang, H., Rose, N. L., Zawisza, E., Pérez, L., & Schwalb, A. (2022). Imprints of the Little Ice Age and the severe earthquake of AD 2001 on the aquatic ecosystem of a tropical maar lake in El Salvador. *Holocene, 32*, 1065–1080. https://doi.org/10.1177/09596836221106965

Xu, C., Xu, X., Zhou, B., & Yu, G. (2013). Revisions of the M 8.0 Wenchuan earthquake seismic intensity map based on co-seismic landslide abundance. *Natural Hazards, 69*, 1459–1476. https://doi.org/10.1007/s11069-013-0757-0

Yakupoğlu, N., Uçarkuş, G., Kadir Eriş, K., Henry, P., & Namık Çağatay, M. (2019). Sedimentological and geochemical evidence for seismoturbidite generation in the Kumburgaz Basin, Sea of Marmara: Implications for earthquake recurrence along the central high segment of the North Anatolian Fault. *Sedimentary Geology, 380*, 31–44. https://doi.org/10.1016/j.sedgeo.2018.11.002

Zhong, N., Jiang, H., Li, H., Su, D., Xu, H., Liang, L., & Fan, J. (2022). The potential of using soft-sediment deformation structures for quantitatively reconstructing paleo-seismic shaking intensity: Progress and prospect. *Environment and Earth Science, 81*, 1–22. https://doi.org/10.1007/s12665-022-10504-8

MIX
Papier aus verantwortungsvollen Quellen
Paper from responsible sources
FSC® C105338

If you have any concerns about our products,
you can contact us on
ProductSafety@springernature.com

In case Publisher is established outside the EU,
the EU authorized representative is:
Springer Nature Customer Service Center GmbH
Europaplatz 3, 69115 Heidelberg, Germany

Printed by Libri Plureos GmbH
in Hamburg, Germany